ISNM 104:
International Series of Numerical Mathematics
Internationale Schriftenreihe zur Numerischen Mathematik
Série Internationale d'Analyse Numérique
Vol. 104

Edited by
K.-H. Hoffmann, Augsburg; H. D. Mittelmann, Tempe;
J. Todd, Pasadena

Springer Basel AG

Bifurcation and Symmetry

Cross Influence between Mathematics and Applications

Edited by

E. Allgower
K. Böhmer
M. Golubitsky

1992

Springer Basel AG

Editors

Prof. Dr. Eugene L. Allgower
Dept. of Mathematics
Colorado State University
Fort Collins, CO 80523
USA

Prof. Dr. Klaus Böhmer
Fachbereich Mathematik
Universität Marburg
Hans-Meerwein-Str., Lahnberge
D–W–3550 Marburg
Germany

Prof. Dr. Martin Golubitsky
Dept. of Mathematics
University of Houston
University Park
Houston, TX 77204–3476
USA

Deutsche Bibliothek Cataloging-in-Publication Data

Bifurcation and symmetry: cross influence between
mathematics and applications / ed. by E. Allgower ... – Basel ;
Boston ; Berlin : Birkhäuser, 1992
  (International series of numerical mathematics ; Vol. 104)
  ISBN 978-3-0348-7538-7   ISBN 978-3-0348-7536-3 (eBook)
  DOI 10.1007/978-3-0348-7536-3
NE: Allgower, Eugen L. [Hrsg.]; GT

Contents

Foreword

Symmetry is a property which occurs throughout nature and it is therefore natural that symmetry should be considered when attempting to model nature. In many cases, these models are also nonlinear and it is the study of nonlinear symmetric models that has been the basis of much recent work. Although systematic studies of nonlinear problems may be traced back at least to the pioneering contributions of Poincaré, this remains an area with challenging problems for mathematicians and scientists. Phenomena whose models exhibit both symmetry and nonlinearity lead to problems which are challenging and rich in complexity, beauty and utility.

In recent years, the tools provided by group theory and representation theory have proven to be highly effective in treating nonlinear problems involving symmetry. By these means, highly complex situations may be decomposed into a number of simpler ones which are already understood or are at least easier to handle. In the realm of numerical approximations, the systematic exploitation of symmetry via group representation theory is even more recent.

In the hope of stimulating interaction and acquaintance with results and problems in the various fields of applications, bifurcation theory and numerical analysis, we organized the conference and workshop *Bifurcation and Symmetry: Cross Influences between Mathematics and Applications* during June 2-7, 8-14, 1991 at the Philipps-University of Marburg, Germany.

In this conference an effort was made to achieve a mixture of scientists who study phenomena exhibiting symmetry, mathematicians who apply the methods of group theory to the analysis of such problems, and numerical analysts who are exploring the systematic use of symmetry in the numerical approximation of solutions. Another aim was to propagate the application of group theoretical methods to new areas and particularly to numerical analysis. We made an effort to bring together established researchers and newcomers, particularly recent PhDs and advanced graduate students.

The present volume contains most of the papers presented at the conference. A list of the speakers and the titles of their talks appears below. In all, there were 55 participants representing 14 countries. The papers appearing in this volume have been refereed. In order to accomodate as many papers as possible, we requested that the authors limit the length of their papers. We wish to express our thanks to the authors for their efforts and to all of the participants for their contributions to the success of the conference.

The papers appearing in this volume fall roughly into four areas: dynamical systems, bifurcation theory, numerical techniques and applications. In each of these areas the papers focus on aspects of the analysis involving symmetry.

We are pleased to acknowledge the support of several organizations: The *Deutsche Forschungsgemeinschaft*, The State of Hessen, The Ursula Kuhlmann Foundation (Marburg), and The Philipps-University Marburg. Without the support of these organizations this conference would not have been possible. Finally, we wish to thank some of those who had an essential impact on the efficient conduct of the conference, Mrs. Olga Kehraus, Drs Bai Yin (Y. Bai), Mei Zhen (Z. Mei) and several graduate students of the *Fachbereich Mathematik* of the University of Marburg.

International Series of Numerical Mathematics, Vol. 104, © 1992 Birkhäuser Verlag Basel

1

Exploiting Equivariance in the Reduced Bifurcation Equations

Eugene L. Allgower[1,4], Klaus Böhmer[2], Mei Zhen[2,3,5]

1. Introduction

Our aim here is to study and exploit equivariance properties of the reduced bifurcation equations for a class of elliptic boundary value problems. To illustrate the approach we consider specifically the problem

$$(1.1) \qquad G(u, \lambda) := \Delta u + \lambda f(u) = 0$$

with doubly periodic boundary conditions

$$(1.2) \qquad u(x+1, y) = u(x, y+1) = u(x, y) \text{ for } (x, y) \in \mathbf{R}^2.$$

The assumptions concerning the function f are that it is smooth, odd, and normalized so that

$$f'(0) = 1, \ f'''(0) \neq 0.$$

We regard G as a map $G : X \times \mathbf{R} \to Y$ where $Y := \{u \in C^{0,\bullet}(\mathbf{R}^2) | u \text{ satisfies } (1.2)\}$, and $X := Y \cap C^{2,\bullet}(\mathbf{R}^2)$. It is easily checked that G is equivariant for the symmetry group $\Gamma := D_4 \times T_2 \times Z_2$, i.e.

$$G(\gamma u, \lambda) = \gamma G(u, \lambda) \qquad \text{for all } \gamma \in \Gamma, \ (u, \lambda) \in X \times \mathbf{R}.$$

The group Γ acting on any $u \in Y$ is generated by the following elements:

(1.3) **Reflections:** $\qquad S_1 u(x, y) := u(1 - x, y), \quad S_2 u(x, y) := u(y, x).$

(1.4) **Translations:** $\qquad T_{(\theta_1, \theta_2)}(x, y) := u(x - \theta_1, y - \theta_2), \quad \theta_1, \theta_2 (\text{mod } 1) \in (0, 1).$

(1.5) **Sign factors:** $\qquad (\sigma u)(x, y) := \sigma u(x, y), \quad \sigma \in Z_2 := \{-1, 1\}.$

In addition, for $u \in X$, it is easy to verify for the inner product

$$(u, v) := \int_\Omega uv \, dxdy \qquad \text{with } \Omega := [0, 1]^2$$

that the invariance

$$(1.6) \qquad (\gamma u, \gamma v) = (u, v) \text{ for } \gamma \in \Gamma \text{ holds.}$$

[1] Colorado State University, Department of Mathematics, Fort Collins, CO 80523, USA.
[2] Fachbereich Mathematik, Universität Marburg, 3550 Marburg/Lahn, F. R. Germany.
[3] Department of Mathematics, Xi'an Jiaotong University, Xi'an 710049, PRC.
[4] Partially supported by NSF Grant DMS-9104058.
[5] Supported by Deutsch Forschungsgemeinschaft.

Since the reduced bifurcation equations [7], [1] differ somewhat from the classical bifurcation equations, which are known to be equivariant, [3], [8], [5], we verify their equivariance and illustrate this on the example (1.1) (1.2). This equivariance is in turn exploited to factor out linear factors and to establish further invariance and equivariance of the remaining quotient factors.

2. Eigenstructures

Since f is an odd function, $u(x,y) \equiv 0$ is a trivial solution of (1.1) (1.2). We want to study bifurcations from the trivial solution curve $\{(0,\lambda) \mid \lambda \in \mathbf{R}\}$. To this end, we note that the eigenvalues of the linearization of (1.1) (1.2) are given by

$$(2.1) \qquad \Lambda := \{\lambda_{mn} := 4(m^2 + n^2)\pi^2 \neq 0 \mid m,n \text{ non-negative integers}\}.$$

Corresponding to the eigenvalue λ_{mn}, we have the eigenfunctions, see also (2.5),

$$\phi_{mn}(x,y) := \phi_m(x) \otimes \phi_n(y)$$

$$(2.2) \qquad \text{where } \phi_m(x) := \sqrt{2}\begin{pmatrix} \sin 2m\pi x \\ \cos 2m\pi x \end{pmatrix} \text{ and } \otimes \text{ denotes the tensor product.}$$

If $m \neq n$, then $\lambda_{nm} = \lambda_{mn}$ and there are also the corresponding eigenfunctions

$$(2.3) \qquad \phi_{nm}(x,y) := \phi_n(x) \otimes \phi_m(y).$$

If the eigenvalue $\lambda_0 := (m^2 + n^2)\pi^2$ has M representations, then the null space

$$(2.4) \qquad N(D_u(G(0,\lambda_0))) =: N(D_u G_0) = R(D_u G_0)^\perp$$

has dimension $4M$. The most frequent case is $M = 2$, i.e., $m \neq n$ and $m^2 + n^2 \neq p^2 + q^2$ for all $p,q \in \mathbf{N}$ with $(p,q) \neq (m,n)$ and $\neq (n,m)$. A basis for $N(D_u G_0)$ is given by

$$(2.5) \qquad \begin{pmatrix} \phi_1(x,y) \\ \phi_2(x,y) \\ \phi_3(x,y) \\ \phi_4(x,y) \end{pmatrix} := 2\begin{pmatrix} \sin 2m\pi x \sin 2n\pi y \\ \sin 2m\pi x \cos 2n\pi y \\ \cos 2m\pi x \sin 2n\pi y \\ \cos 2m\pi x \cos 2n\pi y \end{pmatrix}, \quad \begin{pmatrix} \phi_5(x,y) \\ \phi_6(x,y) \\ \phi_7(x,y) \\ \phi_8(x,y) \end{pmatrix} := 2\begin{pmatrix} \sin 2n\pi x \sin 2m\pi y \\ \sin 2n\pi x \cos 2m\pi y \\ \cos 2n\pi x \sin 2m\pi y \\ \cos 2n\pi x \cos 2m\pi y \end{pmatrix}.$$

Of course, $M = 1$ occurs when $m = n$ and $2m^2 \neq p^2 + q^2$ for $(p,q) \in \mathbf{N} \times \mathbf{N}\setminus\{(m,m)\}$. However, M can also be arbitrarily large [6]. In any case, when the eigenfunctions $\{\phi_1,\ldots,\phi_{4M}\}$ corresponding to the eigenvalue $\lambda_0 = \lambda_{4M}$ with multiplicity M are enumerated as above, we have $(\phi_i,\phi_j) = \delta_{ij}$ for $i,j = 1,\ldots,4M$.

The actions of Γ in the Γ-invariant subspace $N(D_u G_0)$ as defined by the generating elements in (1.3) - (1.5) induce corresponding actions on the vectors

$$\vec{\phi}^* := (\phi_1,\ldots,\phi_{4M})^*$$

which are represented by corresponding orthogonal $4M \times 4M$ matricies A_γ. We list the A_γ for (1.3) and (1.4). For (1.5) the matricies A_γ are merely $\pm I_{4M}$.

(2.6) $\qquad S_1$ induces $S_1\bar{\phi} = A^{(1)}\bar{\phi}$ where $A^{(1)} := diag(A_1^{(1)}, \ldots, A_1^{(1)})$
$$\text{and } A_1^{(1)} = diag(-1, -1, 1, 1).$$

(2.7)

$$S_2 \text{ induces } S_2\bar{\phi} = A^{(2)}\bar{\phi} \text{ where}$$

$$A^{(2)} = \begin{pmatrix} A_1^{(2)} & 0 & & & \\ 0 & 0 & A_1^{(2)} & & \\ & A_1^{(2)} & \ddots & \ddots & \\ & & \ddots & \ddots & A_1^{(2)} \\ & & A_1^{(2)} & 0 & \end{pmatrix}, \quad A^{(2)} = \begin{pmatrix} 0 & A_1^{(2)} & & & \\ A_1^{(2)} & 0 & \ddots & & \\ & \ddots & \ddots & \ddots & \\ & & \ddots & \ddots & A_1^{(2)} \\ & & & A_1^{(2)} & 0 \end{pmatrix}$$

$(M$ odd the upper left block corresponds to $m = n)$ $\qquad (M$ even$)$

$$\text{and } A_1^{(2)} := \begin{pmatrix} 1 & 0 & 0 & 0 \\ 0 & 0 & 1 & 0 \\ 0 & 1 & 0 & 0 \\ 0 & 0 & 0 & 1 \end{pmatrix}.$$

$$T_{(\theta_1,0)} \text{ induces } T_{(\theta_1,0)}\bar{\phi} = A^{(\theta_1,0)}\bar{\phi} \text{ where } A^{(\theta_1,0)} = diag(A_1^{(\theta_1,0)}, \cdots, A_1^{(\theta_1,0)})$$

$$\text{and } A_1^{(\theta_1,0)} = \begin{pmatrix} \cos 2m\pi\theta_1 & \sin 2m\pi\theta_1 \\ -\sin 2m\pi\theta_1 & \cos 2m\pi\theta_1 \end{pmatrix} \otimes \begin{pmatrix} 1 & 0 \\ 0 & 1 \end{pmatrix}$$

The matrix $A^{(0,\theta_2)}$ induced by $T_{(0,\theta_2)}$ is defined similiarly. For $M = 1$, we list a subgroup of $\{A_\gamma \mid \gamma \in \Gamma\}$ which will be useful below. It is generated by the elements in Table 2.9.

3. The reduced bifurcation equations

At a prospective bifurcation point $(0, \lambda_0)$ $(\lambda_0 = \lambda_{mn})$, we seek a solution curve $(u(t), \lambda(t)) \in X \times \mathbf{R}$ such that:

(3.1) $\quad$ (i) $\quad (u(0), \lambda(0)) =: (u_0, \lambda_0) = (0, \lambda_{mn})$,
$\qquad$ (ii) $\quad (\dot{u}(0), \dot{\lambda}(0)) \neq (0, 0)$,
$\qquad$ (iii) $\quad G(u(t), \lambda(t)) = 0$ on an open interval I with $0 \in I$.

Letting $N(D_u G_0) = \text{span}[\phi_1, \ldots, \phi_{4M}]$, we may paraphrase a theorem given in [2], [7] as follows.

Theorem 3.2. *A nontrivial solution of (1.1) (1.2) satisfying (3.1) is necessarily of the form*

$$(u(t), \ \lambda(t)) = \left(t \sum_{i=1}^{4M} \alpha_i(t)\phi_i + t^3 v(t), \lambda_0 + t^2\beta(t) \right)$$

with $\sum |\alpha_i(0)| > 0$, $v(t) \in R(D_u G_0)$, $t \in I$, $\text{sign}(\beta(0)) = -\text{sign}(f'''(0)) \neq 0$.

As a consequence of the above theorem, we have $G(u(t), \lambda(t)) = 0$ if and only if

Table 2.9

γ	A_γ	γ	$A_\gamma = $ permutation matrix
I	I_4	S_2	(1234)
S_1	$\begin{pmatrix} -1 & & & \\ & -1 & & \\ & & 1 & \\ & & & 1 \end{pmatrix}$	$S_1'T(0,\frac{1}{4m})$	$(2143) \leftrightarrow \begin{pmatrix} 0 & 1 & 0 & 0 \\ 1 & 0 & 0 & 0 \\ 0 & 0 & 0 & 1 \\ 0 & 0 & 1 & 0 \end{pmatrix}$
S_1'	$\begin{pmatrix} 1 & & & \\ & -1 & & \\ & & 1 & \\ & & & -1 \end{pmatrix}$	$S_1S_2T(0,\frac{1}{4m})$	(2413)
S_1S_1'	$\begin{pmatrix} 1 & & & \\ & -1 & & \\ & & 1 & \\ & & & -1 \end{pmatrix}$	$S_1T(\frac{1}{4m},0)$	(3412)
		$S_1'S_2T(\frac{1}{4m},0)$	(3142)
		$S_1S_1'T(\frac{1}{4m},\frac{1}{4m})$	(4321)
		$S_1S_1'S_2T(\frac{1}{4m},\frac{1}{4m})$	(4231)

$$(3.3) \qquad G\Big(\sum_{i=1}^{4M} t\,\alpha_i(t)\phi_i + t^3 v(t), \lambda_0 + t^2\beta(t)\Big)/t^3 = 0$$

for $t \neq 0$, and for $t = 0$, a limit of (3.3) holds

$$(3.4) \qquad D_u G_0 v + \Big[\operatorname{sign}\beta(0) \sum_{i=1}^{4M} \alpha_i\phi_i + \lambda_0 f'''(0)\Big(\sum_{i=1}^{4M} \alpha_i\phi_i \Big)^3 \Big]/6 = 0 (\text{see, e.g. } [2]).$$

Now let $P : X \to R(D_u G_0)$ be the projection defined by

$$(3.5) \qquad Pu := u - \sum_{i=1}^{4M} (\phi_i, u)\phi_i$$

Following the usual ideas of a Lyapunov-Schmidt reduction, we have $(u(t), \lambda(t))$ is a nontrivial solution curve of (1.1) (1.2) if and only if

$$(3.6) \qquad PG(\sum_{i=1}^{4M} t\,\alpha_i(t)\phi_i + t^3 v, \lambda_0 + t^2\beta(t))/t^3 =: G_1(v, \alpha_1, \ldots, \alpha_{4M}, t) = 0,$$

$$(I - P)G\,(\sum_{i=1}^{4M} t\,\alpha_i(t)\phi_i + t^3 v, \lambda_0 + t^2\beta(t))/t^3 =: G_2(v, \alpha_1, \ldots, \alpha_{4M}, t) = 0.$$

Since $P : X \to R(D_u G_0)$, the first equation allows v to be implicitly defined as a function of $\bar{\alpha} := (\alpha_1, \ldots, \alpha_{4M})$, and t in a neighborhood of $t = 0$, and the second equation leads to our *reduced bifurcation equations* (see e.g. [2], [7])

$$(3.7) \qquad (\phi_j, G_2(v(\alpha_1, \ldots, \alpha_{4M}), \alpha_1, \ldots, \alpha_{4M}, t)) = 0, \qquad j = 1, \ldots, 4M.$$

Our aim now is to verify the equivariance of the equation (3.6) (hence also of (3.7)), and to exploit this in the next section. We first check that P is equivariant. Since for $\gamma \in \Gamma$ we have

$$\begin{aligned}
P(\gamma u) &= \gamma u - \sum_{\gamma=1}^{4M} (\phi_j, \gamma u)\phi_j \\
&= \gamma u - \sum (\gamma^{-1}\phi_j, \gamma^{-1}\gamma u)\phi_j \qquad \text{(by (1.6))} \\
&= \gamma(u - \sum (\gamma^{-1}\phi_j, u)\gamma^{-1}\phi_j) = \gamma P u
\end{aligned}$$

since $\{\gamma^{-1}\phi_j\}_{j=1}^{4M}$ is also a basis for $D_u G_0$.

The action of Γ on X also induces a corresponding action on the coefficient vector $\bar{\alpha}$ in the following natural way. For $u \in N(D_u G_0)$ we have $u = \bar{\phi}^* \bar{\alpha}$ and so $\gamma u = (A_\gamma \bar{\phi})^* \bar{\alpha} = \bar{\phi}^*(A_\gamma^* \bar{\alpha})$. So we define

$$(3.8) \qquad \gamma \bar{\alpha} := A_\gamma^* \bar{\alpha} \qquad \text{where } \gamma \bar{\phi} = A_\gamma \bar{\phi}$$

and A_γ is defined as in the previous section. To see that G_1 is equivariant, we check

$$\begin{aligned}
G_1(\gamma v, \gamma \bar{\alpha}, t) &= PG(t\bar{\phi}^* \gamma \bar{\alpha} + t^3 \gamma v, \lambda_0 + \beta(t)t^2)/t^3 \\
&= PG(t\bar{\phi}^* A_\gamma^* \bar{\alpha} + t^3 \gamma v, \lambda_0 + \beta(t)t^2)/t^3 \\
&= PG(\gamma(t\bar{\phi}^* \bar{\alpha} + t^3 v, \lambda_0 + \beta(t)t^2)/t^3 = \gamma G_1(v, \bar{\alpha}, t),
\end{aligned}$$

since P and G are equivariant.

Since $G_1(v, \bar{\alpha}, t) = 0$ and $G_1(v(\gamma\bar{\alpha}), \bar{\alpha}, t) =)$ uniquely define $v(\bar{\alpha})$ and $v(\gamma\bar{\alpha})$, respectively, we have by the equivariance of G_1,

$$G_1(\gamma v(\bar{\alpha}), \gamma \bar{\alpha}, t) = \gamma G_1(v(\bar{\alpha}), \bar{\alpha}, t) = 0 = G_1(v(\gamma\bar{\alpha}), \gamma\bar{\alpha}, t)$$

implies v is equivariant as a function of $\bar{\alpha}$. From this we easily check that G_2 is equivariant too, since

$$\begin{aligned}
G_2(v(\gamma\bar{\alpha}), \gamma\bar{\alpha}, t) &= (I - P)G(t\bar{\phi}^* \gamma\bar{\alpha} + t^3 v(\gamma\bar{\alpha}), \lambda_0 + \beta(t)t^2)/t^3 \\
&= (I - P)G(\gamma(t\bar{\phi}^* \bar{\alpha} + t^3 v(\bar{\alpha}), \lambda_0 + \beta(t)t^2)/t^3 \\
&= \gamma G_2(v(\bar{\alpha}), \bar{\alpha}, t).
\end{aligned}$$

From this we can now also conclude that the reduced bifucation equations (3.7) are also equivariant.

4. Exploiting equivariance in the reduced bifurcation equations

Let us denote the reduced bifurcation equations (3.7) by

$$(4.1) \qquad \bar{g}(\bar{\alpha}, t) = (g_1(\bar{\alpha}, t), \ldots, g_{4M}(\bar{\alpha}, t)) := \left(\phi_j, G_2(v(\bar{\alpha}), \bar{\alpha}, t) \right)_{j=1}^{4M} = 0.$$

Recently it has been noted in the case of polynomial systems of equations having equivariance properties, it is possible to extract linear factors out of components of the equations, see e.g. [4]. In this section we obtain similar results for the reduced bifurcation equations and examine the properties which are further inherited by the remaining quotients. We recall from the discussions in section 2, that the equivariance of (4.1) implies

$$(4.2) \qquad \bar{g}(\bar{\alpha}, t) = A_\gamma^{*-1} \bar{g}(A_\gamma^* \bar{\alpha}, t) \quad \text{for all} \quad \gamma \in \Gamma, \ \bar{\alpha} \in \mathbf{R}^{4M}.$$

To make the following discussion simpler, we take $M = 1$. For $M > 1$, analogous results follow similarly. The subgroup of $\{A_\gamma \mid \gamma \in \Gamma\}$ which we use for splitting factors from $\bar{g}(\bar{\alpha}, t) = 0$ is generated by the elements of Table 2.9. From (4.2) with $\gamma = S_1 S_1' S_2 T(\frac{1}{4m}, \frac{1}{4m})$ we have

$$(4.3) \qquad \bar{g}(\bar{\alpha}, t) = \begin{pmatrix} g_4(\alpha_4, \alpha_2, \alpha_3, \alpha_1, t) \\ g_2(\alpha_4, \alpha_2, \alpha_3, \alpha_1, t) \\ g_3(\alpha_4, \alpha_2, \alpha_3, \alpha_1, t) \\ g_1(\alpha_4, \alpha_2, \alpha_3, \alpha_1, t) \end{pmatrix} \quad \text{for } \bar{\alpha} \in \mathbf{R}^4, \ t \in I.$$

For simplicity we supress the t–dependency. From (4.3) we have $\bar{g}(\bar{\alpha}) = 0$ implies

$$0 = g_1(\bar{\alpha}) - g_4(\bar{\alpha}) \equiv g_1(\alpha_1, \alpha_2, \alpha_3, \alpha_4) - g_1(\alpha_4, \alpha_2, \alpha_3, \alpha_1)$$
$$= g_1(\alpha_1, \alpha_2, \alpha_3, \alpha_4) - g_1(\alpha_4, \alpha_2, \alpha_3, \alpha_4) + g_1(\alpha_4, \alpha_2, \alpha_3, \alpha_4) - g_1(\alpha_4, \alpha_2, \alpha_3, \alpha_1).$$

By the mean value theorem we have

$$
\begin{aligned}
(4.4) \qquad g_1(\bar{\alpha}) - g_4(\bar{\alpha}) &= (\alpha_1 - \alpha_4) \int_0^1 \partial_1 g_1(\alpha_4 + \tau(\alpha_1 - \alpha_4), \alpha_2, \alpha_3, \alpha_4) \, d\tau \\
&\quad + (\alpha_4 - \alpha_1) \int_0^1 \partial_4 g_1(\alpha_4, \alpha_2, \alpha_3, \alpha_1 + \tau(\alpha_4 - \alpha_1)) \, d\tau \\
&= (\alpha_1 - \alpha_4) \int_0^1 [\partial_1 g_1(\alpha_4 + \tau(\alpha_1 - \alpha_4), \alpha_2, \alpha_3, \alpha_4) \\
&\quad - \partial_4 g_1(\alpha_1, \alpha_2, \alpha_3, \alpha_4 + \tau(\alpha_1 - \alpha_4))] \, d\tau \\
&=: (\alpha_1 - \alpha_4) Q_1(\alpha_1, \alpha_2, \alpha_3, \alpha_4).
\end{aligned}
$$

From (4.2) with $\gamma = -S_2 T(\frac{1}{4m}, \frac{1}{4m})$ we have

$$(4.5) \qquad \bar{g}(\bar{\alpha}) = \begin{pmatrix} -g_4(-\alpha_4, \alpha_2, \alpha_3, -\alpha_1) \\ g_2(-\alpha_4, \alpha_2, \alpha_3, -\alpha_1) \\ g_3(-\alpha_4, \alpha_2, \alpha_3, -\alpha_1) \\ -g_1(-\alpha_4, \alpha_2, \alpha_3, -\alpha_1) \end{pmatrix}.$$

From (4.5) we have $\bar{g}(\bar{\alpha}) = 0$ implies

$$
\begin{aligned}
0 = g_1(\bar{\alpha}) + g_4(\bar{\alpha}) &\equiv g_1(\alpha_1, \alpha_2, \alpha_3, \alpha_4) - g_1(-\alpha_4, \alpha_2, \alpha_3, -\alpha_1) \\
&= g_1(\alpha_1, \alpha_2, \alpha_3, \alpha_4) - g_1(-\alpha_4, \alpha_2, \alpha_3, \alpha_4) + g_1(-\alpha_4, \alpha_2, \alpha_3, \alpha_4) - g_1(-\alpha_4, \alpha_2, \alpha_3, -\alpha_1).
\end{aligned}
$$

By the mean value theorem we have

$$
\begin{aligned}
(4.6) \qquad g_1(\bar{\alpha}) + g_4(\bar{\alpha}) &= (\alpha_1 + \alpha_4) \int_0^1 [\partial_1 g_1(-\alpha_4 + \tau(\alpha_1 + \alpha_4), \alpha_2, \alpha_3, \alpha_4) \\
&\qquad\qquad + \partial_4 g_1(-\alpha_4, \alpha_2, \alpha_3, -\alpha_1 + \tau(\alpha_1 + \alpha_4))] \, d\tau \\
&=: (\alpha_1 + \alpha_4) Q_4(\alpha_1, \alpha_2, \alpha_3, \alpha_4)
\end{aligned}
$$

By (4.2) with $A_\gamma = \begin{pmatrix} 0 & 0 & 0 & -1 \\ 0 & 1 & 0 & 0 \\ 0 & 0 & -1 & 0 \\ 1 & 0 & 0 & 0 \end{pmatrix}$ corresponding to $\gamma = S_1 S_2 T(\frac{1}{4m}, \frac{1}{4m})$ we have

$$
\bar{g}(\bar{\alpha}) = \begin{pmatrix} -g_4(\alpha_4, \alpha_2, -\alpha_3, -\alpha_1) \\ g_2(\alpha_4, \alpha_2, -\alpha_3, -\alpha_1) \\ -g_3(\alpha_4, \alpha_2, -\alpha_3, -\alpha_1) \\ g_1(\alpha_4, \alpha_2, -\alpha_3, -\alpha_1) \end{pmatrix}.
$$

By $\bar{g}(\bar{\alpha}) = 0$ we have

$$
\begin{aligned}
(4.7) \qquad 0 = g_1(\bar{\alpha}) + g_4(\bar{\alpha}) &= g_1(\alpha_4, \alpha_2, -\alpha_3, -\alpha_1) - g_4(\alpha_4, \alpha_2, -\alpha_3, -\alpha_1)] \\
&= (\alpha_4 + \alpha_1) Q_1(\alpha_4, \alpha_2, -\alpha_3, -\alpha_1) \qquad \text{(by (4.4))}.
\end{aligned}
$$

By (4.6) and (4.7) we have

$$
(4.8) \qquad Q_4(\alpha_1, \alpha_2, \alpha_3, \alpha_4) = Q_1(\alpha_4, \alpha_2, -\alpha_3, -\alpha_1).
$$

With $\gamma = S_1' T(0, \frac{1}{4m})$ and arguing in the same way as above, we obtain

$$
(4.9) \qquad g_2(\bar{\alpha}) - g_3(\bar{\alpha}) = (\alpha_2 - \alpha_3) Q_1(\alpha_2, \alpha_1, \alpha_4, \alpha_3).
$$

Also from

$$
\begin{aligned}
(4.10) \qquad g_2(\bar{\alpha}) + g_3(\bar{\alpha}) &= g_1(\alpha_2, \alpha_1, \alpha_4, \alpha_3) + g_4(\alpha_2, \alpha_1, \alpha_4, \alpha_3) \\
&= (\alpha_2 + \alpha_3) Q_4(\alpha_2, \alpha_1, \alpha_4, \alpha_3) \\
&= (\alpha_2 + \alpha_3) Q_1(\alpha_3, \alpha_1, -\alpha_4, -\alpha_2) \qquad \text{(by (4.8))}.
\end{aligned}
$$

Summarizing the results (4.4) (4.7) (4.9) (4.10) we have

$$
(4.11) \qquad B\bar{g}(\bar{\alpha}) = \begin{pmatrix} (\alpha_1 - \alpha_4) Q_1(\alpha_1, \alpha_2, \alpha_3, \alpha_4) \\ (\alpha_2 - \alpha_3) Q_1(\alpha_2, \alpha_1, \alpha_4, \alpha_3) \\ (\alpha_2 + \alpha_3) Q_1(\alpha_3, \alpha_1, -\alpha_4, -\alpha_2) \\ (\alpha_1 + \alpha_4) Q_1(\alpha_4, \alpha_2, -\alpha_3, -\alpha_1) \end{pmatrix} = 0
$$

where $B = \begin{pmatrix} 1 & 0 & 0 & -1 \\ 0 & 1 & -1 & 0 \\ 0 & 1 & 1 & 0 \\ 1 & 0 & 0 & 1 \end{pmatrix}$ and Q_1 is the quotient factor defined in (4.4).

To simplify our notation, let us set $Q := Q_1$. From (4.11) we have

$$(4.12) \qquad \bar{g}(\bar{\alpha}) = \frac{1}{2} \begin{pmatrix} (\alpha_1 - \alpha_4)Q(\alpha_1, \alpha_2, \alpha_3, \alpha_4) + (\alpha_1 + \alpha_4)Q(\alpha_4, \alpha_2, -\alpha_3, -\alpha_1) \\ (\alpha_2 - \alpha_3)Q(\alpha_2, \alpha_1, \alpha_4, \alpha_3) + (\alpha_2 + \alpha_3)Q(\alpha_3, \alpha_1, -\alpha_4, -\alpha_2) \\ -(\alpha_2 - \alpha_3)Q(\alpha_2, \alpha_1, \alpha_4, \alpha_3) + (\alpha_2 + \alpha_3)Q(\alpha_3, \alpha_1, -\alpha_4, -\alpha_2) \\ -(\alpha_1 - \alpha_4)Q(\alpha_1, \alpha_2, \alpha_3, \alpha_4) + (\alpha_1 + \alpha_4)Q(\alpha_4, \alpha_2, -\alpha_3, -\alpha_1) \end{pmatrix}.$$

Using (4.12) and the equivariance of $\bar{g}$ with $A_\gamma = diag(1, 1, -1, -1)$, equating the coefficients leads to

$$(4.13) \qquad Q(\bar{\alpha}) = Q(-\alpha_4, \alpha_3, \alpha_2, -\alpha_1), \quad Q(-\alpha_4, \alpha_2, -\alpha_3, \alpha_1) = Q(\alpha_1, \alpha_2, -\alpha_3, -\alpha_4).$$

Other equivariant properties of Q can be derived similarly by using other group elements. However, without further specific information regarding the coefficients in the reduced bifurcation equations (3.6), (3.7), it seems that merely formal exploitation of equivariance does not yield further results beyond the structure given by (4.11) with (4.13), and so on. The difficulty stems from the fact that explicit information regarding the remainder function v is not available in this general case. In the next section we consider the special case $t = 0$.

5. The special case $t = 0$, $M = 1$

When $t = 0$, the remainder function v in (3.6) (3.7) plays no role. The coefficients in (3.7) can be computed similarly as Lemma 2.1 in [2], and lead to the equation

$$(5.1) \qquad \bar{g}(\bar{\alpha}, 0) = \begin{pmatrix} \alpha_1[1 - b(3\alpha_1^2 + 3\alpha_2^2 + 3\alpha_3^2 + \alpha_4^2)] - 2b\alpha_2\alpha_3\alpha_4 \\ \alpha_2[1 - b(3\alpha_1^2 + 3\alpha_2^2 + \alpha_3^2 + 3\alpha_4^2)] - 2b\alpha_1\alpha_3\alpha_4 \\ \alpha_3[1 - b(3\alpha_1^2 + \alpha_2^2 + 3\alpha_3^2 + 3\alpha_4^2)] - 2b\alpha_1\alpha_2\alpha_4 \\ \alpha_4[1 - b(\alpha_1^2 + 3\alpha_2^2 + 3\alpha_3^2 + 3\alpha_4^2)] - 2b\alpha_1\alpha_2\alpha_3 \end{pmatrix} = 0,$$

where $b := \lambda_0 \mid f'''(0)\mid/8$ and $\lambda(t)$ is locally parameterized as $\lambda_0 + at^2$, $a := -\text{sign}(f'''(0))$. Once again suppressing the t–dependence we the form (4.11)

$$B\bar{g}(\bar{\alpha}) = \begin{pmatrix} (\alpha_1 - \alpha_4)[1 - 3b(\alpha_1^2 + \alpha_2^2 + \alpha_3^2 + \alpha_4^2) + 2b(\alpha_2\alpha_3 - \alpha_1\alpha_4)] \\ (\alpha_2 - \alpha_3)[1 - 3b(\alpha_1^2 + \alpha_2^2 + \alpha_3^2 + \alpha_4^2) - 2b(\alpha_2\alpha_3 - \alpha_1\alpha_4)] \\ (\alpha_2 + \alpha_3)[1 - 3b(\alpha_1^2 + \alpha_2^2 + \alpha_3^2 + \alpha_4^2) + 2b(\alpha_2\alpha_3 - \alpha_1\alpha_4)] \\ (\alpha_1 + \alpha_4)[1 - 3b(\alpha_1^2 + \alpha_2^2 + \alpha_3^2 + \alpha_4^2) - 2b(\alpha_2\alpha_3 - \alpha_1\alpha_4)] \end{pmatrix}$$

$$(5.2)$$

$$=: \begin{pmatrix} (\alpha_1 - \alpha_4)Q_1(\bar{\alpha}) \\ (\alpha_2 - \alpha_3)Q_2(\bar{\alpha}) \\ (\alpha_2 + \alpha_3)Q_1(\bar{\alpha}) \\ (\alpha_1 + \alpha_4)Q_2(\bar{\alpha}) \end{pmatrix} = 0.$$

There are three essential cases to consider:

I. $Q_1(\bar{\alpha}) \neq 0 \neq Q_2(\bar{\alpha})$. In this case $diag(Q_1(\bar{\alpha})^{-1}, Q_2(\bar{\alpha})^{-1}, Q_1(\bar{\alpha})^{-1}, Q_2(\bar{\alpha})^{-1})Bg(\bar{\alpha}) = (\alpha_1 - \alpha_4, \alpha_2 - \alpha_3, \alpha_2 + \alpha_3, \alpha_1 + \alpha_4)^* = 0$. This obviously yields the trivial solution $\bar{\alpha} = 0$.

II. $Q_1(\bar{\alpha}) = 0 \neq Q_2(\bar{\alpha})$. Then $\alpha_2 - \alpha_3 = 0 = \alpha_1 + \alpha_4$, and for a nontrivial solution we require $\alpha_1 - \alpha_4 \neq 0$ or $\alpha_2 + \alpha_3 \neq 0$.

There are two essential subcases to consider:

(a) $\alpha_1 - \alpha_4 = 0 \quad \alpha_2 + \alpha_3 \neq 0$. Then we have $\alpha_1 = \alpha_4 = 0$ and $Q_1(0, \alpha_2, \alpha_2, 0) = 0$. Hence

$$1 - 3b(2\alpha_2^2) + 2b\alpha_2^2 = 0 \text{ or } \alpha_2 = \pm\frac{1}{2\sqrt{b}}$$

corresponding to the element $\sin 2m\pi(x + y)$ in $N(D_u G_0)$. Now we have a bifurcating branch with the properties $u = 0$ on $x + y = \frac{k}{2m}, k = 0, \pm 1, \ldots$ and $\frac{\partial u}{\partial \nu} = 0$ for $\nu = (1, -1)$. A similar result holds for $(\alpha_1 - \alpha_4) \neq 0$ and $(\alpha_2 + \alpha_3) = 0$. In this case we have bifurcating branches having so-called hidden symmetries.

(b) $(\alpha_1 - \alpha_4)(\alpha_2 + \alpha_3) \neq 0$. Then we have $Q_1(\alpha_1, \alpha_2, \alpha_2, -\alpha_1) = 0$ or $\alpha_1^2 + \alpha_2^2 = \frac{1}{4b}$. This defines a 1-manifold of solutions for which there is a hidden symmetry defined by $u = 0$ on $x + y = \frac{k}{2m} \pm \tan^{-1}\frac{\alpha_1}{\sqrt{1/4b - \alpha_1^2}}$.

III. $Q_1(\bar{\alpha}) = 0 = Q_2(\bar{\alpha})$. These equations yield a 2-manifold which is equivalently described by

$$\alpha_1\alpha_4 - \alpha_2\alpha_3 = 0.$$

$$\alpha_1^2 + \alpha_2^2 + \alpha_3^2 + \alpha_4^2 = \frac{1}{3b}.$$

For $t \neq 0$ it is however, not yet established if the corresponding 2-manifold of bifurcating branches exists, except in some special cases in which (1.1) is restricted to subspaces of X.

The following example shows that a map may have the same equivariance as g in (5.1), but not necessarily the same elimination of quotient factors as in (5.2),

$$\begin{pmatrix} h_1(\bar{\alpha}) \\ h_2(\bar{\alpha}) \\ h_3(\bar{\alpha}) \\ h_4(\bar{\alpha}) \end{pmatrix} := \begin{pmatrix} \alpha_1 \\ \alpha_2 \\ \alpha_3 \\ \alpha_4 \end{pmatrix} - \begin{pmatrix} \alpha_1(\alpha_1^2 + 2\alpha_2^2 + 2\alpha_3^2 + \alpha_4^2) \\ \alpha_2(2\alpha_1^2 + \alpha_2^2 + \alpha_3^2 + 2\alpha_4^2) \\ \alpha_3(2\alpha_1^2 + \alpha_2^2 + \alpha_3^2 + 2\alpha_4^2) \\ \alpha_4(\alpha_1^2 + 2\alpha_2^2 + 2\alpha_3^2 + \alpha_4^2) \end{pmatrix} + \begin{pmatrix} \alpha_2\alpha_3\alpha_4 \\ \alpha_1\alpha_3\alpha_4 \\ \alpha_1\alpha_2\alpha_4 \\ \alpha_1\alpha_2\alpha_3 \end{pmatrix}.$$

Now

$$g_1(\bar{\alpha}) - g_4(\bar{\alpha}) = (\alpha_1 - \alpha_4)[1 - (\alpha_1^2 + 2\alpha_2^2 + 2\alpha_3^2 + \alpha_4^2) - \alpha_2\alpha_3],$$

$$g_2(\bar{\alpha}) + g_3(\bar{\alpha}) = (\alpha_2 + \alpha_3)[1 - (2\alpha_1^2 + \alpha_2^2 + \alpha_3^2 + 2\alpha_4^2) + \alpha_1\alpha_4].$$

6. An example

Based on the information of solutions of the reduced bifurcation equations at $t = 0$, one may determine the expected solution manifolds of (1.1), (1.2) by considering the reduced problem in the subspaces of X. For example, for $M = 2$ and solutions of (4.2) in the form

$(\alpha_1, 0, 0, \alpha_1, 0)$, we choose the subgroup $\Sigma := \{(\theta_1, \theta_2) \in T_2 \mid m\theta_1 - n\theta_2 \ (\bmod \ 1/2) = 0\}$ of Γ and consider the reduced problem

$$(6.1) \qquad Find \quad (u, \lambda) \in X^\Sigma \times \mathbf{R}, \quad such \ that \quad G^\Sigma(u, \lambda) = 0.$$

From (2.4), (2.5) one sees directly that $(0, \lambda_0)$ is a simple bifurcation point of (6.1) and

$$(6.2) \qquad N(D_u G_0^\Sigma) = \mathrm{span}\,[\phi_1 + \phi_4].$$

Thus, the reduced problem (6.1) has exactly one nontrivial solution branch $(u(t), \lambda(t))$ passing through $(0, \lambda_0)$ and

$$(6.3) \qquad (u(t),\ \lambda(t)) = (t\alpha_1(t)(\phi_1 + \phi_4) + t^3 v(t), \lambda_0 + at^2)$$

with $\alpha_1(0) \neq 0$, $v(t) \in R(D_u G_0^\Sigma)$, $t \in I$, $a := -\mathrm{sign}\,(f'''(0))$. Similarly, if we choose the above translation subgroup with negative sign factor, i.e. $\tilde{\Sigma} := -\Sigma$, Then

$$(6.4) \qquad N(D_u G_0^{\tilde{\Sigma}}) = \mathrm{span}\,[\phi_2 - \phi_3].$$

The correspnonding nontrivial solution branches $(u(t), \lambda(t))$ passing through $(0, \lambda_0)$ is in the form

$$(6.5) \qquad (u(t),\ \lambda(t)) = (t\alpha_2(t)(\phi_2 - \phi_3) + t^3 v(t), \lambda_0 + at^2)$$

with $\alpha_2(0) \neq 0$, $v(t) \in R(D_u G_0^{\tilde{\Sigma}})$, $t \in I$, $a := -\mathrm{sign}\,(f'''(0))$.

Other symmetric solution branches can be discussed analoguously by using the equivariant branching lemma or its extensions, see e.g. [1], [3], [5], [8].

References

1 E. L. Allgower, K. Böhmer, and Z. Mei. An extended equivariant branching theory. Preprint, University of Marburg, Fed. Rep. Germany, to appear in *Math. Meth. Appl. Sci.*, 1991.

2 E. L. Allgower, K. Böhmer, Z. Mei. A complete bifurcation scenario for the 2d-nonlinear Laplacian with Neumann boundary conditions on the unit square, in: *Bifurcation and Chaos: Analysis, Algorithms, Applications*, R. Seydel, F. W. Schneider, T. Küpper, H. Troger (Eds.), ISNM 97, pp. 1-18, Birkhäuser Verlag, Basel 1991.

3 G. Cicogna. Symmetry breakdown from bifurcation. *Lett. Nuovo Cimento*, 31:600–602, 1981.

4 K. Gatermann. Symbolic solution of polynomial equation systems with symmetry, in: *Proceedings of ISSAC-90*, S. Watanabe, M. Nagata (Eds.), pp. 112-119, ACM, New York 1990.

5 M. Golubitsky, I. Stewart, and D. G. Schaeffer. *Singularities and Groups in Bifurcation Theory, 2*, Springer-Verlag, Berlin, Heidelberg, New York, 1988.

6 J. R. Kuttler and V. G. Sigillito. Eigenvalues of the Laplacian in two dimensions. *SIAM Rev.*, 26:163–193, 1984.

7 Z. Mei. Bifurcations of a simplified buckling problem and the effect of discretizations. *Manuscripta Math.*, 71:225–252, 1991.

8 A. Vanderbauwhede. *Local Bifurcation Theory and Symmetry*. Pitman, London, 1982.

The homoclinic twist bifurcation point

D. G. Aronson
School of Mathematics
University of Minnesota
Minneapolis
MN 55455
USA

S.A. van Gils
Faculty of Applied Mathematics
University of Twente
P.O. Box 217
7500 AE Enschede
The Netherlands

M. Krupa
Department of Mathematics
State University of Groningen
P.O. Box 800
9700 AV Groningen
The Netherlands

January 28, 1992

Abstract

We analyze bifurcations occurring in the vicinity of a homoclinic twist point for a generic two parameter family of $\mathbf{Z}_2$ equivariant ODE's in four dimensions. The results are compared with numerical results for a system of two coupled Josephson junctions with pure capacitive load.

1980 Mathematics Subject Classification: 58F22, 58F14, 34C25.

1 Introduction

In a recent article Aronson, Golubitsky and Krupa [AGK91] have studied a class of differential equations describing the dynamics of large series arrays of Josephson junctions coupled through an external load. An important feature of the equations is $\mathbf{S}_n$ symmetry realized through permutation of the junctions. The problem addressed in the article is the existence, stability and bifurcations of periodic solutions invariant under the $\mathbf{S}_n$ symmetry. A numerical study indicates that such solutions exist in a region of the two parameter space bounded on one side by a curve of homoclinic solutions. This curve contains an interesting codimension two point which can be roughly characterized in the following way. Consider a unique solution of the linearized flow around the homoclinic orbit whose direction limits in backward time on the most unstable direction of the saddle. In the forward time limit this solution approaches precisely the same direction or exactly the opposite direction. [AGK91] show numerically that as parameters vary along the curve of homoclinic solutions a transition occurs between these two cases. They call a point of such transition a *homoclinic twist point* and show that in every neighborhood of this point there must be points at which the symmetric periodic orbit has a Floquet multiplier $+1$ and a Floquet multiplier -1.

In this article we analyze bifurcations occurring in the vicinity of a homoclinic twist point for a generic two parameter family of $\mathbf{Z}_2$ equivariant ODE's in four dimensions.

Twist point bifurcations for generic vector fields (no symmetry) have received recently a fair amount of attention. They were studied in particular by Yanagida [Yan87], who asserted that they provide a route to homoclinic doubling. Homoclinic doubling is also considered in Chow, Deng and Fiedler [CDF89]. Deng [Den92] has recently described a scenario by which twist point bifurcations can lead to cusp horseshoes, hence complex dynamical behavior. We believe that similar phenomena are also present in the symmetric context.

Consider

$$\dot{x} = F(x, \alpha, \lambda), \tag{1.1}$$

where $x \in \mathbf{R}^4$ and $F : \mathbf{R}^{\times} \mathbf{R}^2 \to \mathbf{R}^4$ is a C^∞ smooth function.

H1 F commutes with the action of $\mathbf{Z}_2$ given by $\kappa(x_1, y_1, x_2, y_2) = (x_1, y_1, -x_2, -y_2)$.

Let $S = \mathbf{R}^2 \times \{0\}$, and $R = \{0\} \times \mathbf{R}^2$. In S, κ acts trivially and in R, κ acts as $-\text{Id}$. Equivariance of F implies that S is invariant under the flow of F. We consider families with the following additional property:

H2 For $\alpha = 0$ (1.1) has a smooth family of homoclinic orbits $\gamma_\lambda(t)$ to a saddle s_λ and $\gamma_\lambda(t) \in S$.

Let

$$\Gamma_\lambda = \{\gamma_\lambda(t) : t \in \mathbf{R}\}. \tag{1.2}$$

Symmetry implies that $DF(s_\lambda, 0, \lambda)$ has the form

$$DF(s_\lambda, 0, \lambda) = \begin{pmatrix} A_\lambda & 0 \\ 0 & B_\lambda \end{pmatrix}, \tag{1.3}$$

where A_λ, B_λ are 2×2 blocks. Similarly

$$DF(\gamma_\lambda(t), 0, \lambda) = \begin{pmatrix} A_\lambda(t) & 0 \\ 0 & B_\lambda(t) \end{pmatrix}. \tag{1.4}$$

Clearly A_λ must have one negative and one positive eigenvalue. We make the following additional hypothesis.

H3 B_λ has one positive and one negative eigenvalue.

Suppose F satisfies $H1$-$H3$. Let ν_+ (ν_-) be the eigenvector corresponding to positive (negative) eigenvalue of B_λ. Consider the linearized equation

$$\dot{y} = DF(\gamma_\lambda(t), 0, \lambda)y. \tag{1.5}$$

For $|t|$ large the behavior of solutions of (1.5) is qualitatively the same as the behavior of the solutions of

$$\dot{y} = DF(s_\lambda, 0, \lambda)y. \tag{1.6}$$

It follows that there exists a unique solution of (1.5), $y_\lambda(t)$ such that $\lim_{t\to-\infty}\frac{y_\lambda(t)}{\|y_\lambda(t)\|} = v_+$ and $\|y_\lambda(0)\| = 1$. Let $y_\infty(\lambda) = \lim_{t\to\infty}\frac{y_\lambda(t)}{\|y_\lambda(t)\|}$. Generically $y_\infty(\lambda) = \pm v_+$. We say that the point $(\alpha,\lambda) = (0,\lambda_0)$ is a *homoclinic twist point* if $y_\infty(\lambda)$ points in the opposite directions for $\lambda > \lambda_0$ and $\lambda < \lambda_0$.

Remark. There exists also a unique solution of (1.5), $z_\lambda(t)$ such that $\lim_{t\to\infty}\frac{z_\lambda(t)}{\|z_\lambda(t)\|} = v_-$ and $\|z_\lambda(0)\| = 1$. Let $z_{-\infty}(\lambda) = \lim_{t\to-\infty}\frac{z_\lambda(t)}{\|z_\lambda(t)\|}$. Generically $z_{-\infty}(\lambda) = \pm v_-$. The invariance of S and the fact that a flow has to preserve orientation imply that $z_{-\infty}(\lambda)$ and $y_\infty(\lambda)$ point to the same side of v_- and v_+ respectively. It follows that if $(0,\lambda_0)$ is a twist point then $z_{-\infty}(\lambda)$ must point to the opposite directions of v_- for $\lambda > \lambda_0$ and $\lambda < \lambda_0$.

It is well known that under mild genericity conditions (1.1) has for each (α,λ) on one side of the line $\alpha = 0$ a unique periodic orbit $P_{(\alpha,\lambda)} \subset S$. The orbits $P_{(\alpha,\lambda)}$ limit on Γ_λ as $\alpha \to 0$. In our analysis we will also encounter periodic orbits lying outside of S. Such orbits can many times enter and leave a neighborhood of the saddle s_λ before coming back to the initial condition. Let V_λ be a small neighborhood of s_λ. We say that a periodic orbit $p(t)$ with minimal period T is a *1-periodic orbit* if the set

$$PV = \{p(t) : t \in [0,T]\} \cap V_\lambda$$

is connected. We say that $p(t)$ is a *2-periodic orbit* if PV has precisely two connected components. We now state the two main theorems of the notes.

Theorem 1.1 *Suppose F satisfies H1-H3 and $(\alpha,\lambda) = 0$ is a twist point. Then, generically, the following statements hold.*

(i). $P_{(\alpha,\lambda)}$ has a pitchfork bifurcation along a continuous curve
$\{(\tau_1(\lambda),\lambda) : \lambda \in [0,\lambda^*), \tau_1(0) = 0\}$.

(ii). $P_{(\alpha,\lambda)}$ has a period doubling bifurcation along a continuous curve
$\{(\tau_2(\lambda),\lambda) : \lambda \in [0,\lambda^*), \tau_2(0) = 0\}$.

(iii). (1.1) has a nonsymmetric homoclinic orbit (not contained in S) along a smooth curve
$\{(\tau_3(\lambda),\lambda) : \lambda \in [0,\lambda^*), \tau_3(0) = 0\}$. *This orbit limits on Γ_0 as $(\alpha,\lambda) \to (0,0)$.*

Theorem 1.2 *For a generic F satisfying the assumptions of Theorem 1.1 there exist regions D_i, $i \in \{1,2\}$ whose boundaries are given by τ_i, $i \in \{1,2\}$ with the following properties*

(i). For each (α,λ) in D_1 there exists a unique 1-periodic orbit and for each (α,λ) in D_2 there exists a unique 2-periodic orbit.

(ii). The periods of the periodic orbits in the regions D_1 and D_2 depend continuously on (α,λ) and tend to ∞ as $(\alpha,\lambda) \to \tau_3$.

Figure 1 shows the four possible relative positions of the curves τ_i and the regions λ_i.

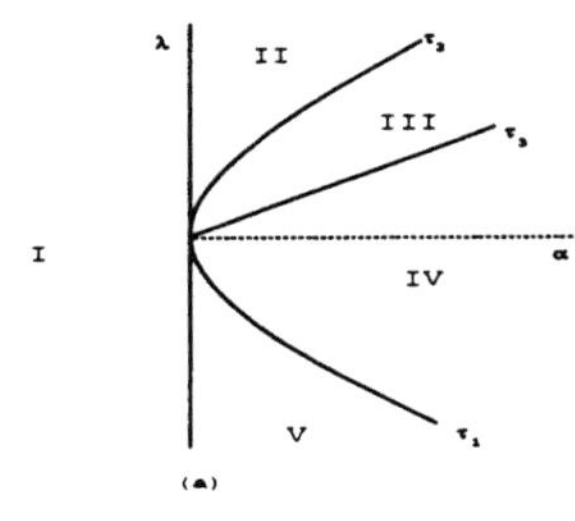

	$(P_{(\alpha,\lambda)},0,0)$	1-per	2-per
I	0	0	0
II	1	0	0
III	1	0	1
IV	1	1	0
V	1	0	0

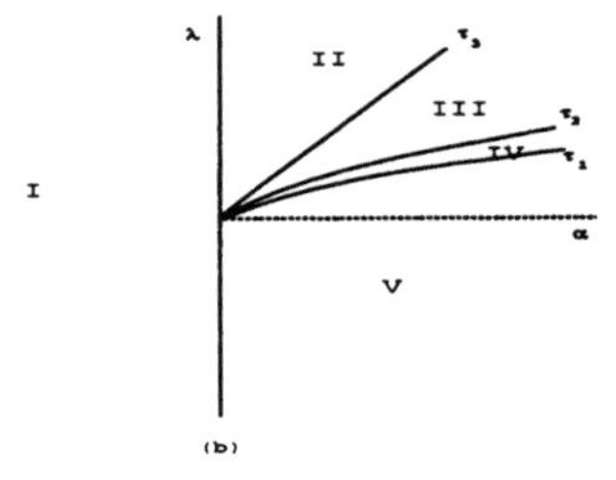

	$(P_{(\alpha,\lambda)},0,0)$	1-per	2-per
I	0	0	0
II	1	0	0
III	1	1	1
IV	1	1	0
V	1	0	0

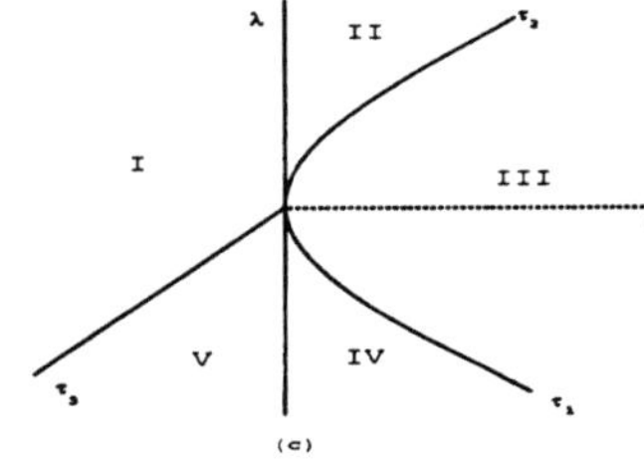

	$(P_{(\alpha,\lambda)},0,0)$	1-per	2-per
I	0	0	1
II	1	0	1
III	1	0	0
IV	1	1	0
V	0	1	0

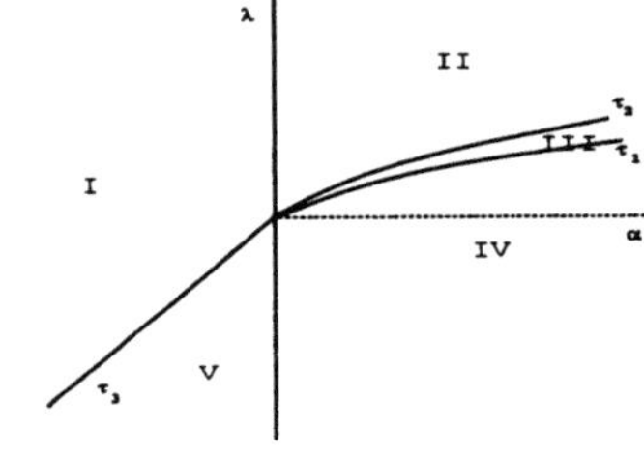

	$(P_{(\alpha,\lambda)},0,0)$	1-per	2-per
I	0	1	1
II	1	1	1
III	1	1	0
IV	1	0	0
V	0	0	0

Figure 1.

Remark. The precise genericity conditions in Theorems 1 and 2 will be stated in Section 3.

The article is organized as follows. In section 2 we describe the procedure of finding a return map for the flow near the homoclinic orbit. We demonstrate how this map is affected by symmetry and explain the role of the parameters. In Section 3 we reformulate and prove Theorems 1 and 2 as results pertaining to the return map. The rest of Section 3 and Section 4 contain the proofs of these results. Section 5 compares the predictions derived from our analysis with the numerical results for the Josephson junction equations.

2 Return map near the homoclinic orbit Γ

Let e_i, $i = 1,\ldots,4$ be the standard basis vectors in $\mathbf{R}^4$. We assume that $s_\lambda \equiv 0$ and that e_i, $i = 1,\ldots,4$ are the eigendirections for $dF(0,(0,\lambda))$. Let (x_1, y_1) be coordinates in R. We assume that x_1, x_2 correspond to the stable directions and y_1, y_2 to the unstable directions. We now make the following hypothesis:

Hypothesis H4 The eigenvalues of $dF(0,(0,0))$ are nonresonant to order k, with $k \geq 3$.

It follows from from (H4) and a theorem of Sternberg (see [Ste57]) that for (α, λ) near $(0,0)$ the flow of (1.1) can be linearized by a C^k smooth change of coordinates with k large. Hence we assume that there exists a neighborhood V of 0 in $\mathbf{R}^4$ where the flow of (1.1) is given by the flow of (1.2). We define hyperplanes

$$\Sigma^{\text{out}} = \{(x_1, \delta, x_2, y_2)\}$$

$$\Sigma^{\text{in}} = \{(\delta, y_1, x_2, y_2)\}.$$

If we choose δ small enough so that $(0,\delta,0,0) \in V$ and $(\delta,0,0,0) \in V$ then $(0,\delta,0,0) \in \Gamma$ and $(\delta,0,0,0) \in \Gamma$. By scaling the variables we can achieve that $\delta = 1$.

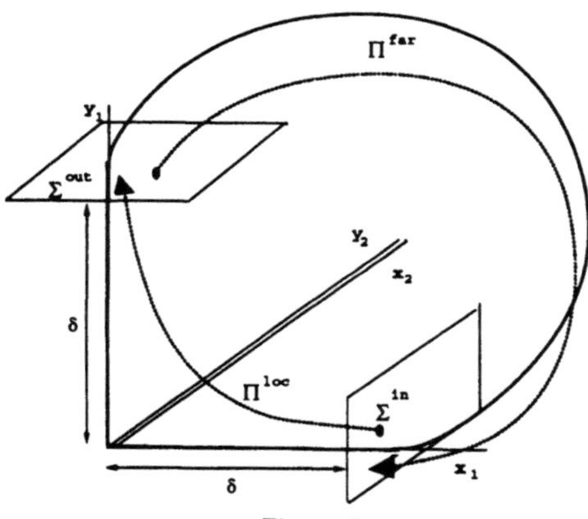

Figure 2

We assume that the stable eigenvalue in S has been scaled to -1. Let ν_1, ν_2 denote the eigenvalues corresponding to e_2 and e_4 and $-\mu$ the eigenvalue corresponding to e_3. The linear flow on Σ^{in} is given by

$$(1, y_1^{\text{in}}, y_2^{\text{in}}, y_2^{\text{in}}) \overset{\varphi_s}{\mapsto} (e^{-s}, e^{\nu_1 s} y_1^{\text{in}}, e^{-\mu s} x_2^{\text{in}}, e^{\nu_2 s} y_2^{\text{in}}).$$

We let Π^{loc} be the first hit map from Σ^{in} to Σ^{out}. The range of Π^{loc} is given by

$$\mathcal{R}(\Pi^{\text{loc}}) = \{\underline{x} \in \Sigma^{\text{out}} \mid x_1^{\text{out}} \geq 0\}.$$

Lemma 2.1 *The inverse of* Π^{loc} *on* $\mathcal{R}(\Pi^{\text{loc}})$ *is given by*

$$(\Pi^{\text{loc}})^{-1} \begin{pmatrix} u \\ 1 \\ x_2^{\text{out}} \\ y_2^{\text{out}} \end{pmatrix} = \begin{pmatrix} 1 \\ u^{\nu_1} \\ u^{-\mu} x_2^{\text{out}} \\ u^{\nu_2} y_2^{\text{out}} \end{pmatrix}. \tag{2.1}$$

We now define the part of the return map given by the first hit map from Σ^{out} to Σ^{in}. Let Π^{far} denote this map. The domain of this mapping is given by $\Sigma^{\text{out}} \cap V$, where V is a neighborhood of the origin containing $(0,1,0,0)$ and $(1,0,0,0)$. Note that Σ^{in} and Σ^{out} are invariant under κ. It follows that Π^{far} commutes with the action of κ. Also for $\alpha = 0$, $\Pi^{\text{far}}(0,1,0,0) = (1,0,0,0)$. Hence Π^{far} has the form

$$\Pi^{\text{far}} \begin{pmatrix} x_1^{\text{out}} \\ 1 \\ x_2^{\text{out}} \\ y_2^{\text{out}} \end{pmatrix} = \begin{pmatrix} 1 \\ \alpha + a \\ b_1 x_2^{\text{out}} + b_2 y_2^{\text{out}} \\ b_3 x_2^{\text{out}} + b_4 y_2^{\text{out}} \end{pmatrix} \tag{2.2}$$

where: $a, b_1, b_2, b_3, b_4 : \Sigma^{\text{out}} \times \mathbf{R}^2 \to \mathbf{R}$ and are invariant under the action of κ. Without loss of generality we assume that b_i, $i = 1..4$, are C^∞-smooth functions of u, x^2, y^2.

Let $z = (u, x_2^{\text{out}}, y_2^{\text{out}})$ denote the coordinates in Σ^{out} and $w = (r, x_2^{\text{in}}, y_2^{\text{in}})$ the coordinates in Σ^{in}. Hence $a, b_i, i = 1, \ldots, 4$, depend on (z, μ), $\mu = (\alpha, \lambda)$. The existence of a symmetric homoclinic for $\alpha = 0$ implies that $a(0, (0, \lambda)) = 0$. We choose the parameter α such that $a(\underline{0}, \alpha, \lambda) = 0$.

The action of symmetry on the coordinates z and w is given by:

$$\kappa \begin{pmatrix} u \\ x_2^{\text{out}} \\ y_2^{\text{out}} \end{pmatrix} = \begin{pmatrix} u \\ -x_2^{\text{out}} \\ -y_2^{\text{out}} \end{pmatrix},$$

$$\kappa \begin{pmatrix} r \\ x_2^{\text{in}} \\ y_2^{\text{in}} \end{pmatrix} = \begin{pmatrix} r \\ -x_2^{\text{in}} \\ -y_2^{\text{in}} \end{pmatrix}.$$

Let $\Pi = \Pi^{\text{loc}} \circ \Pi^{\text{far}}$ and let

$$\begin{pmatrix} \hat{u} \\ \hat{x}_2^{\text{out}} \\ \hat{y}_2^{\text{out}} \end{pmatrix} = \Pi \begin{pmatrix} u \\ x_2^{\text{out}} \\ y_2^{\text{out}} \end{pmatrix}. \tag{2.3}$$

We rewrite (2.3) as

$$(\Pi^{\text{loc}})^{-1} \begin{pmatrix} \hat{u} \\ \hat{x}_2^{\text{out}} \\ \hat{y}_2^{\text{out}} \end{pmatrix} = \Pi^{\text{far}} \begin{pmatrix} u \\ x_2^{\text{out}} \\ y_2^{\text{out}} \end{pmatrix}.$$

Hence the map Π can be implicitly expressed by the equations

$$\begin{aligned}
\hat{u}^{\nu_1} &= \alpha + a \\
\hat{x}_2^{\text{out}} &= u^\mu (b_1 x_2^{\text{out}} + b_2 y_2^{\text{out}}) \\
\hat{u}^{\nu_2} \hat{y}_2^{\text{out}} &= b_3 x_2^{\text{out}} + b_4 y_2^{\text{out}}.
\end{aligned} \tag{2.4}$$

The domain consists of those $\underline{x} \in \Sigma^{out} \cap V$ such that $\alpha + a \geq 0$.

Recall the definition of the vectors v_+ and v_- and the trajectory $y_\lambda(t)$ given in section 1. Under the assumptions of Section 2, $v_- = e_3$ and $v_+ = e_4$. Let T_0, T_1 be such that $\gamma(T_0) = (1, 0, 0, 0)$ and $\gamma(T_1) = (0, 1, 0, 0)$. It follows that $y_\lambda(T_0) = e_4$. From (2.2) we infer that

$$y_\lambda(T_1) = \begin{pmatrix} 0 \\ 0 \\ b_3(\underline{0}, 0, \lambda) \\ b_4(\underline{0}, 0, \lambda) \end{pmatrix},$$

Consequently $\lim_{t \to \infty} y_\lambda(t) = (\operatorname{sgn} b_4(\underline{0}, 0, \lambda)) v_-$. Therefore we have the following

Lemma 2.2 *The point $(\alpha, \lambda) = (0, 0)$ in parameter space is a twist point if $b_4(\underline{0}, 0, \lambda)$ changes sign as λ passes through* 0.

We assume that

H5 $\frac{\partial}{\partial \lambda} b_4(\underline{0}, 0, 0) \neq 0$.

By redefining the parameters α, λ we can write b_4 as $b_4 = \lambda + f$, where $f(\underline{0}, \alpha, \lambda) = 0$. We now rewrite (2.4) as

$$\begin{aligned} \hat{u}^{\nu_1} &= \alpha + a \\ \hat{x} &= u^\mu(b_1 x + b_2 x) \\ \hat{u}^{\nu_2}\hat{y} &= b_3 x + (\lambda + f)y. \end{aligned} \tag{2.5}$$

Note that writing Π as in in (2.4) *hides* the singularities. The expression (2.4) is most suitable for finding equilibria and periodic points of Π.

3 Periodic points in S

The periodic orbits $P_{(\alpha, \lambda)}$ correspond to fixed points of (2.5) with $x = y = 0$. This leads to the equation. (we suppress the dependence of a on (α, λ)):

$$u^{\nu_1} = \alpha + a(u, 0, 0). \tag{3.1}$$

We assume that

H6 $a_u = \frac{\partial}{\partial u} a(0, 0, 0) \neq 0$.

It is clear that (2.5) has a unique fixed point for each $\alpha > 0$ or $\alpha < 0$.

Let $(u^*, 0, 0)$ denote the fixed point of (3.1). Let $b_i^* = b_i(u^*, 0, 0)$, $i = 1, 2, 3$ $f^* = f(u^*, 0, 0)$ and $a_u^* = \frac{\partial}{\partial u} a(u^*, 0, 0)$. We have the following lemma:

Lemma 3.1

$$d\Pi((u^*, 0, 0), (\alpha, \lambda)) = \begin{pmatrix} \frac{1}{\nu_1}(u^*)^{1-\nu_1} Q_u^* & 0 & 0 \\ 0 & (u^*)^\mu b_1^* & (u^*)^\mu b_2^* \\ 0 & (u^*)^{-\nu_2} b_3^* & (u^*)^{-\nu_2}(\lambda + f^*) \end{pmatrix}.$$

Proof The formula is obtained by implicit differentiation of (2.5) using the fact that a, b_i, $i = 1, 2, 3$ and f are invariant under the action of x and the fact that $(u^*, 0, 0)$ is a fixed point of Π.   ∎

Let

$$A = \begin{pmatrix} (u^*)^{\mu} b_1^* & (u^*)^{\mu} b_2^* \\ (u^*)^{-\nu_2} b_3^* & (u^*)^{-\nu_2}(\lambda + f^*) \end{pmatrix}. \tag{3.2}$$

The eigenvalues of A correspond to the eigenvalues of $d\Pi(u^*, 0, 0, \alpha, \lambda)$ in the nonsymmetric directions (directions contained in R). The following proposition describes the asymptotic behavior of these eigenvalues. In [AGK91] an analogous result is proved for the system of ODE's describing large series arrays of Josephson junctions. See also section 5.

Proposition 3.2 *For (α, λ) small enough A has real eigenvalues μ_+ and μ_- with $|\mu_+| > 1$ and $|\mu_-| < 1$. For fixed $\lambda \neq 0$ these eigenvalues have the following limiting behavior:*

$$\lim_{u^* \to 0} \mu_- = 0,$$

$$\lim_{u^* \to 0} \mu_+ = \begin{cases} \infty & \text{if} \quad \lambda > 0 \\ -\infty & \text{if} \quad \lambda < 0. \end{cases}$$

4   Proofs of Theorem 1.1 and 1.2

We begin by restating Theorems 1 and 2 as results for the map Π defined by (2.5). Note that the periodic orbit $P_{(\alpha, \lambda)}$ corresponds to the fixed point $(u^*, 0, 0)$ and a nonsymmetric homoclinic orbit of (1.1) corresponds to a fixed point of Π of the form $(0, 0, y)$. In the sequel we will use the notation $u = \frac{\partial}{\partial u}(0, 0, 0)$, etc.

Proposition 4.1 *Let Π be the map defined by (2.5). Let H1-H6 hold and assume that $f_{y^2} \neq 0$. Then the following hold:*

(i). *$(u^*, 0, 0)$ has a pitchfork bifurcation along a continuous curve $\alpha = \tau_1(\lambda)$, $\tau_1(0) = 0$.*

(ii). *$(u^*, 0, 0)$ has a period doubling bifurcation along a continuous curve $\alpha = \tau_2(\lambda)$, $\tau_2(0) = 0$.*

(iii). *Π has a fixed point of the form $(0, 0, y^*)$ along a differentiable curve $\alpha = \tau_3(\lambda)$, $\tau_3(0) = 0$. $y^* \to 0$ as $\lambda \to 0$.*

Proposition 4.2 *Let H1-H5 hold. Assume in addition that $a_u f_{y^2} - a_{y^2} f_u \neq 0$, $a_{y^2} \neq 0$, $f_{y^2} \neq 0$, $a_{y^2} b_{20} b_{30} \neq 0$, and $f_{y^2} \neq 0$. Then there exist regions D_i, $i = 1, 2$, whose boundaries are given by the curve τ_3 and the curve τ_i, $i = 1, 2$, with the following properties:*

(i). *For each (α, λ) in D_1, there exists a unique fixed point of the form (u_1, x_1, y_1), $u_1 > 0$, $(x_1, y_1) \neq 0$.*
   For each (α, λ) in D_2, Π has a unique $\mathbb{Z}_2$ invariant period 2 orbit of the form (u_2, x_2, y_2), $u_2 > 0$, $(x_2, y_2) \neq 0$.

(ii). *The fixed point and the period 2 point depend continuously on (α, λ) and $u_i \to 0$, $i = 1, 2$, as $(\alpha, \lambda) \to (\tau_3(\lambda), \lambda)$.*

Proof of Proposition 4.1

(i). Π has a symmetry breaking pitchfork bifurcation if one of the eigenvalues of A, $\mu_\pm$, equals 1. The equation $\mu_\pm = 1$ is equivalent to

$$\operatorname{tr} A - 1 = \det A. \tag{4.1}$$

Using (3.2) we obtain

$$\lambda + f^* - (u^*)^{\nu_2} + (u^*)^\mu(b_2^* b_3^* - b_1^*(\lambda + f^*)) + (u^*)^{\mu+\nu_2} b_1^* = 0. \tag{4.2}$$

The curve τ_1 is obtained as the solution of (4.2) and (3.1).

(ii). Analogously $(u^*, 0, 0)$ has a period doubling bifurcation if $\mu_+ = -1$ or $\mu_- = -1$. Using (3.2) we arrive at the equation

$$\lambda + f^* + (u^*)^{\nu_2} - (u^*)^\mu(b_2^* b_3^* - b_1^*(\lambda + f^*)) + (u^*)^{\mu+\nu_2} b_1^* = 0 \tag{4.3}$$

The curve τ_2 is obtained as the solution of (4.3) and (3.1).

(iii). We look for fixed points of (2.5) of the form $(0, 0, y)$. This leads to the equations

$$\begin{cases} \alpha + a &= 0 \\ x &= 0 \\ y(\lambda + f) &= 0 \end{cases} \tag{4.4}$$

The dependence of a and f on (α, λ) is of higher order and we can, with no loss of generality, neglect it. We rewrite (4.4) as

$$\begin{aligned} \alpha &= -a(0, 0, y^2, \alpha, \lambda) \\ \lambda &= -f(0, 0, y^2, \alpha, \lambda). \end{aligned} \tag{4.5}$$

By the Implicit Function Theorem we can solve for $\alpha = \alpha^*(y^2)$, $\lambda = \lambda^*(y^2)$, such that $\alpha^*(0) = \lambda^*(0) = 0$. This determines a half curve $(\tau_3(\lambda), \lambda)$ if $f^{y^2} \neq 0$. $\blacksquare$

Proof of Proposition 4.2. We suppress the dependence of a, b_1, b_2, b_3 and f on (α, λ). This dependence is of higher order and does not affect the analysis.

We find the region D_1 by solving for fixed points of (2.5). Such fixed points must satisfy the equation

$$\begin{aligned} u^{\nu_1} &= \alpha + a \tag{4.6a} \\ x &= u^\mu(b_1 x + b_2 y) \tag{4.6b} \\ u^{\nu_2} y &= b_3 x + y(\lambda + f). \tag{4.6c} \end{aligned}$$

By the contraction mapping principle we solve (4.6b) for x as a function of u, y, α, λ. By symmetry there exists a function φ such that

$$x = u^\mu y \varphi(u, y^2).$$

Substituting into (4.6a) and (4.6c) and canceling the factor y we get

$$u^{\nu_1} = \alpha + \tilde{a} \tag{4.7a}$$

$$u^{\nu_2} = u^\mu \varphi \tilde{b}_3 + \lambda + \tilde{f}, \tag{4.7b}$$

where

$$\tilde{a}(u, y^2) = a(u, u^\mu y \varphi(u, y^2), y)$$
$$\tilde{b}_3(u, y^2) = b_3(u, u^{2\mu} y^2 \varphi^2(u, y^2), y^2)$$
$$\tilde{f}(u, y^2) = f(u, u^{2\mu} y^2 \varphi^2(u, y^2), y^2).$$

A straightforward computation gives that

$$\begin{vmatrix} \frac{\partial \alpha^\bullet}{\partial u} & \frac{\partial \alpha^\bullet}{\partial y^2} \\ \frac{\partial \lambda^\bullet}{\partial u} & \frac{\partial \lambda^\bullet}{\partial y^2} \end{vmatrix} = a_u f_{y^2} - f_u a_{y^2} - \nu_1 f_2 u^{\nu_1 - 1} + \nu_2 a_{y^2} u^{\nu_2 - 1} - \mu a_{y^2} b_{20} b_{30} u^{\mu - 1}$$
$$+ \mathcal{O}\left(u^{\min\{\mu, 2\mu - 1, \mu + \nu_2 - 1, 1, 1, \mu + \nu_1 - 1, \nu_2\}} + y^2 + u^{\min\{\mu - 1, \nu_1 - 1\}} y^2 \right).$$

From the nondegeneracy assumptions it follows that for each pair (u, y^2), with $u > 0$ and $y^2 > 0$ their corresponds a unique pair (α, λ) satisfying (4.7a) and (4.7b). The boundaries of the region D_1 are obtained from (4.7a) and (4.7b) by putting $u = 0$ and varying y^2, and putting $y^2 = 0$ while varying u (positive). The last boundary reduces to the pitchfork bifurcation line, see (??) and the sentence below. The first boundary is described by the set of equations

$$\alpha + a(0, 0, y^2) = 0 \tag{4.8}$$
$$\lambda + f(0, 0, y^2) = 0.$$

If $f_{y^2} \neq 0$ then this set of equations reduces to $\alpha - \tau_3(\lambda) = 0$ where $\tau_3(\lambda) = \frac{a_{y^2}}{f_{y^2}}\lambda$. To find the region D_2 we look for points (u, x, y) such that under the transformation (2.5) $(\hat{u}, \hat{x}, \hat{y}) = (u, -x, -y)$. Substituting this relation into (2.5) we get

$$\begin{cases} u^{\nu_1} = \alpha + a \\ x = -u^\mu(b_1 x + b_2 x) \\ u^{\nu_2} y = -(b_3 x + y(\lambda + f)). \end{cases} \tag{4.9}$$

Through a procedure analogous to the one applied for D_1 we find the region D_2. One of its boundaries is given by the period doubling bifurcation variety. The other boundary is obtained by solving (4.9) with $u = 0$. This leads to the set of equations (4.8), and hence D_2 has this boundary in common with D_1.   ∎

A straightforward computation shows that we must distinguish between
i: $\min\{\nu_1, 1\} > \min\{\mu, \nu_2\}$, ii: $\min\{\nu_1, 1\} < \min\{\mu, \nu_2\}$, iii: $f_{y^2} < 0$ and iv: $f_{y^2} > 0$. Figure $1.a$ corresponds to $i + iii$, Figure $1.b$ to $ii + iii$, Figure $1.c$ to $i + iv$ and Figure $1.d$ to $ii + iv$.

5   Application to arrays of Josephson junctions

In [AGK91] the dynamics of current biased Josephson junctions with global coupling is considered. Using the software package AUTO [Doe81] numerical results have been obtained for a system of two oscillators. There are two parameters in the problem: β (measuring the capacitence) and I (the applied biased current). In the case of pure capacitive load a codimension-two bifurcation is observed where, in parameter space, a curve of period doubling bifurcations and a curve of fixed point bifurcations terminate on a curve of homoclinic orbits in a point, say P. On one side of this curve of homoclinic orbits, in phase periodic solutions exist. In [AGK91] it is shown analytically that there is a change in the orientation of the tangent space of the homoclinic orbits at P. The bifurcation picture computed with AUTO is shown schematically below, see also [AGK91]

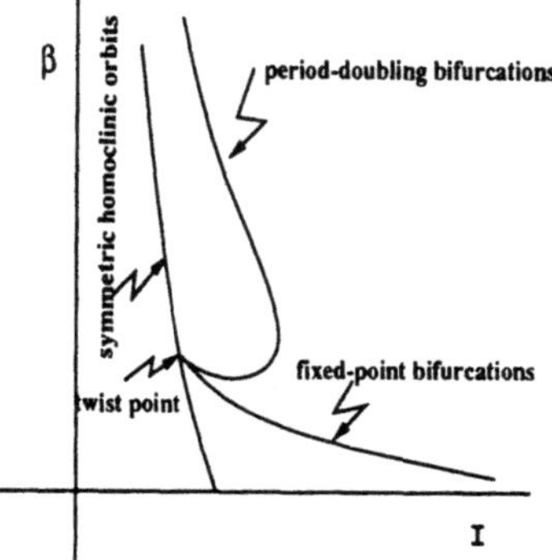

Figure 1: Homoclinic twist in the Josephson system.

Theorem 4.1 promises a third curve. For parameter values on this curve asymmetric homoclinic orbits exist, and the period of both the period doubled solutions and the asymmetric periodic solutions approaches infinity as the parameters approach this third curve. We have found in section 4 that there are four qualitatively different bifurcation pictures. From the computations in [AGK91] we know the numerical values of I and β at the twist point. Thus we are able to compute the eigenvalues at that point. It turns out that the symmetric eigenvalues are 0.350 and -0.642. The eigenvalues in the asymmetric direction are 0.612 and -2.999. This implies that we are in case (b) or (d) of Figure 1 (compare the last remark of the previous section). We have done numerical calculations with AUTO to find the the curve τ_i, $i = 1..3$. The results are in agreement with the theory. Unfortunately we have not yet numerically calculated the coefficient f_{y^2}. This would further restrict the possible bifurcation diagrams.

More bifurcations occur at the homoclinic twist bifurcation point. For instance, we have proved that a line of 2-homoclinic orbits ends at the twist point. For a three dimensional mapping with a twist point n-homoclinic orbits exist in the vicinity of the twist point for arbitrary $n \in \mathbb{N}$. These and other results will appear in a forthcoming paper [Aa92].

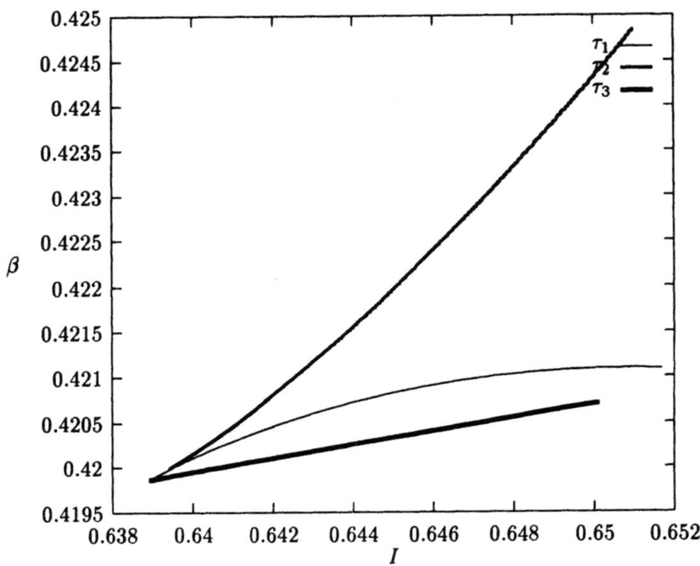

Figure 2: Bifurcations at the homoclinic twist point.

References

[Aa92]   D.G. Aronson and al. The dynamics of coupled current-biased josephson junctions, part ii. Technical report, University of Minnesota, 1992.

[AGK91]   D.G. Aronson, M. Golubitsky, and M. Krupa. Coupled arrays of josephson juncties, and bifurcation of maps with s_n symmetry. *Nonlinearity*, 4:861–902, 1991.

[CDF89]   S-N. Chow, B. Deng, and B. Fiedler. Homoclinic bifurcation at resonant eigenvalues. Technical report, Konrad-Zuse-Zentrum für Informationstechnik Berlin, 1989.

[Den92]   B. Deng. Homoclinic twisting bifurcations and cusp horseshoe maps. Technical report, University of Nebraska, 1992.

[Doe81]   E.J. Doedel. Auto: A program for the automatic bifurcation analysis of autonomous systems. *Cong. Num.*, 30:265–284, 1981.

[Ste57]   S. Sternberg. Local contractions and a theorem of poincaré. *Amer. J. Math.*, pages 809–824, 1957.

[Yan87]   E. Yanagida. Branching of double pulse solutions from single pulse solutions in nerve axon equations. *J. Diff. Eqs.*, pages 243–262, 1987.

International Series of Numerical Mathematics, Vol. 104, © 1992 Birkhäuser Verlag Basel

High Corank Steady-State Mode Interactions on a Rectangle

Peter Ashwin

Mathematics Institute, Warwick University, Coventry CV4 7AL, U.K.[*]

Abstract

We consider bifurcation of steady states from a trivial solution of PDEs. The existence of arbitrarily high corank bifurcation points is shown to be generic in restrictions of PDEs with Euclidean equivariance to systems periodic on rectangular (or rhombic) lattices under certain rationality assumptions. As an example, we examine the Kuramoto-Sivashinsky equation on a rectangular domain with Neumann boundary conditions, using Liapunov-Schmidt reduction in a two parameter setting.

1   Introduction

As observed by Kuttler and Sigillito [12] the Laplacian on a square can have arbitrarily high multiplicities of eigenvalues.

For a partial differential equation (PDE) on a square, such behaviour is not generic in the class of bifurcation problems with D_4 (square) symmetry, as the only irreducible representations of this group are one or two dimensional, and generically the action of the symmetry group on the kernel will be irreducible [8, 9], giving a corank of two at most. In this paper, we show that high corank

---

[*]Present address: FB Mathematik, Universität Marburg, 3550 Marburg, Germany

behaviour is generic in the restriction of Euclidean equivariant problems to rectangular or rhombic domains with simple boundary conditions.

As an example, we consider Liapunov-Schmidt reduction of the Kuramoto-Sivashinsky equation [15]. Mode interactions give us an opportunity to examine the crossover between bifurcation in small and large scale domains. For more details, see [5, 4].

2   Number theoretic degeneracies

The arbitrarily high order mode interactions are generated by the linearised partial differential operator being equivariant under the (non-compact) Euclidean group of all translations and rotations in the plane, in other words, for a problem can be posed on the whole of $\mathbf{R}^2$:

$$G(u(x))(y) = 0$$

with $u(x)$ in some Banach space X of functions from $\mathbf{R}^2$ to $\mathbf{R}$. Assume dG is a Fredholm linear operator from X to some space Y. This is equivariant under the action of $\gamma \in E(2)$ defined by:

$$dG(u(\gamma(x)))(\gamma(y)) = dG(u(x))(y).$$

This equivariance makes it easy to verify that the operator acts on complex exponentials $e_k(x) = \exp(ik.x)$ $(k \in \mathbf{R}^2)$ thus: $dG(e_k) = P(|k|)e_k$ with $P : \mathbf{R}^+ \to \mathbf{C}$ a polynomial in $|k|^2$ for a partial differential operator (Ian Melbourne, pers. comm.).

For a nonlinear parabolic PDE bifurcation problem $u_t = G(u, \lambda)$, the steady state bifurcation points correspond to solutions of $P(|k|) = 0$. For the problem on $\mathbf{R}^2$, there will be a *critical circle* of wavevectors that bifurcate simultaneously, and so an infinite dimensional kernel.

Restricting the above problem to one with Neumann or Dirichlet boundary conditions (NBC, DBC resp.) on a *rectangular domain* $[0, a\pi] \times [0, b\pi]$, combinations of complex exponentials will be in the kernel at bifurcation for values of the wavevector $k = (l/a, m/b)$ with $(l, m) \in \mathbf{Z}^2$; e.g. $cos(lx_1/a)cos(mx_2/b)$ for NBC. Thus, under the assumption that all eigenspaces are spanned by a finite set of Fourier modes, the dimension of the kernel is equal to the number of solutions of:

$$\frac{l^2}{a^2} + \frac{m^2}{b^2} = |k|^2 = (P^{-1}(0))^2 \tag{1}$$

for $(l, m) \in \mathbf{Z}^2$. The number of solutions of this depend on the rationality of a/b:

2.1 Rectangles with $a^2/b^2 \in \mathbf{Q}$

This includes the square $(a = b)$ discussed in [12, 11];

Writing $a/b = p/(q\sqrt{d})$ with integers p, q and some square-free integer $d > 0$, means we can rewrite equation 1 as $q^2l^2 + dp^2m^2 = n$, and thus the number of solutions is at least $R(n/(pq)^2)$ for any n such that $pq|\sqrt{n}$, where $R(n)$ is the number of solutions of the equation

$$x^2 + dy^2 = n$$

with $d \in \mathbf{N}$ square-free and $(x, y) \in \mathbf{Z}^2$. For all d, $R(n)$ can be shown to be unbounded using the following theorem (which is not sharp).

Theorem 2.1 $S(n) = sup_{r \leq n} R(r) > O((\log n)^\epsilon)$ for $0 < \epsilon < 1$. In particular, it is unbounded in n.

Proof For any $m \in \mathbf{N}$, define $\alpha = 2md + i\sqrt{d}$. then $\arg(\alpha) < \tan(\arg(\alpha)) < 2m$, and the set of numbers $S = \{\alpha^l \bar{\alpha}^{2m-l}; 0 \leq l < m\}$ has m distinct elements in $\mathbf{Z}(\sqrt{-d})$, with $\beta \in S$ implies $\beta\bar{\beta} = \alpha^m\bar{\alpha}^m = n \in \mathbf{N}$. Thus we have shown that for any m, d there exists an n with at least m solutions to $x^2 + dy^2 = n$. Therefore, $(\alpha\bar{\alpha})^m = n$ implies $m \log m > O(\log n)$ and the result follows. $\square$

One consequence of this lemma is that for large $R(n)$, the arguments of the critical wavevectors fill the interval $[0, 2\pi)$. We define

$$\mathcal{R}(n) = \{x + iy\sqrt{d} : x^2 + dy^2 = n\}$$

(so $R(n) = |\mathcal{R}(n)|$) and the following lemma comes from considering S defined in theorem 2.1.

Lemma 2.1 *For any $\epsilon > 0$, there exists an infinite sequence $\{n_i\}$ with $n_i \to \infty$ such that an ϵ-neighbourhood of $\{\arg z : z \in \mathcal{R}(n_i)\}$ covers $[0, 2\pi)$.*

2.2 Rectangles with $a^2/b^2 \notin \mathbf{Q}$

These cannot have arbitrarily large null spaces [13]. There is at most one solution in $\mathbf{Z}_+^2$ of the equation $b^2 l^2 + a^2 m^2 = n$ if $\{a^2, b^2\}$ are independent over the rationals. This is because in this case, the set of n lie within a vector space over $\mathbf{Q}$ of dimension 2.

The results above generalise easily to domains of higher dimension, and permit combinations of NBC, DBC, PBC in different dimensions.

Rhombic and hexagonal lattices   The arguments above also apply to the case when we have periodic boundary conditions on a rhombus, triangle or hexagon, as by considering an $E(2)$ equivariant problem with periodicity on a rhombic lattice with generating vectors $v_1 = (1, 0)$ and $v_2 = (a, b)$ $(b \neq 0)$, we have eigenfunctions of the form $\exp(ilx)\exp(im(ax + by))$ with eigenvalues $\lambda = (l + am)^2 + b^2 m^2$. The number of solutions of this can be unbounded in the same way as for the rectangle if both $(a/b)^2$ and a are rational. Thus, in particular a hexagonal lattice $(a, b) = (1/2, \sqrt{3}/2)$ can have unbounded corank at bifurcation.

3   The Kuramoto-Sivashinsky equation

The Kuramoto-Sivashinsky (KS) equation upon suitable scaling [6] can be written:

$$u_t + \nabla^4 u + \nabla^2 u + (\nabla u)^2 = 0. \tag{2}$$

Consider this equation on the rectangle $[0, a\pi] \times [0, b\pi]$ with NBC $u_n = (\nabla u)_n = 0$ on the edges. The system is equivariant [9] under the symmetry group of the rectangle $\mathbf{D}_2$ or the square $\mathbf{D}_4$.

Defining: $e(k, l) = \cos\left(\frac{kx}{a}\right)\cos\left(\frac{ly}{b}\right)$ with $(k, l) \in \mathbf{N}^2$, we note that this is a basis of eigenvectors of the linearised equation. By substituting $u(x, y, t) =$

$\sum_{k=0}^{\infty} \sum_{l=0}^{\infty} y(k,l) \, e(k,l)$ into equation 2, multiplying by $e(k,l)$ and integrating over all space, we get a weak form of the equation;

$$y_t(k,l) = \omega(k,l) \, y(k,l) + \sum_{i,j,m,n} r_{ijmn} \, y(i,j) \, y(m,n)$$

with steady solutions obeying $y_t(k,l) = 0$ for all (k,l).

3.1 Steady-state bifurcations from $u = 0$.

With NBC, the eigenfunction $e(k,l)$ of the linearised equation has a steady-state bifurcation when

$$\left(\frac{k^2}{a^2} + \frac{l^2}{b^2} \right) = 1. \tag{3}$$

We regard the two numbers $(a,b) \in \mathbf{R}^2$ as bifurcation parameters. The set of bifurcation points in the (a,b) plane is shown in figure 1 for $(a,b) \in [0,20]^2$. At a constant value of (k,l), the curves can be parameterised by $(a,b) = (k \csc(\theta), l \sec(\theta))$ for $\theta \in (0,2\pi)$.

As discuss previously, there are multiple bifurcations caused by *number-theoretic degeneracies* in this equation, and these appear as multiple intersections of bifurcation lines in figure 1.

3.2 Liapunov-Schmidt reduction

Looking closer at the multiple bifurcations, we can investigate analytically the creation of branches of mixed-mode steady solutions from the trivial solution using Liapunov-Schmidt reduction [8]. We project the full equations onto the null space at bifurcation, and thus reduce the bifurcation problem from infinite dimension to the dimension of the null space. Consider the equation for steady solutions:

$$\Phi(u, \lambda) = 0$$

with $\lambda = (a,b) \in \mathbf{R}^2$ and write $\Phi \equiv d\Phi_{(0,0)} + \phi$;

$$\Phi \; : \; C^{4,\bullet}(\Omega) \times \mathbf{R}^2 \to C^{0,\bullet}(\Omega)$$

$$\Phi(u, \lambda) \; \equiv \; \nabla^4 u + \nabla^2 u + (\nabla u)^2 \; (= -u_t),$$

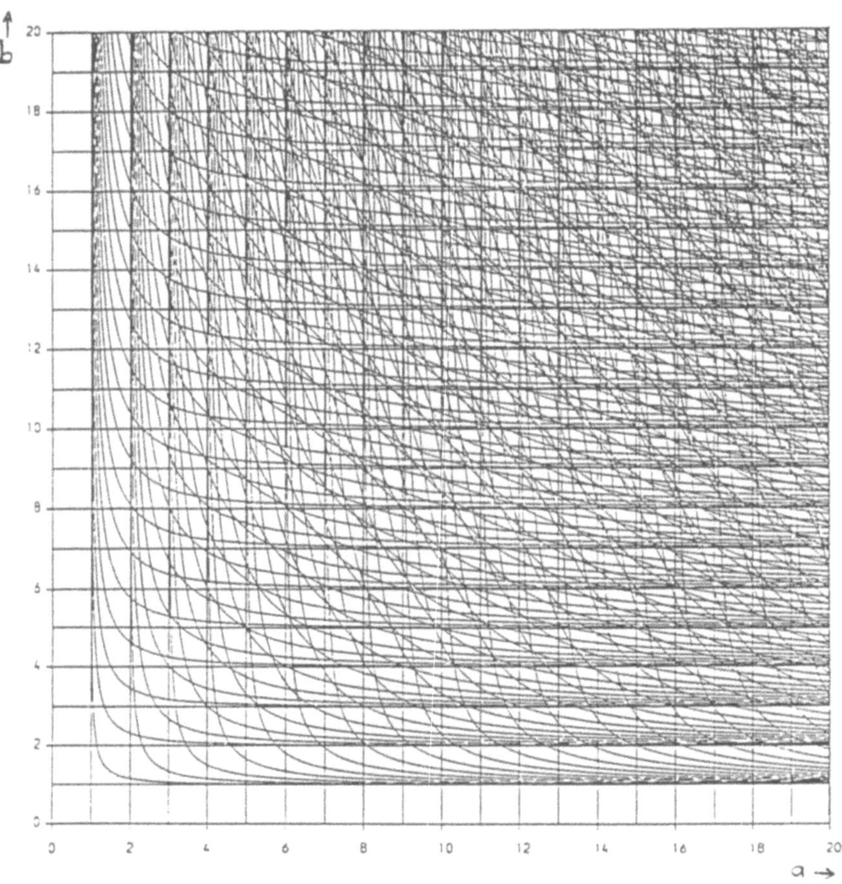

Figure 1:

The locations of the bifurcations in the (a,b) plane for $(a,b) \in [0,20]^2$. Note the crossings of the primary mode bifurcations are mode interactions, and mixed mode solutions may also bifurcate from such points. Note that the whole diagram has the structure of the monoid $(\mathbf{N}^2, \times)$, as described by Aston in [6].

where $\Omega = [0, a\pi] \times [0, b\pi]$ and $C^{k,\bullet}(\Omega)$ is the space of functions with k continuous derivatives $\overline{\Omega} \to \mathbf{R}$ satisfying NBC (i.e. $u_n = (\nabla^2 u)_n = 0$) and a Hölder condition to ensure $d\Phi$ is Fredholm.

When $d\Phi_{(0,0)}$ has a non-trivial null space we can perform a bifurcation analysis by projecting onto the null space. Assume that the bifurcation takes place at $\lambda = 0$.

Let $(N, 0) \oplus (0, \mathbf{R}^2)$ be the null space of the operator $d\Phi_{(0,0)}$ (note the trivial action on λ). Using the Fredholm condition, we construct a splitting of $C^{4,\bullet}(\Omega)$ into $N \oplus N^\perp$. Define the inverse of the linear part thus: $(d\Phi_{(0,0)})^{-1} : \text{Range}(d\Phi)^\perp \to N^\perp$ and invert the projection onto the kernel by considering

$$u_1 \in N, \ \lambda \in \mathbf{R}^2 \tag{4}$$

Define sucessive approximations, (suppressing the dependence of ϕ on λ) by:

$$u_i = (d\Phi_{(0,0)})^{-1}(-\phi(u_{i-1})) + u_1 \tag{5}$$

and at each stage let $E_i = d\Phi_{(0,0)} u_{i-1} + \phi(u_{i-1}) = \Phi(u_{i-1})$. By construction the projection onto the null space of u_i is u_1. The projection of E_∞ onto $\text{Range}(d\Phi)^\perp$ are the reduced bifurcation equations.

Theorem 3.1 *[4] For the iteration defined by equation 5, there exists $\epsilon > 0$ such that for $||u_1||$ and $||\lambda|| < \epsilon$, the iteration $u_i \to u_\infty$ converges.*

This gives convergence to some u_∞ with $E_\infty = \Phi(u_\infty, \lambda) \in \text{Range}(d\Phi)^\perp$. If $E_\infty = 0$, u_∞ is a solution with projection u_1 onto the null space N.

In order to move the parameter dependence on domain size to a dependence in the equations, we set $\Omega = [0, 2\pi]^2$ and rescale the lengths. The nonlinear terms at $(a_0/\sqrt{1-\bar{a}}, b_0/\sqrt{1-\bar{b}})$ are polynomial functions of u, its spatial derivatives up to fourth order, and the perturbations of the parameters $\bar{a}$ and $\bar{b}$. We truncate all equations at first order in the parameters and at a fourth order in the null space coordinates. The iteration is performed [14] using the computer algebra package MAPLE by WATCOM.

3.3   (8,1):(1,8):(7,4):(4,7)

This is an example of a four-mode interaction on a square domain. The interaction
happens at $(\bar{a},\bar{b}) = (0,0)$ where $(a,b) = (\sqrt{65}/\sqrt{1-\bar{a}}, \sqrt{65}/\sqrt{1-\bar{b}})$, and if the
null space coordinates are $u\,e(8,1) + v\,e(1,8) + w\,e(7,4) + x\,e(4,7)$, then the
equations are found to be:

$$0 = -\frac{u(64\bar{a} - \bar{b})}{65} + \frac{106907u^3}{559248} + \frac{6064w^2u}{15873} - \frac{128v^2u}{3201} + \frac{173x^2u}{759}$$

$$0 = -\frac{w(49\bar{a} + 16\bar{b})}{65} + \frac{26276w^3}{6288} + \frac{6064u^2w}{15873} + \frac{173v^2w}{759} + \frac{1568x^2w}{8319}.$$

and those obtained by permuting $(u,v,w,x) \mapsto (v,u,x,w)$. Upon solving these
truncated equations, there are pure, two-mode and three-mode branches of solutions, but no four-mode. We do not list them all (up to reflection, there are 41
possible branches counting all permutations of $+$ and $-$ signs, of which only 24
are real). As examples, some representative branches have (u,v,w,x) approx.:

$$(0,0,0,0) \qquad\qquad\qquad\qquad\qquad\qquad \text{trivial}$$
$$(\pm\sqrt{5.15\bar{a} + 0.08\bar{b}}, 0, 0, 0) \qquad\qquad\qquad \text{pure } u \text{ mode}$$
$$(\pm\sqrt{5.40\bar{a} + 1.21\bar{b}}, \pm\sqrt{1.21\bar{a} + 5.40\bar{b}}, 0, 0) \qquad \text{mixed } u - v \text{ mode}$$
$$(\pm\sqrt{6.2\bar{a} + .3\bar{b}}, 0, \pm\sqrt{-.4\bar{a} + .1\bar{b}}, \pm\sqrt{.3\bar{a} + .2\bar{b}}) \quad \text{mixed } u - w - x \text{ mode}$$

The only three mode solutions that occur are $v - w - x$ and $u - w - x$, and
figure 2 shows how these modes fit together in a small circuit about the mode
interaction.

4   Conclusions and discussion

For $E(d)$ problems restricted to d dimensional rectangles with certain boundary
conditions, it is generic in certain cases to have *hidden symmetries* in the problem
which enlarge the null space. Perturbing the equation to break the Euclidean
symmetry (e.g. by discretizing on a rectangular lattice) would generically destroy
these high order mode interactions, as would applying more general boundary
conditions.

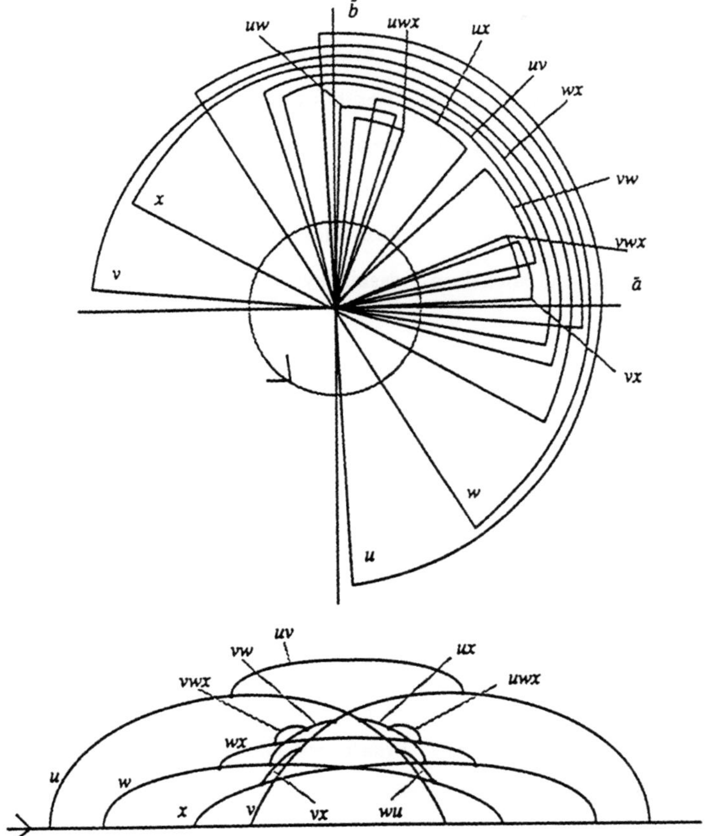

Figure 2:

The four mode interaction between $(8,1):(1,8):(7,4):(4,7)$. Note there are pure modes, two mode and three mode branches in any neighbourhood of the origin of the perturbed parameter plane $(\bar{a}, \bar{b})$.

Liapunov-Schmidt reduction finds a rich structure of bifurcating branches from the trivial solution at these points. Although the branches found in the above examples are all unstable, the resulting bifurcation equations can be used to determine branch switching for continuation algorithms [1]. The singularity strutures agree with those obtained theoretically by [2, 3, 10, 7] and their theory would give more efficient ways to perform these reductions. The method described is also applied to the Michelson-Sivashinsky equation, a nonlinear integro-differential equation in [4].

Acknowledgements   I would like to thank Ian Stewart, Jacques Furter, Ian Melbourne, Klaus Böhmer, Mei Zhen and Gabriela Gomes for many helpful discussions; Gene Allgower for drawing my attention to [12] and Mike Harrison and Charles Matthews for help with number theory. This work was done with the support of a British Gas Research Scholarship.

References

[1] G. Allgower, K. Böhmer, and Mei Z. Branch switching at a corank-4 bifurcation point of semi-linear elliptic problems with symmetry. Preprint, Mathematics Dept, University of Marburg, 1990.

[2] D. Armbruster and G. Danglmayr. Corank-two bifurcations for the brusselator with non-flux boundary conditions. *Dyn. and Stab. Systems*, 1:187–200, 1986.

[3] D. Armbruster and G. Danglmayr. Coupled stationary bifurcations in non-flux boundary value problems. *Math. Proc. Camb. Phil. Soc.*, 101:167–192, 1987.

[4] P. Ashwin. Applications of dynamical systems with symmetry. PhD Thesis, Maths Inst., University of Warwick, 1991.

[5] P. Ashwin. Steady state mode interactions in the Kuramoto Sivashinsky equation on a rectangle. Preprint, University of Warwick, Mathematics Institute, 1991.

[6] P.J. Aston. Scaling laws and bifurcation. In *Singularity theory and its applications, Warwick 1989*, volume 1463 of *LNM*, pages 1–21. Springer, Berlin, 1991.

[7] J.D. Crawford, M. Golubitsky, M.G.M. Gomes, E. Knobloch, and I.N. Stewart. Boundary conditions as symmetry constraints. In *Proceedings of symposium on singularities and their applications, Warwick, 1989*. Springer, New York, 1991.

[8] M. Golubitsky and D. Schaeffer. *Groups and singularities in bifurcation theory volume 1*, volume 51 of *App. Math. Sci.* Springer, New York, 1986.

[9] M. Golubitsky, I.N. Stewart, and D. Schaeffer. *Groups and singularities in bifurcation theory volume 2*, volume 69 of *App. Math. Sci.* Springer, New York, 1988.

[10] M.G.M. Gomes. Steady-state mode interactions in rectangular domains. MSc thesis, Maths Inst., University of Warwick, 1989.

[11] G.H. Hardy and Wright. *An introduction to the theory of numbers.* Cambridge University Press, 1956.

[12] J.R. Kuttler and V.G. Sigillito. Eigenvalues of the Laplacian in two dimensions. *SIAM Review*, 26:163–193, 1984.

[13] W-M. Ni and I. Takagi. On the Neumann problem for some semilinear elliptic equations and systems of activator-inhibitor type. *Trans AMS*, 297:351–368, 1986.

[14] R.H. Rand and D. Armbruster. *Perturbation methods, bifurcation theory and computer algebra*, volume 65 of *Applied Math. Sciences.* Springer, New York, 1987.

[15] G.I. Sivashinsky. Nonlinear analysis of hydrodynamic instability in flames- I. Derivation of the basic equations. *Acta Astronautica*, 4:1177–1206, 1977.

International Series of Numerical Mathematics, Vol. 104, © 1992 Birkhäuser Verlag Basel

Numerical Investigation of the Bifurcation from Travelling Waves to Modulated Travelling Waves

P. Aston[1], A. Spence[2], W. Wu[3]

P. Aston[1], A. Spence[2], W. Wu[3]

[1] Department of Mathematical
and Computing Sciences
University of Surrey
Guildford GU2 5XH

[2] School of Mathematical Sciences
University of Bath
Bath BA2 7AY

[3] Department of Mathematics
University of Jilin
Changchun
China

November 5, 1991

Abstract

We consider the numerical aspects of a Hopf bifurcation which occurs on a branch of travelling wave solutions in equations with O(2) symmetry. The Jacobian at every travelling wave solution is singular due to the O(2) symmetry which precludes the use of standard Hopf theory. Our approach is to add a spatial phase condition which removes the degeneracy in the Jacobian and allows standard theory to be applied. The numerical implications of this approach are also considered. The methods are applied to the Kuramoto-Sivashinsky equation. Numerical results are obtained which confirm a conjecture by Kevrekidis, Nicolaenko and Scovel (*SIAM J. Appl. Math.* **50**, 760-790) and that are in agreement with results obtained by Armbruster, Guckenheimer and Holmes (*SIAM J. Appl. Math.* **49**, 676-691) based on a centre-unstable manifold reduction which is not formally valid.

1　Introduction

In this paper, we consider nonlinear time dependent problems which have O(2) symmetry and our interest centres on the numerical aspects of a particular type of Hopf bifurcation which occurs in such systems. Travelling wave solutions can be found as steady state solutions of a modified system and arise either from Hopf bifurcations on an O(2) symmetric (trivial) solution or from steady state bifurcations on a branch of non-trivial solutions. The numerical aspects of these two types of bifurcation have been studied recently by Wu, Aston and Spence (1991) and Aston, Wu and Spence (1992). The complication which arises in these problems is that the Lie group O(2) forces the Jacobian of the system to be singular at every steady state solution so that standard bifurcation theory does not apply. We now consider the situation when a Hopf bifurcation occurs on a branch of travelling wave solutions giving rise to a branch of *modulated travelling waves* which can be represented as motion on a torus.

Krupa (1990) and Vanderbauwhede, Krupa and Golubitsky (1989) have considered this type of problem when the symmetry consists of any compact Lie group. Two vector fields are constructed, one which is tangent to the group orbit of solutions and the other which is normal to the group orbit. This approach enables bifurcation problems to be analysed successfully for a wide range of Lie groups. However, no attention is paid to the numerical aspects of the problem. We restrict attention to the Lie group O(2) and show how, by a careful choice of spatial phase condition, the problem can be considered using standard symmetry breaking Hopf bifurcation theory. This enables standard codes to be used for the detection of such Hopf bifurcations. More care is required for following a branch of modulated travelling wave solutions but we show how this can be done using the integral phase constraints which are implemented in the AUTO continuation package (Doedel (1981)).

The plan of the paper is as follows. In Section 2 we set the scene by introducing the problem and the basic results we require. In Section 3 we introduce the new spatial phase condition and show how the problem then fits into the standard Hopf theory. The numerical implementation is considered in Section 4 followed by a numerical example in Section 5 which confirms a conjecture made by Kevrekidis, Nicolaenko and Scovel (1990) regarding solutions of the Kuramoto-Sivashinsky equation. The numerical results are also in agreement with results of Armbruster, Guckenheimer and Holmes (1989) which are based on a reduction to a centre-unstable manifold but are not formally valid.

2　Background Theory

In this Section, we review the main results we require from our previous paper (Aston, Spence and Wu (1992) which we henceforth refer to as ASW) without proof.

We consider the general equation

$$\dot{X} = g(X, \lambda), \qquad g : \mathbf{R}^n \times \mathbf{R} \to \mathbf{R}^n \qquad (2.1)$$

and we assume that g is O(2) equivariant, where O(2) is the group generated by a rotation r_α, $\alpha \in [0, 2\pi)$ and a reflection s. We consider solutions of the form

$$X(t) = r_{ct} x(t). \qquad (2.2)$$

Such solutions with x independent of time correspond to rotating or travelling wave solutions of (2.1) and were considered in our previous paper (ASW). We now consider the case when $x(t)$ is time periodic and $c \neq 0$ which corresponds to *modulated travelling waves* that consist of time periodic solutions drifting with constant velocity along the group orbit. Substituting the required form of the solution (2.2) into the equation (2.1) gives

$$\dot{x} = g(x, \lambda) - cAx \qquad (2.3)$$

where $A := r_0' = \frac{dr_\alpha}{d\alpha}\big|_{\alpha=0}$. The derivation of (2.3) also requires the result that $r_\alpha' = r_\alpha A$ (see ASW). Steady state solutions of (2.3) correspond to steady state solutions of (2.1) if $c = 0$ or travelling wave solutions of (2.1) if $c \neq 0$. For the sake of convenience, we define

$$\tilde{g}(x, c, \lambda) := g(x, \lambda) - cAx.$$

Differentiating the equivariance condition $r_\alpha \tilde{g}(x, c, \lambda) = \tilde{g}(r_\alpha x, c, \lambda)$ with respect to α leads to the following result.

Lemma 2.1

If (x_0, c_0, λ_0) is a solution of the steady state equation $\tilde{g}(x, c, \lambda) = 0$, then

$$\tilde{g}_x(x_0, c_0, \lambda_0) A x_0 = 0.$$

This result shows that $\tilde{g}_x(x, c, \lambda)$ has a one-dimensional null space at every steady state solution of (2.3). Clearly, standard bifurcation theory cannot be applied in this case since a basic assumption is that the Jacobian is only singular at isolated points. However, this singularity can be eliminated by the addition of a phase condition of the form

$$p(x, c, \lambda) = 0.$$

Combining $\tilde{g}(x, c, \lambda)$ and the phase condition leads to the larger system

$$G(y, \lambda) := \left\{ \begin{array}{l} \tilde{g}(x, c, \lambda) \\ p(x, c, \lambda) \end{array} \right. \qquad (2.4)$$

where $y = (x, c)$. Conditions for the Jacobian of G to be non-singular are given in the next result.

Lemma 2.2

If (x_0, c_0, λ_0) is a solution of $\tilde{g}(x, c, \lambda) = 0$ at which

(i) Null $\tilde{g}_x(x_0, c_0, \lambda_0) = \text{span}\{Ax_0\}$,

(ii) Range $\tilde{g}_x(x_0, c_0, \lambda_0) = \{z \in \mathbf{R}^n \;:\; < \psi_0, z >= 0\}$,

where $< \cdot \,, \cdot >$ denotes the inner product on $\mathbf{R}^n$, then $G_y(y_0, \lambda_0)$ is non-singular provided that

(iii) $< \psi_0, A x_0 > \neq 0$,

(iv) $< p_x(x_0, c_0, \lambda_0), A x_0 > \neq 0$.

Condition (iv) is a non-degeneracy condition which the phase function p must satisfy so that the degeneracy in $\tilde{g}_x(x, c, \lambda)$ is not reproduced in $G_y(y, \lambda)$. If condition (iii) is violated at a particular value of λ, then generically, bifurcation to a branch of travelling waves occurs at that point if $c = 0$ and x is a steady state solution of (2.1), or alternatively, there is a turning point on the branch of travelling waves if $c \neq 0$.

An important feature of the system (2.4) is that it is equivariant with respect to a *discrete* group of symmetries if the phase function p is chosen carefully.

Lemma 2.3

If the phase function p satisfies

$$\begin{aligned}
p(sx, -c, \lambda) &= -p(x, c, \lambda), \\
p(r_{2\pi/n} x, c, \lambda) &= p(x, c, \lambda), \quad n \in \mathbf{Z}^+
\end{aligned}$$

then $G(y, \lambda)$ is equivariant with respect to the dihedral group D_n generated by $R_{2\pi/n}$ and S which act on $y = (x, c)$ by

$$Sy := \begin{bmatrix} sx \\ -c \end{bmatrix}, \tag{2.5a}$$

$$R_{2\pi/n} y := \begin{bmatrix} r_{2\pi/n} x \\ c \end{bmatrix}. \tag{2.5b}$$

We refer to solutions of (2.1) which satisfy $sx = x$ as *symmetric solutions*. Clearly, any solution of $G(y, \lambda) = 0$ which satisfies $Sy = y$ must have $c = 0$ and so consists of a symmetric steady state solution. If the solution also satisfies $R_{2\pi/n} y = y$ for some $n \in \mathbf{Z}^+$, then the steady state solution (branch) has D_n symmetry. Breaking the reflectional symmetry S leads to branches of non-symmetric travelling wave solutions of (2.1) with Z_n symmetry.

3   Hopf Bifurcations on Travelling Wave Solutions

We now consider the situation when there is a Hopf bifurcation on a branch of Z_n-symmetric travelling wave solutions resulting in a bifurcating branch of modulated travelling waves. There is always a two-parameter family of such solutions in that any solution can be translated in time and translated (or rotated) in space using r_α.

Our approach to the problem is to add an appropriate phase condition to the original equation in order to fix the *spatial* phase of the solution and then apply *standard* theory to this larger system.

We first of all define a system which admits the type of solutions we seek based on the system $G(y, \lambda)$ defined in (2.4) with $y = (x, c)$. In order to apply standard Hopf theory, we need a system of the form

$$\dot{y} = G(y, \lambda) \tag{3.1}$$

and so we must consider how to get the $\dot{y}$ term. Clearly the first equation in (3.1) is simply (2.3) which involves an $\dot{x}$ term. The second equation of (3.1) is

$$\dot{c} = p(x, c, \lambda). \tag{3.2}$$

Now we want time periodic solutions of (3.1) to correspond to modulated travelling wave solutions which have *constant* velocity c. Thus, we must choose the phase function $p(x, c, \lambda)$ to ensure that a time periodic solution of (3.2) gives rise to a constant value of c. This is achieved by making p independent of both the velocity c and time t and so we use

$$\dot{c} = P(\bar{x}, \lambda) \tag{3.3}$$

where $\bar{x}$ is the average of the time periodic function $x(t)$ over one period (T) defined by

$$\bar{x} = \frac{1}{T} \int_0^T x(t) \, dt.$$

The solution of the differential equation (3.3) is then given by

$$c(t) = P(\bar{x}, \lambda)t + k$$

for some constant k. Applying periodic boundary conditions to this solution implies that

$$c(t) = \text{constant}, \tag{3.4a}$$

$$P(\bar{x}, \lambda) = 0. \tag{3.4b}$$

Equation (3.4a) gives us the type of solution that we require while equation (3.4b) gives us a phase condition to fix the spatial phase of the solution. Thus, there is only a one parameter family of solutions of (3.1), corresponding to time translation, as in the standard Hopf context. As in the previous case however (see Lemma 2.2 (iv)), the phase condition must satisfy a non-degeneracy condition.

Lemma 3.1

The phase condition (3.4b) will fix the spatial phase of a solution (x_0, c_0, λ_0) of (2.3) if the non-degeneracy condition

$$< P_{\bar{x}}(\bar{x}_0, \lambda_0), A\bar{x}_0 > \neq 0 \tag{3.5}$$

is satisfied.

Proof

Linearising (2.3) about (x_0, c_0, λ_0) gives

$$\dot{\phi} = g_x(x_0, \lambda_0)\phi - c_0 A\phi.$$

This linear equation is satisfied by $\phi = \dot{x}_0$ and $\phi = Ax_0$ due to the translational invariance of the solution in space and time. Linearising the phase condition (3.4b) also gives

$$< P_{\bar{x}}(\bar{x}_0, \lambda_0), \bar{\phi} >= 0. \tag{3.6}$$

The spatial phase will be fixed by the phase condition provided that (3.6) is *not* satisfied by $\phi = Ax_0$, that is

$$< P_{\bar{x}}(\bar{x}_0, \lambda_0), \overline{Ax_0} > \neq 0.$$

Condition (3.5) follows immediately from this using the relation $\overline{Ax_0} = A\bar{x}_0$. $\qquad\square$

We note that equation (3.6) is not violated when $\phi = \dot{x}_0$ as $\overline{(\dot{x}_0)} = 0$ since the derivative of a time periodic function always has zero mean. Thus the phase condition (3.4b) fixes the spatial phase but not the temporal phase as required.

We now consider the symmetry properties of G. In order to apply the standard theory, we need to show that G is equivariant with respect to the group $Z_n \times S^1$ where the S^1 action corresponds to translation in time.

Theorem 3.2

If the phase function $P(\bar{x}, \lambda)$ satisfies the condition

$$P(r_{2\pi/n}\bar{x}, \lambda) = P(\bar{x}, \lambda), \tag{3.7}$$

then $G(y, \lambda)$ is equivariant with respect to $Z_n \times S^1$ where $R_{2\pi/n} \in Z_n$ acts as in (2.5b) and $\theta \in S^1$ acts by

$$\theta \begin{bmatrix} x(t) \\ c(t) \end{bmatrix} = \begin{bmatrix} x(t + \frac{\theta T}{2\pi}) \\ c(t + \frac{\theta T}{2\pi}) \end{bmatrix}.$$

Proof

The Z_n symmetry follows immediately from the equivariance of $\tilde{g}(x, c, \lambda)$ with respect to r_α and (3.7). The S^1 equivariance follows from the fact that $\bar{x}$ and $P(\bar{x}, \lambda)$ are invariant under translation in time. $\qquad\square$

The situation we are interested in is when $\tilde{g}_x(x, c, \lambda)$ has a pair of complex eigenvalues which cross the imaginary axis. However, we want to deal with the eigenvalues of $G_y(y, c)$. The next result, due to Dellnitz (1991) establishes the relationship between these eigenvalues.

Theorem 3.3

Suppose that (x_0, c_0, λ_0) is a solution of $G((x, c), \lambda) = 0$. If the eigenvalues of $\tilde{g}_x(x_0, c_0, \lambda_0)$ are σ_i, $i = 1, \ldots, n$ with $\sigma_n = 0$, then the eigenvalues of $G_y((x_0, c_0), \lambda_0)$

are σ_i, $i = 1, \ldots, n - 1$ and $\pm\delta$ where

$$\delta \ = \ < P_x(x_0, \lambda_0), Ax_0 >^{\frac{1}{2}} \ .$$

We note that the non-degeneracy condition (3.5) is precisely $\delta \neq 0$. Clearly then, if $\tilde{g}_x(x_0, c_0, \lambda_0)$ has eigenvalues $\pm i\omega$ then so has $G_y((x_0, c_0), \lambda_0)$. Thus, standard Hopf bifurcation theory can be applied to $G(y, \lambda)$ taking into account the symmetry group $Z_n \times S^1$ of the system (see Golubitsky, Stewart and Schaeffer (1988), Ch XVI). This theory predicts three possible types of Hopf bifurcation related to the different irreducible representations of the group Z_n which are

- symmetry preserving Hopf bifurcation associated with the trivial irreducible representation $R = I$;

- symmetry breaking Hopf bifurcation associated with the one-dimensional absolutely irreducible representation $R = -I$ for n even resulting in a bifurcating branch on which the spatial period has doubled;

- symmetry breaking Hopf bifurcation associated with the two-dimensional irreducible representation of complex type given by

$$R = \left[\begin{array}{cc} \cos 2\pi k/n & \sin 2\pi k/n \\ -\sin 2\pi k/n & \cos 2\pi k/n \end{array} \right], \quad k = 1, \ldots, m$$

where $m = (n - 1)/2$ if n is odd or $m = (n - 2)/2$ if n is even. These irreducible representations exist for $n \geq 3$.

4 Numerical Implementation

There are a number of aspects of this problem which need careful consideration with regard to the numerical implementation. Firstly, we consider the choice of the phase function P. The simplest possibility is

$$P(\bar{x}, \lambda) = < \ell, \bar{x} > \tag{4.1}$$

for some fixed $\ell \in \mathbf{R}^n$. Other more sophisticated functions could be used based on the work of Jepson and Keller (1984) but we will stay with this simple choice. The additional eigenvalues of $G_y((x_0, c_0), \lambda_0)$ are then $\pm\delta$ where

$$\delta = < \ell, Ax_0 >^{\frac{1}{2}} \ .$$

If $< \ell, Ax_0 >$ is negative, then the eigenvalues $\pm\delta$ will be *on* the imaginary axis. In practice, the real part of these eigenvalues is likely to be small of arbitrary sign and if the sign changes from one step to the next along a path of solutions, then a package would recognise this as a (spurious) Hopf bifurcation point. Thus, it is important to choose ℓ so that $< \ell, Ax_0 >$ is positive to ensure that the additional eigenvalues lie on the real axis. This is achieved by choosing $\ell = Ax_0$ where x_0 is the last solution found.

Detection of this type of Hopf bifurcation can be achieved by using a package such as AUTO (Doedel (1981)) on the system $G(y, \lambda) = 0$. We note that on a travelling wave solution of $G(y, \lambda) = 0$, x is independent of time and so $\bar{x} = x$. Thus the phase condition reduces to $< \ell, x > = 0$ which is a simple algebraic equation that is easily implemented.

A more efficient method of detection is to use the block diagonalisation of $G_y(y, \lambda)$ which arises due to the symmetry of the problem. This block structure is based on the isotypic components of the space which are related to the irreducible representations of the group Z_n in this case (see Aston (1991)). Now the eigenvector Ax_0 associated with the zero eigenvalue of $\tilde{g}_x(x_0, c_0, \lambda_0)$ is in the isotypic component associated with the trivial irreducible representation $R = I$ and so only the corresponding block of $\tilde{g}_x((x_0, c_0), \lambda_0)$ will be singular at every solution. Thus, the part of $G_y((x_0, c_0), \lambda_0)$ which arises from the phase condition can be added just to this block to make it generally non-singular and detection algorithms can then be applied to this augmented block matrix. The other blocks are not singular at every solution and so standard detection algorithms can be applied to these blocks without any amendment. Once a Hopf bifurcation has been detected, standard methods can again be employed for locating it precisely.

This procedure entails extra work in separating the Jacobian into the different blocks but results in a considerable saving in the computation of the eigenvalues which is the part that is often prohibitively expensive computationally.

To compute the branch of modulated travelling waves, we do not use the form of the phase condition (3.3) since it is not easy to implement and it is wasteful in that we would then be computing the constant value of c at every mesh point. Thus, we use the form of the phase condition given by (3.4b) as the additional scalar equation to go with the scalar variable c. Using the phase function defined by (4.1) gives the phase condition

$$< \ell, \bar{x} > \; = \; \frac{1}{T} \int_0^T \; < \ell, x(t) > \; dt \; = \; 0. \tag{4.2}$$

The continuation package AUTO allows the use of integral phase constraints such as (4.2). The system to be solved for the modulated travelling waves thus consists of equation (2.3) with the spatial phase condition (4.2) together with a standard temporal phase condition. This system is solved for the function x and the scalar variables c and T.

Once a Hopf bifurcation has been detected, a starting solution for the above system can be obtained in the usual way for the x variable by using the information contained in the eigenvectors. This first order approximation to x has zero mean since it only has terms which involve $\sin \omega t$ and $\cos \omega t$. This then implies that $c = c_0$ to first order, by (3.3) with P defined by (4.1). This information is sufficient to obtain an initial approximation for the computation of the branch of modulated travelling waves.

5   Numerical Results

As a numerical example of the methods we have developed, we consider the Kuramoto-Sivashinsky (KS) equation

$$U_t + 4U_{xxxx} + \lambda(U_{xx} + UU_x) = 0 \tag{5.1}$$

where U is 2π periodic in x and of zero mean. This equation is O(2) equivariant with respect to the following group action :

$$r_\alpha U(x,t) = U(x+\alpha,t), \quad \alpha \in [0,2\pi)$$

$$sU(x,t) = -U(-x,t).$$

For this group action, the linear operator A is defined by

$$AU = U_x$$

and so substituting $U = r_{ct}u$ into (5.1) gives

$$u_t + 4u_{xxxx} + \lambda(u_{xx} + uu_x) + cu_x = 0 \tag{5.2}$$

A spectral Galerkin method is used to solve this equation where u is approximated by

$$u(x,t) \simeq \sum_{k=1}^{N} a_k(t)\sin kx + b_k(t)\cos kx. \tag{5.3}$$

Note that there is no constant term since u has zero mean. Substituting the form of u given by (5.3) into the equation (5.2) leads to a system of $2N$ ordinary differential equations for the coefficients a_k and b_k. These equations can then be solved using AUTO. To compare with the literature, we present our results in terms of $v(x,t)$ (of zero mean) defined by the relation $v_x = u$.

Recently, many authors have studied the KS equation. In particular, Kevrekidis, Nicolaenko and Scovel (1990) solved the initial value problem numerically over a range of the parameter λ to investigate the stable solutions which occur and compared these results with numerically obtained bifurcation results. A branch of travelling wave solutions bifurcates from the first primary branch of steady state solutions at $\lambda = 13.005$ and on this branch, there is a Hopf bifurcation point of the type we are interested in at $\lambda = 17.399$ (see Fig. 5.1). These solutions are stable as far as the Hopf bifurcation at which point they lose stability. Kevrekidis, Nicolaenko and Scovel did not find a stable modulated travelling wave solution for $\lambda > 17.399$ and so conjectured that there was a subcritical, unstable branch arising from the Hopf bifurcation. They also observed that on the second primary branch, the 2 conjugate steady state solution branches are connected by heteroclinic cycles between 2 secondary bifurcation points which occur at $\lambda = 16.1399$ and $\lambda = 22.557$. However, these heteroclinic cycles are only attracting for λ greater than approximately 16.8. This lead them to conjecture further that the subcritical branch of modulated travelling waves arising from the Hopf

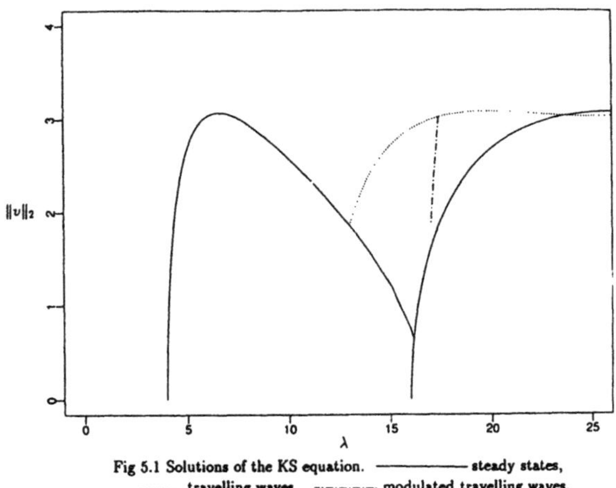

Fig 5.1 Solutions of the KS equation. —————— steady states,
————— travelling waves, ————— modulated travelling waves.

bifurcation at $\lambda = 17.399$ may terminate in a heteroclinic cycle at $\lambda = 16.8$ with the steady state solutions being primary branch 2 solutions.

A different approach to this problem is given by Armbruster, Guckenheimer and Holmes (1989) who perform a reduction to a 4 dimensional centre-unstable manifold in the neighbourhood of the second bifurcation from the trivial solution which occurs at $\lambda = 16$ in our framework. This reduction is formally valid only locally to the bifurcation point. However, they consider non-local solutions and obtain some interesting results. They first consider a third order truncation of the reduced equations and find a *supercritical* branch of modulated travelling waves arising from the Hopf bifurcation on the travelling wave branch which then terminates on the second primary branch in a heteroclinic cycle at $\lambda = 17.0909$ which is in close agreement with the conjectures of Kevrekidis, Nicolaenko and Scovel (1990). They then proceeded to take a fourth order truncation of the reduced equations and found almost identical results except that the modulated travelling wave branch became *subcritical* from the Hopf bifurcation thus giving exact agreement with the conjectures. The modulated travelling wave branch was found to terminate at $\lambda = 17.0794$ in this case.

While these results do look convincing, they are not of course rigorous and so we investigate the same branch of modulated travelling waves numerically. If a heteroclinic cycle is approached on the branch of modulated travelling waves, then the period T should tend to infinity and the velocity c should approach zero.

The first primary branch was computed followed by the branch of travelling waves which bifurcates from it using the methods described in ASW. The Hopf bifurcation on the travelling wave branch was calculated to be at $\lambda = 17.3979$ in good agreement with the results of Kevrekidis, Nicolaenko and Scovel (1990). The branch of modulated travelling waves was then computed using the methods described in the previous

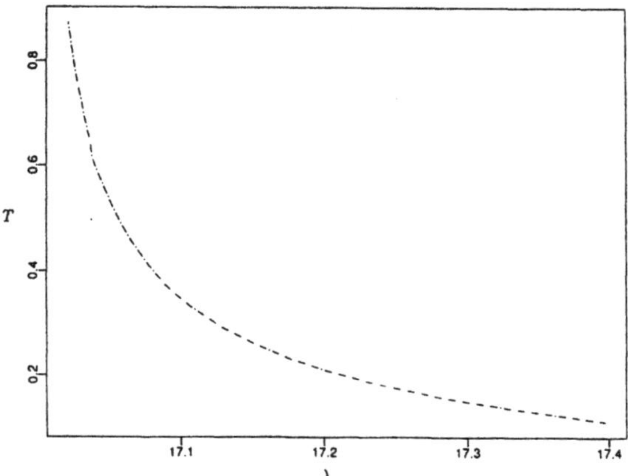

Fig 5.2 Variation in period along the modulated TW branch.

section. The value of N used was 15 and an increasing number of mesh points in the time variable were taken as the computation proceeded. The results are shown in Fig. 5.1. As expected, the period increases as λ falls as shown in Fig. 5.2. AUTO automatically adjusts the position of the mesh points which enables the solution branch to be followed some distance as the period increases although a limit was reached beyond which it became too expensive to continue the computations. However, approximating the curve in Fig. 5.2 with a function of the form $1/(a\lambda - b)$ which passes through 2 points gives values of $a = 23.907$ and $b = 405.788$ resulting in a value of $\lambda = 16.9736$ at which the period becomes infinite. Also, approximating the graph of c against λ, as shown in Fig. 5.3, with a quartic polynomial passing through 5 points and finding

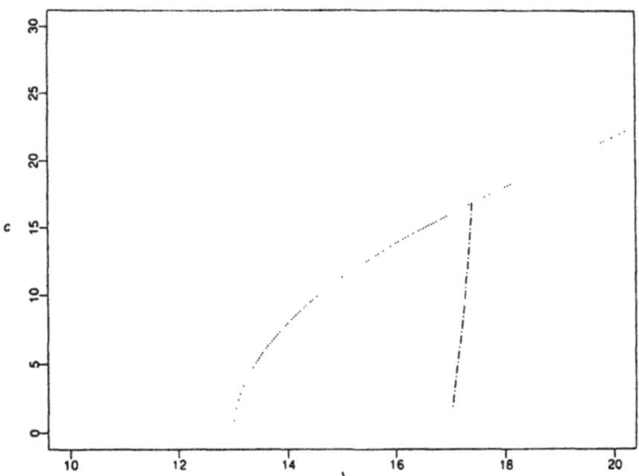

Fig 5.3 Variation in velocity along the modulated TW branch.

where the polynomial crosses the λ axis (ie. when $c = 0$) gives a value of $\lambda = 16.9785$ which is very close to the estimated value of λ at which the period becomes infinite, thus indicating that the branch of modulated travelling waves does indeed terminate at a heteroclinic cycle. In Fig. 5.4, c is plotted against the reciprocal of the period and it is clear that the extension of this path would go through the origin. Thus, our numerical results support the conjectures of Kevrekidis, Nicolaenko and Scovel (1990) and provide a more accurate estimate of the point at which the branch of modulated travelling waves terminates.

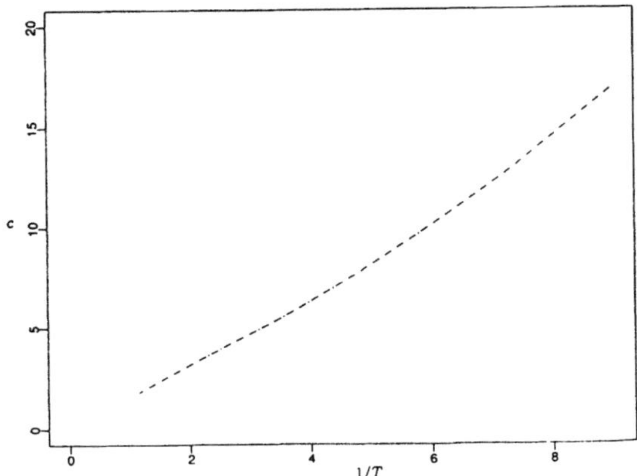

Fig. 5.4 Velocity against the reciprocal of the period on the modulated TW branch.

Finally, we mention that all of the 3 possible types of Hopf bifurcation associated with the different types of irreducible representation occur in this problem. The branch of travelling waves we have been considering can be rescaled by a factor of 2 or 3 to give branches which have solutions of period π or $2\pi/3$ respectively. The Hopf bifurcation already found also scales up onto these branches, but in addition, there are also Hopf bifurcations on the second branch at $\lambda = 63.391$ and on the third branch at $\lambda = 130.228$.

Acknowledgements

The authors acknowledge a correction made to an earlier version of this work by E. Knobloch.

References

Armbruster, D., Guckenheimer, J. and Holmes, P. (1989). Kuramoto-Sivashinsky dynamics on the center-unstable manifold. *SIAM J. Appl. Math.* **49**, 676-691.

Aston P.J. (1991). Analysis and computation of symmetry-breaking bifurcation and scaling laws using group theoretic methods. *SIAM. J. Math. Anal.*, **22**, 181-212.

Aston P.J., Spence A. and Wu W. (1992). Bifurcation to rotating waves in equations with O(2)-symmetry. To appear in *SIAM J. Appl. Math.*

Dellnitz. M. (1991). Computational bifurcation of periodic solutions in systems with symmetry. To appear in *IMA J. Num. Anal.*

Doedel, E.J. (1981). AUTO : A program for the automatic bifurcation analysis of autonomous systems. *Congressus Numerantium* **30**, 265-284.

Golubitsky M., Stewart I. and Schaeffer D.G. (1988). *Singularities and Groups in Bifurcation Theory, Vol. II*. Appl. Math. Sci. **69**, Springer, New York.

Jepson, A.D. and Keller, H.B. (1984). Steady state and periodic solution paths : their bifucations and computations. In *Numerical Methods for Bifurcation Problems*, Eds. T. Küpper, H.D. Mittelmann and H. Weber, ISNM **70**, Birkhäuser.

Kevrekidis, I.G., Nicolaenko, B. and Scovel, J.C. (1990). Back in the saddle again : A computer assisted study of the Kuramoto-Sivashinsky equation. *SIAM J. Appl. Math.* **50**, 760-790.

Krupa, M. (1990). Bifurcations of relative equilibria. *SIAM J. Math. Anal.* **21**,1453-1486.

Vanderbauwhede, A., Krupa, M. and Golubitsky, M. (1989). Secondary bifurcations in symmetric systems. *Proceedings of the EQUADIFF Conference*, eds. C.M. Dafermos, G. Ladas and G. Papanicolaou, 709-716.

Wu,W, Aston, P.J. and Spence, A. (1991). Rotating waves from Hopf bifurcations in equations with O(2) symmetry. Submitted to *SIAM J. Sci. Stat. Comp.*.

International Series of Numerical Mathematics, Vol. 104, © 1992 Birkhäuser Verlag Basel

Mode Interactions of an Elliptic System on the Square

Klaus Böhmer

Fachbereich Mathematik, Universität Marburg, 3550 Marburg/Lahn, FRG

Mei Zhen*

Fachbereich Mathematik, Universität Marburg, 3550 Marburg/Lahn, FRG and
Department of Mathematics, Xi'an Jiaotong University, Xi'an 710049, PRC

Abstract

Mode interactions of a semilinear system of second order elliptic differential equations are discussed at the intersection points of two curves of bifurcation points, using a modified Lyapunov-Schmidt method and symmetries.

1. Introduction

We consider bifurcations of a semilinear system of second order elliptic differential equations of the form:

$$G(u, \lambda, d) := \begin{pmatrix} \Delta u_1 + \lambda f_1(u_1, u_2) \\ d\Delta u_2 + \lambda f_2(u_1, u_2) \end{pmatrix} = 0, \tag{1.1}$$

where λ, $d \in \mathbf{R}, d \neq 0$, $u = (u_1, u_2)^T \in X$ and $G : X \times \mathbf{R}^2 \mapsto Y$ with

$$X := \left\{ u = (u_1, u_2)^T \mid u_i \in C^{2,s}(\Omega), \ u_i|_{\partial\Omega} = 0, \ i = 1, 2 \right\}, \tag{1.2}$$

$$Y := \left\{ u = (u_1, u_2)^T \mid u_i \in C^{0,s}(\Omega), \ i = 1, 2 \right\}. \tag{1.3}$$

The domain Ω is chosen as the square $[0, 1] \times [0, 1]$ and the functions $f_i : \mathbf{R}^2 \mapsto \mathbf{R}, i = 1, 2$, are smooth. Let $C^{k,s}$ denote the space of all k-times differentiable functions u, such that u and its derivatives up to the order k are locally Hölder continuous with exponent s.

The system (1.1) describes steady state of the so-called reaction and diffusion equations. These problems occur widely as models for the dynamics of multispecies populations whose individuals are capable of random spatial migration, as models in chemical reaction, biological pattern formations, nerve conductions, see e.g., [4], [15], [16] and the Fitzhugh-Nagumo model in [14].

Let us introduce an L^2- or dual product in X:

$$\langle u, v \rangle = \int_\Omega (u_1 v_1 + u_2 v_2) dx dy \quad \text{for all} \ u = (u_1, u_2)^T, v = (v_1, v_2)^T \in X. \tag{1.4}$$

To simplify the discussion we consider in this paper a special class of (1.1), i.e., we choose f_1, f_2 to be odd functions with $f_i(-x, -y) = -f_i(x, y)$, $i = 1, 2$ for all $x, y \in \mathbf{R}$. Examples for models with odd functions f_i may be found in [6], [10] and in [17]. In this case,

$$f(0) = 0, \qquad D^2 f(0) = 0 \quad \text{for} \ f(u) := (f_1(u), f_2(u)) \tag{1.5}$$

---

* The work was supported by the Deutsche Forschungsgemeinschaft, F. R. Germany.

and $\{(0, \lambda, d); \lambda, d \in \mathbf{R}\}$ is a two dimensional manifold of trivial solutions of (1.1). Assume

$$Df(0) \neq 0, \quad D^3 f(0) \neq 0. \tag{1.6}$$

We want to study solution branches of (1.1) bifurcating from the trivial solution curve. Since the domain Ω has the symmetry $D_4 := \{S_1, S_2; S_1', S_2', I, R, R^2, R^3\}$ with the action

$$S_1(x, y) = (1 - x, y), \quad S_2(x, y) = (y, x) \quad \text{for all } (x, y) \in \Omega \tag{1.7}$$

and the mapping G is autonomous and odd in u, we define a group $\Gamma := Z_2 \times D_4$ with the action of Γ on u as follows:

$$\gamma u(x, y) = \begin{pmatrix} \gamma u_1(x, y) \\ \gamma u_2(x, y) \end{pmatrix} = \begin{pmatrix} \pm u_1(\sigma^{-1}(x, y)) \\ \pm u_2(\sigma^{-1}(x, y)) \end{pmatrix}, \quad \text{for all } \gamma = \pm \sigma \in \Gamma, \ \sigma \in D_4. \tag{1.8}$$

It is easy to verify that the mapping G is Γ-equivariant, i.e.,

$$G(\gamma u, \lambda, d) = \gamma G(u, \lambda, d) \quad \text{for all } \gamma \in \Gamma, \ u \in X, \ \lambda, d \in \mathbf{R}. \tag{1.9}$$

Consequently, equivariant bifurcation theory may be used to study the bifurcations of (1.1), see e.g., [2], [5], [7], [9], [18].

An outline of the paper is as follows. In Section 2 we determine the bifurcation points of (1.1) on the manifold $\{(0, \lambda, d); \lambda, d \in \mathbf{R}\}$ of trivial solutions and all possible intersection points of two curves of bifurcation points. Section 3 describes null spaces of $D_u G$, $D_u G^*$ at the intersection points. In Section 4 we discuss the mode interactions of (1.1) at these points with consideration on the symmetries of (1.1).

2. Intersection Points of Bifurcation Curves

Due to $D_\lambda G(0, \lambda, d) = D_d G(0, \lambda, d) = 0$ for all λ, $d \in \mathbf{R}$, bifurcation points on the manifold of trivial solutions are those points $(0, \lambda, d) \in X \times \mathbf{R}^2$, where $D_u G(0, \lambda, d)u = 0$ admits nontrivial solutions. For simplicity we will suppress the phase coordinate 0 of $(0, \lambda, d)$ in the sequel and show that these points (λ, d) are element of (different) quadratic curves $S(L_c)$, see (2.9). We are interested in the intersection points of these curves.

For any $(\lambda, d) \in \mathbf{R}^2$, simple calculations yield:

$$D_u G(0, \lambda, d) = \begin{pmatrix} \Delta + \lambda f^0_{1u_1} & \lambda f^0_{1u_2} \\ \lambda f^0_{2u_1} & d\Delta + \lambda f^0_{2u_2} \end{pmatrix}. \tag{2.1}$$

Here and below, if no confusion arises, $D_u G^0, DG^0, f^0_{1u_1}, \ldots$ denote the evaluations of $D_u G, DG, \frac{\partial f_1}{\partial u_1}, \ldots$ at the point $(0, \lambda, d)$. Since $\{\sin m\pi x \sin n\pi y, \ m, n \in \mathbf{N}\}$ are eigenfunctions of the Laplacian Δ on Ω with Dirichlet boundary conditions, they determine an orthogonal basis for $L^2(\Omega)$. Consequently,

$$\{(a, b)^T \sin m\pi x \sin n\pi y; \ m, n \in \mathbf{N}, \ (a, b) = (1, 0) \text{ or } (0, 1)\} \tag{2.2}$$

is an orthogonal basis of the space X. At the same time, one sees easily that the subspace

$$\mathbf{R}^2 \sin m\pi x \sin n\pi y = \{(a, b)^T \sin m\pi x \sin n\pi y; \ a, b \in \mathbf{R}\} \tag{2.3}$$

for each $m, n \in \mathbf{N}$ is invariant under $D_u G^0$. Thus, for any given $(\lambda, d) \in \mathbf{R}^2$, the operator $D_u G^0$ is singular if and only if there exist nontrivial $a, b \in \mathbf{R}$ and $m, n \in \mathbf{N}$, such that

$$D_u G^0 \begin{pmatrix} a \\ b \end{pmatrix} \sin m\pi x \sin n\pi y = 0. \tag{2.4}$$

This yields a system for $a, b \in \mathbf{R}$ and $m, n \in \mathbf{N}$:

$$M(\lambda, d, c) \begin{pmatrix} a \\ b \end{pmatrix} := \begin{pmatrix} \lambda f^0_{1u_1} - c & \lambda f^0_{1u_2} \\ \lambda f^0_{2u_1} & \lambda f^0_{2u_2} - dc \end{pmatrix} \begin{pmatrix} a \\ b \end{pmatrix} = 0, \tag{2.5}$$

here and in the sequel,

$$c := (m^2 + n^2)\pi^2, \quad L_c := m^2 + n^2 > 0. \tag{2.6}$$

The system (2.5), and hence (2.4), has nontrivial solutions if and only if the determinant of its coefficient matrix vanishes, i.e. λ, d have to satisfy the equation:

$$\det(M(\lambda, d, c)) = \delta\lambda^2 - c(f^0_{1u_1}d + f^0_{2u_2})\lambda + c^2 d = 0, \tag{2.7}$$

where

$$\delta := \det(Df(0)) = f^0_{1u_1}f^0_{2u_2} - f^0_{2u_1}f^0_{1u_2}. \tag{2.8}$$

For a given c, the subset of bifurcation points $(0, \lambda, d)$ of (1.1) defines a bifurcation curve:

$$S(L_c) := \{(\lambda, d) \in \mathbf{R}^2 \mid \lambda, d \text{ satisfy } (2.7) \text{ and } d \neq 0\}. \tag{2.9}$$

Lemma 2.1: *For any $(\lambda, d) \in \mathbf{R}^2$, the operator $D_u G^0$ is singular if and only if there exist $m, n \in \mathbf{N}$ such that $(\lambda, d) \in S(L_c)$.*

Remark 2.1: $S(L_c)$ represents a conic section, in particular, one or two straight lines, a hyperbola or a parabola. Ellipses do not occur, see e.g., [3], [8], [11]. Hence, $(\lambda, d) \in S(L_c) \cap S(L_{c'}) \cap S(L_{c''})$ with $c \neq c' \neq c'' \neq c$ if and only if $(\lambda, d) = (0, 0)$.

Proof: Since (2.2) represents an orthogonal basis for X and since the subspace in (2.3) is invariant under $D_u G^0$, a nontrivial $v \in N(D_u G^0)$ has to have a nontrivial component in (2.3), hence (2.4), (2.5) have to admit a nontrivial solution, in other words, (2.7) has to be satisfied and vice versa. ∎

The equation (2.7) characterizes, for a given c, those (λ, d) which yield an eigenvalue zero for $M(\lambda, d, c)$. Depending on whether (λ, d) is an element of one or two different bifurcation curves $S(L_c)$, the structure of solution branches of (1.1) changes dramatically. If (λ, d) is an element of only one $S(L_c)$, bifurcations of (1.1) at (λ, d) can be studied by a modified Lyapunov-Schmidt method, see e.g., [1], [3]. In this paper, we want to study the mode interactions of (1.1) at intersection points of two different $S(L_c)$.

Let L_{c_1}, L_{c_2} correspond to $c_1 \neq c_2$, respectively. The characteristic equations of the bifurcation curves $S(L_{c_1})$, $S(L_{c_2})$ are

$$\delta\lambda^2 - c_1(f^0_{1u_1}d + f^0_{2u_2})\lambda + c_1^2 d = 0, \tag{2.10}$$

$$\delta\lambda^2 - c_2(f^0_{1u_1}d + f^0_{2u_2})\lambda + c_2^2 d = 0. \tag{2.11}$$

Obviously, this system has a trivial solution $(\lambda, d) = (0,0)$ which is excluded in (1.1) by the assumption $d \neq 0$. We will solve it in the following steps:

1)   If $\delta = 0$ then due to $c_1 \cdot c_2 \cdot (c_1 - c_2) \neq 0$, (2.10) and (2.11) have merely the trivial solution $(\lambda, d) = (0,0)$ and

$$S(L_{c_1}) \cap S(L_{c_2}) = \emptyset. \tag{2.12}$$

2)   If $\delta \neq 0$, $f^0_{1u_1} = 0$, the difference of (2.10) and (2.11) yields

$$d = f^0_{2u_2}\,\lambda/(c_1 + c_2). \tag{2.13}$$

Substituting this into (2.10),

$$[\delta\lambda + c_1 c_2(c_1 + c_2)^{-1} f^0_{2u_2}]\lambda = 0. \tag{2.14}$$

For $f^0_{2u_2} \neq 0$ this equation has exactly one nontrivial solution:

$$(\lambda_1, d_1) := -\frac{c_1 c_2 f^0_{2u_2}}{\delta(c_1 + c_2)}\left(1, \frac{f^0_{2u_2}}{c_1 + c_2}\right) \quad \Longrightarrow \quad S(L_{c_1}) \cap S(L_{c_2}) = \{(\lambda_1, d_1)\}. \tag{2.15}$$

3)   If, finally $\delta \cdot f^0_{1u_1} \neq 0$, the equations (2.10), (2.11) are equivalent to

$$\frac{\delta}{c_1}\lambda^2 - (f^0_{1u_1}d + f^0_{2u_2})\lambda + c_1 d = 0, \tag{2.16}$$

$$(f^0_{1u_1}d + f^0_{2u_2})\lambda - (c_1 + c_2)d = 0. \tag{2.17}$$

Taking the sum of (2.16) and (2.17) we find $d = \delta\lambda^2/(c_1 c_2)$. Substituting this into (2.17) one gets an equation for λ:

$$[\delta f^0_{1u_1}\lambda^2 - (c_1 + c_2)\delta\lambda + c_1 c_2 f^0_{2u_2}]\lambda = 0.$$

For $f^0_{2u_2} \neq 0$ it has two nontrivial solutions:

$$\lambda_\pm := \{(c_1 + c_2)\delta \pm \sqrt{[\delta(c_1 + c_2)]^2 - 4c_1 c_2\delta f^0_{1u_1} f^0_{2u_2}}\,\}/(2\delta f^0_{1u_1}). \tag{2.18}$$

Thus we have two different intersection points

$$S(L_{c_1}) \cap S(L_{c_2}) = \{(\lambda_+, d_+), \lambda_-, d_-)\}. \tag{2.19}$$

Simple calculations yield

$$N(M(\lambda, d, c_i)) = \mathrm{span}[a_i], \quad N(M(\lambda, d, c_i)^T) = \mathrm{span}[a_i^*], \quad i = 1,2 \tag{2.20}$$

and

$$a_i := \begin{cases} (-\lambda f^0_{1u_2}, \lambda f^0_{1u_1} - c_i)^T & \text{for } (\lambda, d) = (\lambda_1, d_1) \text{ or } (\lambda_\pm, d_\pm) \text{ with } f^0_{1u_2} f^0_{2u_1} \neq 0; \\ (-\lambda f^0_{2u_1}, \lambda f^0_{2u_2} - dc_1)^T & \text{for } (\lambda, d) = (\lambda_+, d_+) \text{ with } f^0_{1u_2} f^0_{2u_1} = 0, \ i = 1; \\ (-\lambda f^0_{1u_2}, \lambda f^0_{1u_1} - c_2)^T & \text{for } (\lambda, d) = (\lambda_+, d_+) \text{ with } f^0_{1u_2} f^0_{2u_1} = 0, \ i = 2; \\ (-\lambda f^0_{1u_2}, \lambda f^0_{1u_1} - c_1)^T & \text{for } (\lambda, d) = (\lambda_-, d_-) \text{ with } f^0_{1u_2} f^0_{2u_1} = 0, \ i = 1; \\ (-\lambda f^0_{2u_1}, \lambda f^0_{2u_2} - dc_2)^T & \text{for } (\lambda, d) = (\lambda_-, d_-) \text{ with } f^0_{1u_2} f^0_{2u_1} = 0, \ i = 2 \end{cases} \tag{2.21}$$

and

$$
\mathbf{a}_i^* := \begin{cases}
(-\lambda f_{2u_1}^0, \lambda f_{1u_1}^0 - c_i)^T & \text{for } (\lambda, d) = (\lambda_1, d_1) \text{ or } (\lambda_\pm, d_\pm) \text{ with } f_{1u_2}^0 f_{2u_1}^0 \neq 0; \\
(-\lambda f_{1u_2}^0, \lambda f_{2u_2}^0 - dc_1)^T & \text{for } (\lambda, d) = (\lambda_+, d_+) \text{ with } f_{1u_2}^0 f_{2u_1}^0 = 0, \ i = 1; \\
(-\lambda f_{2u_1}^0, \lambda f_{1u_1}^0 - c_2)^T & \text{for } (\lambda, d) = (\lambda_+, d_+) \text{ with } f_{1u_2}^0 f_{2u_1}^0 = 0, \ i = 2; \\
(-\lambda f_{2u_1}^0, \lambda f_{1u_1}^0 - c_1)^T & \text{for } (\lambda, d) = (\lambda_-, d_-) \text{ with } f_{1u_2}^0 f_{2u_1}^0 = 0, \ i = 1; \\
(-\lambda f_{1u_2}^0, \lambda f_{2u_2}^0 - dc_2)^T & \text{for } (\lambda, d) = (\lambda_-, d_-) \text{ with } f_{1u_2}^0 f_{2u_1}^0 = 0, \ i = 2.
\end{cases} \tag{2.22}
$$

3. The Null Spaces of $D_u G^0$ and $(D_u G^0)^*$

In this section we first derive $(D_u G^0)^*$ and then determine the null spaces of the operators $D_u G^0$ and $(D_u G^0)^*$. They are closely related and are necessary in our descriptions of the structure of solution manifolds bifurcating from the bifurcation curve $S(L_c)$.

Under the L^2-product (1.4) in X, the adjoint operator for $D_u G^0$ can be derived from

$$
\langle v, D_u G^0 u \rangle = \left\langle \begin{pmatrix} v_1 \\ v_2 \end{pmatrix}, \begin{pmatrix} \Delta u_1 + \lambda f_{1u_1}^0 u_1 + \lambda f_{1u_2}^0 u_2 \\ d\Delta u_2 + \lambda f_{2u_1}^0 u_1 + \lambda f_{2u_2}^0 u_2 \end{pmatrix} \right\rangle
$$

$$
= \int_\Omega [v_1(\Delta u_1 + \lambda f_{1u_1}^0 u_1 + \lambda f_{1u_2}^0 u_2) + v_2(d\Delta u_2 + \lambda f_{2u_1}^0 u_1 + \lambda f_{2u_2}^0 u_2)]dxdy
$$

$$
= \int_\Omega [u_1(\Delta v_1 + \lambda f_{1u_1}^0 v_1 + \lambda f_{2u_1}^0 v_2) + u_2(d\Delta u_2 + \lambda f_{1u_2}^0 v_1 + \lambda f_{2u_2}^0 v_2)]dxdy
$$

$$
= \left\langle \begin{pmatrix} u_1 \\ u_2 \end{pmatrix}, \begin{pmatrix} \Delta v_1 + \lambda f_{1u_1}^0 v_1 + \lambda f_{2u_1}^0 v_2 \\ d\Delta v_2 + \lambda f_{1u_2}^0 v_1 + \lambda f_{2u_2}^0 v_2 \end{pmatrix} \right\rangle \qquad \forall u, v \in X.
$$

Hence

$$
D_u G^{0*} := (D_u G^0)^* = \begin{pmatrix} \Delta + \lambda f_{1u_1}^0 & \lambda f_{2u_1}^0 \\ \lambda f_{1u_2}^0 & d\Delta + \lambda f_{2u_2}^0 \end{pmatrix}. \tag{3.1}
$$

Let $c = (m^2 + n^2)\pi^2$ be an l-fold eigenvalue of the Laplacian Δ on Ω with Dirichlet boundary conditions, i.e., there are l different pairs of (m_i, n_i), $i = 1, \ldots, l$, such that, with $(m_i, n_i) = (n_j, m_j)$ for some $i \neq j$,

$$
c = (m_i^2 + n_i^2)\pi^2, \quad \text{and} \quad \phi_i := 2\sin m_i\pi x \sin n_i\pi y, \quad i = 1, \ldots, l. \tag{3.2}
$$

Theorem 3.1: *Let $(\lambda_0, d_0) \in S(L_{c_1}) \cap S(L_{c_2})$, where $c_1 \neq c_2$ represent l_1 and l_2 fold eigenvalues of the Laplacian Δ on Ω with Dirichlet boundary conditions, respectively. Then the null spaces of $D_u G^0$ and $D_u G^{0*}$ have the following form:*

$$
\begin{cases}
N(D_u G^0) = \operatorname{span}[\Phi_i(L_{c_j}), i = 1, \ldots, l_j, j = 1, 2], \\
N(D_u G^{0*}) = \operatorname{span}[\Phi_i^*(L_{c_j}), i = 1, \ldots, l_j, j = 1, 2], \quad X = N(D_u G^0) \oplus R(D_u G^{0*}),
\end{cases} \tag{3.3}
$$

where the $\Phi_i(L_{c_j}), \Phi_i^*(L_{c_j})$ are defined with the l_j different ϕ_i in (3.2) and $\mathbf{a}_j$, $\mathbf{a}_j^*$ in (2.21) (2.22) as

$$
\Phi_i(L_{c_j}) := \mathbf{a}_j \phi_i, \qquad \Phi_i^*(L_{c_j}) := \mathbf{a}_j^* \phi_i, \quad i = 1, \ldots, l_j, \ j = 1, 2. \tag{3.4}
$$

4. Mode Interactions

Along the bifurcation curve $S(L_{c_1})$ the dimension of the null space $N(D_u G(0, \lambda, d))$ jumps at the intersection points of $S(L_{c_1})$ with another bifurcation curve $S(L_{c_2})$. New bifurcating solution branches, or the so-called mode interactions are expected at these points (cf. [8], [11]). For reasons of simplifications we confine the discussions to the special case $f^0_{1u_2} \cdot f^0_{2u_1} \neq 0$ and to the points $(\lambda_0, d_0) \in S(L_{c_1}) \cap S(L_{c_2})$ with

$$L_{c_1} \longleftrightarrow c_1 = (m_1^2 + n_1^2)\pi^2, \qquad L_{c_2} \longleftrightarrow c_2 = (m_2^2 + n_2^2)\pi^2 \neq c_1, \tag{4.1}$$

where c_1, c_2 are 2-fold eigenvalues of the Laplacian Δ on Ω with Dirichlet boundary conditions. The spaces $N(D_u G^0)$ and $N(D_u G^{0^*})$ are four dimensional and the basis elements Φ_i, $i = 1, \ldots, 4$, are chosen as follows, see (3.4)

$$\Phi_i(x,y) := \Phi_1(L_{c_i})(x,y), \qquad \Phi_{i+2}(x,y) := \Phi_i(y,x), \ i = 1, 2. \tag{4.2}$$

The elements Φ_i^*, $i = 1, \ldots, 4$, are defined similarly. We fix the parameter d as d_0 and consider (1.1) as a problem dependent on only one parameter λ. In fact, for variable d two dimensional bifurcating manifolds are expected. These solutions branches needs special parametrizations and will be discussed at another place. If $(u(t), \lambda(t))$ is a smooth solution curve of (1.1) through the point $(0, \lambda_0)$ and

$$(u(0), \lambda(0)) = (0, \lambda_0), \quad (\dot{u}(0), \dot{\lambda}(0)) \neq 0, \tag{4.3}$$

then one sees from Theorem 3.1 that there are smooth functions $\alpha_i(t)$, $\beta(t) \in \mathbf{R}$ and $w(t) \in R(D_u G^{0^*})$, $i = 1, \ldots, 4$, such that

$$(u(t), \lambda(t)) = \left(t \sum_{i=1}^{4} \alpha_i(t)\Phi_i + tw(t), \ \lambda_0 + t\beta(t) \right). \tag{4.4}$$

To determine the functions $\alpha_i(t), \beta(t)$ and $w(t)$ in (4.4), we differentiate the equation

$$G(u(t), \lambda(t), d_0) = 0 \tag{4.5}$$

with respect to t at $t = 0$ and obtain $D_u G^0 w(0) = 0$. Hence,

$$w(0) = 0. \tag{4.6}$$

The second derivative of (4.5) with respect to t at $t = 0$ yields with $\alpha_i := \alpha_i(0)$, $\beta := \beta(0)$,

$$\beta D_{u\lambda} G^0 \sum \alpha_i \Phi_i + D_u G^0 \dot{w}(0) = 0. \tag{4.7}$$

Applying $\langle \Phi_j^*, \cdot \rangle$ to (4.7), we derive

$$(-\lambda f^0_{2u_1}, \lambda f^0_{1u_1} - c) \begin{pmatrix} f^0_{1u_1} & f^0_{1u_2} \\ f^0_{2u_1} & f^0_{2u_2} \end{pmatrix} \begin{pmatrix} -\lambda f^0_{1u_2} \\ \lambda f^0_{1u_1} - c \end{pmatrix} \left(\phi_j, \sum \alpha_i \phi_i \right) \beta = 0,$$

where $c = c_1$ or c_2 with the corresponding functions ϕ_1, ϕ_2 in (3.2). The orthogonality of ϕ_1, ϕ_2 with $c_1 \neq c_2$ provides

$$(\delta f^0_{1u_1} \lambda^2 - 2c\delta\lambda + c^2 f^0_{2u_2})\alpha_j\beta = 0, \qquad j = 1, \ldots, 4.$$

If

$$a_0 := \delta f^0_{1u_1} \lambda^2 - 2c\delta\lambda + c^2 f^0_{2u_2} \neq 0, \tag{4.8}$$

then we have either

$$\text{(i)} \quad \beta \neq 0, \ \alpha_j = 0, j = 1, \ldots, 4 \quad \text{or} \quad \text{(ii)} \quad \beta = 0 \ \text{ with } \sum |\alpha_j| > 0, \tag{4.9}$$

In both cases the equation (4.7) induces that $D_u G^0 \dot{w} = 0$. Thus at $t = 0$,

$$\dot{w} = 0. \tag{4.10}$$

Consequently, by rescaling the functions $\alpha_i(t)$ and $\beta(t)$, we obtain

Theorem 4.1: *If the conditions (4.1) and (4.8) are satisfied, then the solution curve (4.4) has the form either*

$$(u(t), \lambda(t)) = \left(t^2 \sum_{i=1}^{4} \alpha_i(t)\Phi_i + t^3 w(t), \lambda_0 + t\beta(t) \right) \text{ with } \beta(0) \neq 0, w(t) \in R(D_u G^{0*}) \tag{4.11}$$

or

$$(u(t), \lambda(t)) = \left(t \sum_{i=1}^{4} \alpha_i(t)\Phi_i + t^3 w(t), \lambda_0 + t^2\beta(t) \right), \sum_{i=1}^{4} |\alpha_i(0)| > 0, \ w(t) \in R(D_u G^{0*}). \tag{4.12}$$

The branch (4.11) corresponds usually to the trivial solution curve $\{(0, \lambda, d_0); \lambda \in \mathbf{R}\}$, see [13]. We are interested in the solution branches (4.12) satisfying

$$\beta(0) \neq 0. \tag{4.13}$$

The inequality (4.13) holds for all bifurcating solution branches if the third derivative of f satisfies proper conditions, see [3], [13]. If $\beta(0) > 0$, the function $\lambda(t)$ in (4.12) may be locally rewritten as

$$\lambda(t) = \lambda_0 + t^2. \tag{4.14}$$

If $\beta(0) < 0$, then t^2 in (4.14) should be replaced by $-t^2$. The other functions $\alpha_i(t), i = 1, \ldots, 4, w(t)$ may be determined by the modified Lyapunov-Schmidt method, see e.g., [1], [13]. Nevertheless, the reduced bifurcation equations for $\alpha_1, \ldots, \alpha_4$ are rather complicated and difficult to solve. We want to make use of the symmetries of (1.1) to simplify the discussion. Obviously, symmetries of the basis functions Φ_i in (4.2) depend on the parity of m_i, n_i and their combinations. For example, for

$$(m_1, n_1) = (\text{even, even}) \quad \text{and} \quad (m_2, n_2) = (\text{odd, odd}), \tag{4.15}$$

the basis functions Φ_1, Φ_2 have different symmetries along $S(L_{c_1})$ and $S(L_{c_2})$ and

$$\Sigma_{\Phi_i} = \{-S_1, -S_1'; I, R^2\}, \quad i = 1, 3, \quad \Sigma_{\Phi_i} = \{S_1, S_1'; I, R^2\}, \quad i = 2, 4. \tag{4.16}$$

In this case, all solution branches of (1.1) bifurcating at $S(L_{c_i})$ persist at $(0, \lambda_0, d_0)$. As an example, let us consider solution branches with the symmetry $\Sigma_1 := \Sigma_{\Phi_1}$ and the reduced problem:

$$Find \quad (u, \lambda) \in X^{\Sigma_1} \times \mathbf{R}, \quad such\ that \quad G^{\Sigma_1}(u, \lambda) = 0, \tag{4.17}$$

where $X^{\Sigma_1} := \{u \in X \mid \sigma u = u \text{ for all } \sigma \in \Sigma_1\}$ and $G^{\Sigma_1} := G|_{X^{\Sigma_1} \times \mathbf{R}}$. Now $(0, \lambda_0)$ is a corank-2 bifurcation point of (4.17) and

$$N(D_u G^{\Sigma_1}(0, \lambda_0)) = \text{span}[\Phi_1, \Phi_3], \quad N((D_u G^{\Sigma_1}(0, \lambda_0))^*) = \text{span}[\Phi_1^*, \Phi_3^*]. \tag{4.18}$$

Solution branches of (4.17) corresponding to (4.12) are in the form

$$(u(t),\ \lambda(t)) = (t(\alpha_1(t)\Phi_1 + \alpha_3(t)\Phi_3) + t^3 w(t),\ \lambda_0 + t^2). \tag{4.19}$$

To determine these solution branches, we define an enlarged system:

$$F(w, \alpha_1, \alpha_3, t) := \begin{pmatrix} G^{\Sigma_1}(t(\alpha_1\Phi_1 + \alpha_3\Phi_3) + t^3 w,\ \lambda_0 + t^2)/t^3 \\ \langle \Phi_1, w \rangle \\ \langle \Phi_3, w \rangle \end{pmatrix} = 0, \tag{4.20}$$

where $F : X \times \mathbf{R}^2 \times \mathbf{R} \mapsto Y \times \mathbf{R}^2$ and at $t = 0$, the mapping F, correspondingly, the enlarged system (4.20) is defined by its limit:

$$F(w, \alpha_1, \alpha_3, 0) := \begin{pmatrix} D_u G^{\Sigma_1}0 w + D_{u\lambda}G^{\Sigma_1}0(\alpha_1\Phi_1 + \alpha_3\Phi_3) + D_{uuu}G^{\Sigma_1}0(\alpha_1\Phi_1 + \alpha_3\Phi_3)^3/6 \\ \langle \Phi_1, w \rangle \\ \langle \Phi_3, w \rangle \end{pmatrix} = 0. \tag{4.21}$$

The smoothness of the mapping F follows directly from the smoothness of G. Taking L^2-products of $\Phi_j^*, j = 1, 3$ with the first equation in (4.21), one gets a system for α_1, α_3:

$$(3\alpha_1^2 + 4\alpha_3^2 + b)\alpha_1 = 0,$$

$$(4\alpha_1^2 + 3\alpha_3^2 + b)\alpha_2 = 0 \quad \text{with} \quad b := \frac{8\mathbf{a}_1^{*T}Df(0)\mathbf{a}_1}{\lambda_0\mathbf{a}_1^{*T}D^3f(0)\mathbf{a}_1^3}. \tag{4.22}$$

For $b < 0$, the system (4.22) has eight isolated solutions. Correspondingly, (4.21) has eight nonsingular solutions, which in turn, leads to eight solution branches of (4.20) by the implicit function theorem and eight solution branches of (1.1) passing through $(0, \lambda_0, d_0)$, see e.g., [1], [13]. Due to the oddness of f, exactly four of the eight solution branches are different. We refer to [3] for the details.

The other combinations of (m_i, n_i), $i = 1, 2$ can be studied similarly. In particular, if m_1, m_2 are even and n_1, n_2 are odd, respectively, new solution branches emerge at

$(0, \lambda_0, d_0)$ in the form of mixed mode of $S(L_{c_1})$ and $S(L_{c_2})$. In fact, now the basis functions Φ_1, Φ_2 have the same symmetries along $S(L_{c_1})$ and $S(L_{c_2})$ and

$$\Sigma_{\Phi_i} = \{-S_1, S_1'; I, -R^2\}, \quad i = 1, 2, \quad \Sigma_{\Phi_i} = \{S_1, -S_1'; I, -R^2\}, \quad i = 3, 4. \tag{4.23}$$

The subgroups Σ_{Φ_1} and Σ_{Φ_3} are conjugate. The solution branches with the symmetry $\Sigma_1 := \Sigma_{\Phi_1}$ are in the form

$$(u(t), \lambda(t)) = (t(\alpha_1(t)\Phi_1 + \alpha_2(t)\Phi_2) + t^3 w(t), \ \lambda_0 + t^2). \tag{4.24}$$

and can be determined by an enlarged system similarly as (4.20), (4.21). The reduced bifurcation equations for α_1, α_2 are

$$\frac{\lambda_0}{6}\langle \Phi_j^*, D^3 f(0)(\alpha_1 \Phi_1 + \alpha_2 \Phi_2)^3\rangle + \alpha_j \langle \Phi_j^*, Df(0)\Phi_j\rangle = 0, \quad j = 1, 2. \tag{4.25}$$

This system is usually more complicated than (4.22) and not all solutions of it are real, see e.g., (4.27). Thus the existence of solution branches at $(0, \lambda_0, d_0)$ with the mixed modes of $S(L_{c_1})$ and $S(L_{c_2})$ depends also on the mode numbers m_i, n_i, $i = 1, 2$.

Example 4.1: (cf. Kirchgässner [10]) Consider the model (1.1) with

$$f(u) = \begin{pmatrix} u_1 + u_2 - u_1^3 \\ -u_1 + Au_2 - u_1^2 u_2 \end{pmatrix}, \qquad A > 0, \quad d = 1. \tag{4.26}$$

For any c_1, c_2 in (2.6), choosing $A = (c_1 c_2)^{-1}(c_1 + c_2)^2 - 1$, one gets a intersection point of $S(L_{c_1})$ and $S(L_{c_2})$ as a real solution of (2.10), (2.11):

$$(\lambda_0, d_0) = (c_1 c_2/(c_1 + c_2), 1).$$

For $c_1 = 20\pi^2$, $c_2 = 10\pi^2$, we have $b = -27\pi^{-6}/2000$ in (4.22). Thus all four sloution branches of (1.1) bifurcating along $S(L_{c_1})$ persist at $(0, \lambda_0, d_0) = (0, 20\pi^2/3, 1)$. For $c_1 = 5\pi^2$ and $c_2 = 13\pi^2$, the bifurcation equations (4.25) are

$$\begin{aligned}
(3\alpha_1^2 + 2\alpha_2^2 - 2\alpha_1\alpha_2 - \frac{139968}{3570125\pi^6})\alpha_1 &= 0, \\
(2\alpha_1^2 + 3\alpha_2^2 - \frac{139968}{1373125\pi^6})\alpha_2 &= 0.
\end{aligned} \tag{4.27}$$

It has four real nontriviall solutions $\pm(216\sqrt{15}\pi^{-3}/4225, 0)$, $\pm(0, 216\sqrt{39}\pi^{-3}/4225)$, which lead to two different solution branches of (1.1) bifurcating at $(0, \lambda_0, d_0) = (0, 65\pi^2/18, 1)$.

References

[1] Allgower, E. L., Böhmer, K., Mei, Z.: A complete bifurcation scenario for the 2d-nonlinear Laplacian with Neumann boundary conditions on the unit square, in: *Bifurcation and Chaos: Analysis, Algorithms, Applications*, R. Seydel, F. W. Schneider, T. Küpper, H. Troger (Eds.), ISNM **97**, pp. 1-18, Birkhäuser Verlag, Basel 1991

[2] Allgower, E. L., Böhmer, K., Mei, Z.: An extended equivariant branching theory, to appear in *Math. Meth. in Appl. Sci.*, 1991

[3] Böhmer, K., Mei, Z.: Bifurcations for an elliptic system on the square, Department of Mathemtics, University of Marburg, preprint 1991

[4] Cantrell, R. S.: On coupled multiparameter nonlinear elliptic systems, *Trans. Amer. Math. Soci.* **294**, 263-289 (1986)

[5] Cicogna, G.: Symmetry breakdown from bifurcation, *Lett. Nuovo Cimento* **31**, 600–602 (1981)

[6] Cohen, H., Hoppensteadt, F. C., Miura, R. M.: Slowly-modulated oscillations in nonlinear diffusion process, *SIAM J. Appl. Math.* **33**, 217-229 (1977)

[7] Dellnitz, M., Werner, B.: Computational methods for bifurcation problems with symmetries-with special attention to steady state and Hopf bifurcation points, *J. Comp. Appl. Math.* **26**, 97-123 (1989)

[8] Eilbeck, J. C., Furter, J. E.: Understanding steady-state bifurcation diagrams for a model reaction-diffusion system, in: *Continuation and Bifurcations: Numerical Techniques and Applications*, D. Roose, B. De Dier, A. Spence (Eds.), NATO ASI Series C313, pp. 25-42, Kluwer Academic Publishers 1990

[9] Golubitsky, M., Stewart, I. N., Schaeffer, D. G.: *Singularities and Groups in Bifurcation Theory*, Vol. II, Springer-Verlag, Heidelberg Berlin New York 1988

[10] Kirchgässner, K.: Waves in weakly-coupled parabolic systems, in: *Nonlinear Analysis and Optimization*, C. Vinti (Ed.), Lecture Notes in Math. **1107**, Springer-Verlag, Berlin Heidelberg New York 1984

[11] López-Gómez, J., Duncan, K. N., Eilbeck, J. C., Molina, M.: Structure of solution manifolds in a strongly coupled elliptic system, Preprint 1990

[12] Mei, Z.: Solution branches at corank-2 bifurcation points with symmetry, in: *Bifurcations and Chaos: Analysis, Algorithms, Applications*, R. Seydel, F. W. Schneider, T. Küpper, H. Troger (Eds.), ISNM **97**, pp. 251-255, Birkhäuser Verlag, Basel 1991

[13] Mei, Z.: Bifurcations of a simplified buckling problem and the effect of discretizations, *Manuscripta Mathematica* **71**, 225-252 (1991)

[14] Miura, R. M.: A nonlinear WKB method and slowly-modulated oscillations in nonlinear diffusion processes, in: *Diffusion*, W. E. Fitzgibbon, H. F. Walker (Eds.), Research Notes in Math. **14**, pp. 155-170, Pitman London 1977

[15] Murray, J. D.: *Mathematical Biology*, Biomaths. Texts. **19**, Springer-Verlag, Berlin Heidelberg New York 1989

[16] Nicolis, G., Prigogine, I.: *Self-Organization in Nonequilibrium Systems: From Dissipative Structure to Order through Fluctuation*, Wiley, New York 1977

[17] Price, C. B., Wambacq, P., Oosterlinck, A.: Computing with reaction-diffusion systems: applications in image processing, in: *Continuation and Bifurcations: Numerical Techniques and Applications*, D. Roose, B. De Dier, A. Spence (Eds.), NATO ASI Series C313, pp. 379-388, Kluwer Academic Publishers 1990

[18] Vanderbauwhede, A.: *Local Bifurcation Theory and Symmetry*, Pitman London 1982

International Series of Numerical Mathematics, Vol. 104, © 1992 Birkhäuser Verlag Basel

Secondary, Tertiary and Quarternary States of Fluid Flow

by F.H. Busse and R.M. Clever
Institute of Physics, University of Bayreuth, W-8580 Bayreuth,
and Institute of Geophysics and Planetary Physics,
University of California at Los Angeles, CA 90024

Summary

Symmetry considerations are an important tool for the study of the evolution from simple to complex flows through subsequent bifurcations. For physical as well as mathematical reasons the attention is usually focused on the basic state with the highest degree of symmetry exhibiting the flow phenomena of interest. Only fluid systems that are steady in time and homogeneous with respect to two spatial directions will thus be considered in this paper. While rolls predominate as solutions for secondary states of fluid flow, a large variety of three-dimensional solutions can be realized as tertiary fluid states. Of particular interest are those states that are stationary at least with respect to a moving frame of reference. Even quarternary states can be realized in stationary form in special cases. Some new results will be presented for the case of thermal convection in a layer heated from below in the presence of plane Couette flow.

1. Introduction

The Rayleigh-Bénard convection layer and the Taylor-Couette system have long been used as the prime examples for the application of bifurcation theory in fluid mechanics. These two cases are the most studied and best known ones of a large class of systems in which series of transitions from simple to more complex forms of fluid flow can be observed. For reviews we refer to [1], [2]. Inclined convection layers or differentially heated Taylor-Couette systems already exhibit significant changes and more dramatic variations must be expected when mean flow in new directions are imposed. All of these systems have in common that the fluid is confined between two parallel walls and that they are approximately homogeneous in two spatial dimensions. In this paper we intend to consider this type of systems from a more general point of view and present some new results in particular cases.

Fluid systems that depend essentially only on a single spatial coordinate are of special interest to physicists since they tend to exhibit mechanisms of instability and transitions in their most simple form. Because of the high degree of symmetry, the bifurcation phenomena seen in those systems are also most attractive from an aesthetic point of view. While the realisation of uniformity in two dimensions poses some challenges to experimentalists, it simplifies the task of the numerical analysts who attempt to simulate those systems with the help of a computer. Symmetries play an essential role in such analyses in more than one way. The breaking of symmetries is a characteristic feature of bifurcations which may go unrecognized without it. Symmetries facilitate the mathematical analysis in that group-theoretical methods become applicable which permit a complete survey of the manifold of solutions. In programming computers the use of symmetries is particularly valuable because it can increase the efficiency by a large factor. There are thus many incentives to consider the version of a problem with a maximum degree of symmetries in its external conditions.

The analysis of this paper starts with the observation that the solutions bifurcating from the basic static state typically describe roll like fluid motions. Taylor vortices and convection rolls are but the most wellknown examples [1].[2].A large variety of solutions becomes available through secondary bifurcations. Physically preferred solutions appear to be characterized by at least two broken symmetries. In section 2 and 3 we try to summarize from a symmetry point of view the large body of work on secondary and tertiary states of fluid flow. In sections 4 and 5 we present some new results for the special example of convection in the presence of plane Couette flow.

2. Rolls and their instabilities

We consider fluid systems under steady external conditions that are homogeneous with respect to two spatial dimensions such as the fluid annulus between coaxial cylinders. While the basic state of the fluid reflects the symmetry of the system, the secondary state bifurcating from the basic state is typically characterized by a single wavevector $\underset{\sim}{\ell}$ in the plane spanned by the two homogeneous directions. There are degenerate bifurcation problems in which the preferred bifurcating solutions at the onset of instability is characterized by more than one wavevector $\underset{\sim}{\ell}$. A wellknown example are hexagonal cells in a Rayleigh-Bénard layer. But even in the latter case roll solutions described by a single vector $\underset{\sim}{\ell}$ are preferred under symmetric conditions. Such roll solutions can be written in the form

$$\varphi = \sum_{m,n} a_{mn} \exp\{im\,\underset{\sim}{\ell}\cdot\underset{\sim}{r}\}g_n(z) \qquad \text{with} \qquad a_{-mn} = a_{mn}^{*} \tag{2.1}$$

where φ is a representative variable (order parameter) of the fluid flow and where $\underset{\sim}{r}$ denotes the position vector. The subscript m runs through all integers, while the subscript n runs through all positive ones. The coordinate z is normal to the walls and the real functions g_n represent a complete system of functions satisfying individually the boundary conditions for φ. Besides steady solutions of the form (2.1) time periodic solutions may also occur. But this possibility will not be considered here.

In order to study the secondary solutions bifurcating from roll solutions of the form (2.1), it is convenient to focus the attention on rolls of maximal symmetries. Symmetry about the midplane of the fluid layer can be attained, for example, by the consideration of the small gap limit in the configuration of the cylindrical annulus and by the assumption of nearly the same angular velocities of the cylinders in the case of the Taylor-Couette system. The function $g_n(z)$ can thus be chosen such that the conditions

$$g_n(z) = (-1)^{n+1}g_n(-z) \tag{2.2}$$

are satisfied. In the following we shall use Cartesian coordinates with the y-coordinate in the direction of the preferred wavevector $\underset{\sim}{\ell}$. But the arguments remain the same if, for instance, helical coordinates for a cylindrical problem are used. The symmetry properties of rolls are given in table 1 where we have introduced the wavenumber $\alpha = |\underset{\sim}{\ell}|$.

Table 1: Symmetry properties of roll solutions

translation symmetry in time:	$\dfrac{\partial}{\partial t}\,\varphi = 0$
translation symmetry in longit. direction:	$\dfrac{\partial}{\partial x}\,\varphi = 0$
transverse periodicity:	$\varphi(y + \dfrac{2\pi}{\alpha},z) = \varphi(y,z)$
transverse reflection:	$\varphi(y,z) = \varphi(-y,z)$ or $a_{-mn} = a_{mn}$
inversion about roll axis:	$\varphi(y,z) = -\varphi(\dfrac{\pi}{\alpha} - y, -z)$ or $a_{mn} = 0$ for odd $m + n$

While the first three properties hold for all steady roll solutions, the remaining two properties can be achieved only in special cases. In particular the last condition requires the symmetry about the midplane mentioned above. Infinitesimal disturbances of the roll solutions (2.1) can be written in the form

$$\tilde{\varphi} = \exp\{ibx + idy + \sigma t\}\sum_{m,n} \tilde{a}_{mn} \exp\{im\alpha y\}g_n(z) \tag{2.3}$$

according to Floquet's theory. Solutions of the disturbance equations for which the eigenvalue σ crosses the imaginary axis correspond to instabilities. In general those instabilities are preferred for which at least two of the symmetries of table 1 are broken. The following table lists the symmetries that are broken by the instabilities of convection rolls in a Rayleigh–Bénard layer (see, for example, [3],[4] and earlier papers cited therein).

Table 2: Symmetries Broken by Bifurcations from Convection Rolls

Symmetries	translation in time	longitud. translation	transverse periodicity	transverse reflection	inversion about axis
Properties of disturbance	$\sigma_i \neq 0$	$b \neq 0$	$d \neq 0$	$\tilde{\alpha}_{mn} \neq \tilde{\alpha}_{-mn}$	$\tilde{\alpha}_{mn} \neq 0$ for m+n=odd
Eckhaus Instab.			X	X	
Crossroll Inst.		X			X
Knot-Inst.		X			X
Even Blob-Inst.	X	X			
Odd Blob-Inst.	X	X			X
Oscillatory Inst.	X	X		X	
Skewed Var. Inst.		X	X	X	
Osc.Skewed Var.I.	X	X	X	X	
Zig-Zag Instab.		X		X	

3. Tertiary solutions and their instabilities

Not all of the instabilities of steady rolls lead to three-dimensional solutions. Only the second through sixth instabilities of Table 2 lead to new solutions that are periodic in both directions with respect to which the problem is homogeneous. These tertiary solutions can be represented in the form

$$\varphi = \sum_{l\,m\,n} a_{lmn} \exp\{il\alpha_x x + im\alpha_y y\}g_n(z) \quad \text{with} \quad a_{-l-mn} = a^*_{lmn} \tag{3.1}$$

which includes traveling wave convection if x is the coordinate in a moving system of coordinates, $x = \hat{x} - ct$. The form (3.1) is particularly suitable for a numerical investigation of the tertiary states since the symmetry properties of these states can easily be taken into account through appropriate restrictions on the coefficients a_{lmn}. In table 3 we list a few examples for solutions of the form (3.1) that have already been investigated. In the actual numerical analysis the range of summation in (3.1) must be restricted to a finite interval which must be chosen sufficiently large, such that the properties of the solutions change very little with a variation of the interval.

Table 3: Symmetries of Twice Spatially Periodic Tertiary Flows

Tertiary solution	Reflection Symmetries	Inversion Symmetry	Remarks
bimodal convection,	$a_{l-mn} = a_{lmn}$	$a_{lmn} = 0$ for $l+m+n =$ odd	[5], [6]
knot convection	$a_{-lmn} = a_{lmn}$		[7]
travel. wave conv., wavy rolls with Poiseuille flow	$a_{lmn} = -a_{l-mn}$ for odd l $a_{lmn} = a_{l-mn}$ for even l	$a_{lmn} = 0$ for $m+n =$ odd	$x = \hat{x} - ct$ [8], [9] [10]
wavy rolls in inclined layers, wavy rolls with Couette flow, wavy Taylor vortices	$a_{lmn} = -a_{l-mn}$ for odd l $a_{lmn} = a_{l-mn}$ for even l	$a_{lmn} = -a_{-lmn}$ for odd $m+n$ $a_{lmn} = a_{-lmn}$ for even $m+n$	[11] [12] [13] [14]

Wavy Taylor vortices have also been investigated without the small gap approximation in which case the inversion symmetry disappears [15], [16].
The stability of stationary solutions of the form (3.1) can be studied as in the case of the secondary solutions by the superposition of infinitesimal disturbances,

$$\tilde{\varphi} = \exp\{iby + idx + \sigma t\} \sum_{l\,m\,n} \tilde{a}_{lmn} \exp\{il\alpha_x x + im\alpha_y y\}g_n(z) \tag{3.2}$$

Of particular interest are those disturbances which do not change the x,y-periodicity interval of the stationary tertiary flow, i.e. disturbances with b = d = 0. Depending on the number n of symmetries exhibited by the tertiary solutions, disturbances with b = d = 0 can be separated in 2^n classes, i.e. the disturbances possess either the same or the opposite parity as the stationary solution with respect to a particular symmetry property. Since steady knot convection and bimodal convection exhibit three different symmetries as indicated in table 3, eight different classes of disturbances can be defined [5]. In the other cases listed in the table this number is reduced to four. From the computational point of view this reduction is often essential for stability calculations.

4. Longitudinal Convection Rolls in the Presence of Plane Couette Flow

In order to illustrate the general discussion of the preceding sections we consider the special case of convection in a horizontal layer heated from below which is bounded by isothermal rigid boundaries. The latter are moved parallel and relative to each other in the x-direction. Using the height d of the layer as length scale, d^2/κ as time scale where κ is the thermal diffusivity and $(T_2 - T_1)Ra^{-1}$ as temperature scale where T_1 and T_2 are temperatures of the boundaries, we obtain a dimensionless description of the problem. We introduce the representation

$$\underset{\sim}{u} = RePz\underset{\sim}{i} + U_1^{(x)}\underset{\sim}{i} + U_1^{(y)}\underset{\sim}{j} + \nabla \times (\nabla \times \underset{\sim}{k}\varphi) + \nabla \times \underset{\sim}{k}\psi \equiv \underset{\sim}{U} + \underset{\sim}{\delta}\varphi + \underset{\sim}{\epsilon}\psi , \qquad (4.1)$$

where $\underset{\sim}{i}$, $\underset{\sim}{j}$, $\underset{\sim}{k}$ are the unit vectors in the x, y, z direction and where $\underset{\sim}{U}$ represents the horizontal average of the velocity field, and obtain the equations

$$\nabla^4\Delta_2\varphi - \Delta_2\theta = P^{-1}\{\underset{\sim}{\delta}\cdot[(\underset{\sim}{\delta}\varphi + \underset{\sim}{\epsilon}\psi)\cdot\nabla(\underset{\sim}{\delta}\varphi + \underset{\sim}{\epsilon}\psi)] + (\underset{\sim}{U}\cdot\nabla + \partial_t)\nabla^2\Delta_2\varphi - \partial_{zz}^2\underset{\sim}{U}\cdot\nabla\Delta_2\varphi\} \qquad (4.2a)$$

$$\nabla^2\Delta_2\psi = P^{-1}\{\underset{\sim}{\epsilon}\cdot[(\underset{\sim}{\delta}\varphi + \underset{\sim}{\epsilon}\psi)\cdot\nabla(\underset{\sim}{\delta}\varphi + \underset{\sim}{\epsilon}\psi)] + (\underset{\sim}{U}\cdot\nabla + \partial_t)\Delta_2\psi - \partial_z\underset{\sim}{U}\cdot\underset{\sim}{\epsilon}\Delta_2\varphi\} \qquad (4.2b)$$

$$\nabla^2\theta - Ra\Delta_2\varphi = (\underset{\sim}{\delta}\varphi + \underset{\sim}{\epsilon}\psi)\cdot\nabla\theta + (\underset{\sim}{U}\cdot\nabla + \partial_t)\theta \qquad (4.2c)$$

for the scalar fields φ and ψ and for the deviation θ of the temperature from the state of pure conduction. The operators $\underset{\sim}{\delta}$ and $\underset{\sim}{\epsilon}$ are defined in (4.1). Besides the Reynolds number Re given by the relative motion of the boundaries, the Rayleigh number Ra and the Prandtl number P appear in the equations. They obey their usual definition

$$Ra = \frac{\gamma g(T_2 - T_1)d^3}{\nu \kappa} \quad , \quad P = \frac{\nu}{\kappa}$$

The first bifurcation from the basic solution

$$\theta \equiv \varphi \equiv \psi \equiv U_1^{(x)} \equiv U_1^{(y)} \equiv 0 \tag{4.3}$$

of the problem assumes the form of longitudinal rolls

$$\theta = \theta(y,z), \quad \varphi = \varphi(y,z), \quad \psi = \psi(y,z), \quad U_1^{(y)} \equiv 0 \tag{4.4}$$

It is remarkable that $\varphi(y,z)$ and $\theta(y,z)$ are independent of Re and that ψ and $U_1^{(x)}$ are proportional to Re. In the case $P = 1$, ψ and $U_1^{(x)}$ are given by

$$\partial_y\psi(y,z) = -(\theta - \bar{\theta})Re/Ra, \quad U_1^{(x)} = -\bar{\theta}Re/Ra \tag{4.5}$$

where the bar indicates the horizontal average. The mean shear thus tends to disappear as the convection reaches a nearly isothermal state in the interior of the layer. For $P<1$ the mean shear will be reversed in the interior at elevated Rayleigh numbers as can already be noticed from the case $P = 0.71$ shown in figure 1. For $P>1$ the deviations from the plane Couette profile become less pronounced.

The similarity between the transports of heat and momentum are also evident in figure 2 where the heat Nusselt number N and the momentum Nusselt number S are plotted. The Nusselt number is defined as the ratio of transports with convection and without convection. For $P = 1$, S equals N; in general S is given by $P^{-1}N$. The kinetic energy of the toroidal component of the velocity field

$$E_{tor} \equiv \tfrac{1}{2} <|\nabla \times \underset{\sim}{k}\psi|^2> \tag{4.6a}$$

rapidly increases after the onset of convection, but decreases at higher Rayleigh number because the energy E_{mf} of the mean flow from which E_{tor} is derived decreases with Rayleigh number. The energy of the poloidal component

$$E_{pol} \equiv \tfrac{1}{2} <|\nabla \times (\nabla \times \underset{\sim}{k}\varphi)|^2> \tag{4.6b}$$

is independent of the Reynolds number, of course.

The numerical results presented here and in the following have been obtained with truncated versions of the Galerkin expansions (2.1), (2.3), (3.1), (3.2). For details we refer to [12],[17].

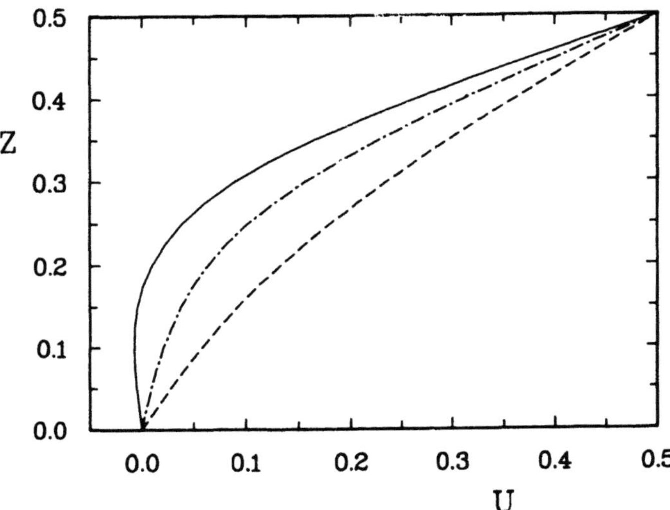

Fig. 1. Profiles of the mean flow in the case of convection in the presence of
plane Couette flow with Ra = 5000, Re = 400, α_y = 3.117. The solid
(dashed) profiles correspond to the case of two dimensional
longitudinal rolls with P = 0.71 (P = 2.5). The dash-dotted profile
corresponds to wavy-roll convection with P = 0.71, α_x = 0.9.

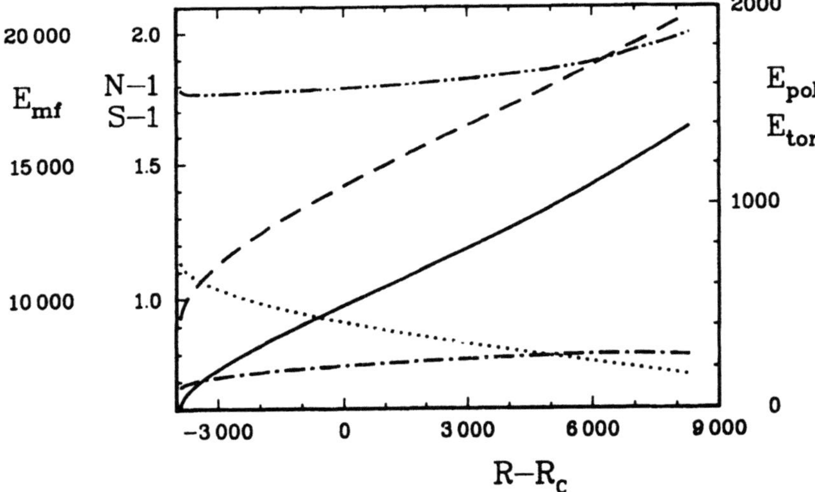

Fig. 2. Properties of longitudinal convection rolls in the presence of plane
Couette flow with Re = 700. The heat and shear Nusselt numbers N
(solid line) and S (dashed), the energies of the poloidal
(dash-dotted), toroidal (dash double-dotted) and mean (dotted)
components of the velocity held are shown as function of the Rayleigh
number for P = 0.71, α_y = 3.117.

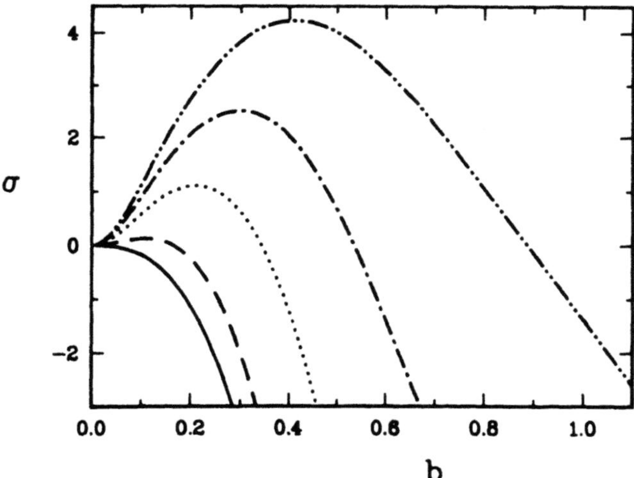

Fig. 3. The growthrate σ of the wavy instability of rolls as function of the
longitudinal wavenumber b for different Rayleigh numbers, Re = 1750,
1800, 1900, 2000, 2100 (from bottom to top) in the case P = 0.71,
Re = 400, α_y = 3.117.

The instabilities of longitudinal rolls are strongly affected by the presence
of the mean shear Of particular importance is the wavy instability which
exhibits the same symmetries as the zig-zag instability of table 1. A typical
dependence of the growthrate σ on parameters of the problem is shown in
figure 3. For more details we refer to [11],[12].

5. Tertiary and Quarternary Solutions in the Presence of Plane Couette Flow

The wavy instability bifurcates supercritically at low Reynolds numbers of the
order Re $\approx$ 200. As it evolves the interaction of the wavy rolls with the mean
shear leads to a decrease of the transport efficiency. At Reynolds numbers of
the order 400 for P $\approx$ 1 the bifurcation occurs subcritically if α_x is
sufficiently high and the mean shear provides an additional energy source for
the three-dimensional motion. The effect becomes dominant at higher Reynolds
numbers such that the heat and shear Nusselt number becomes nearly independent
of the Rayleigh number and solutions can be obtained even for negative

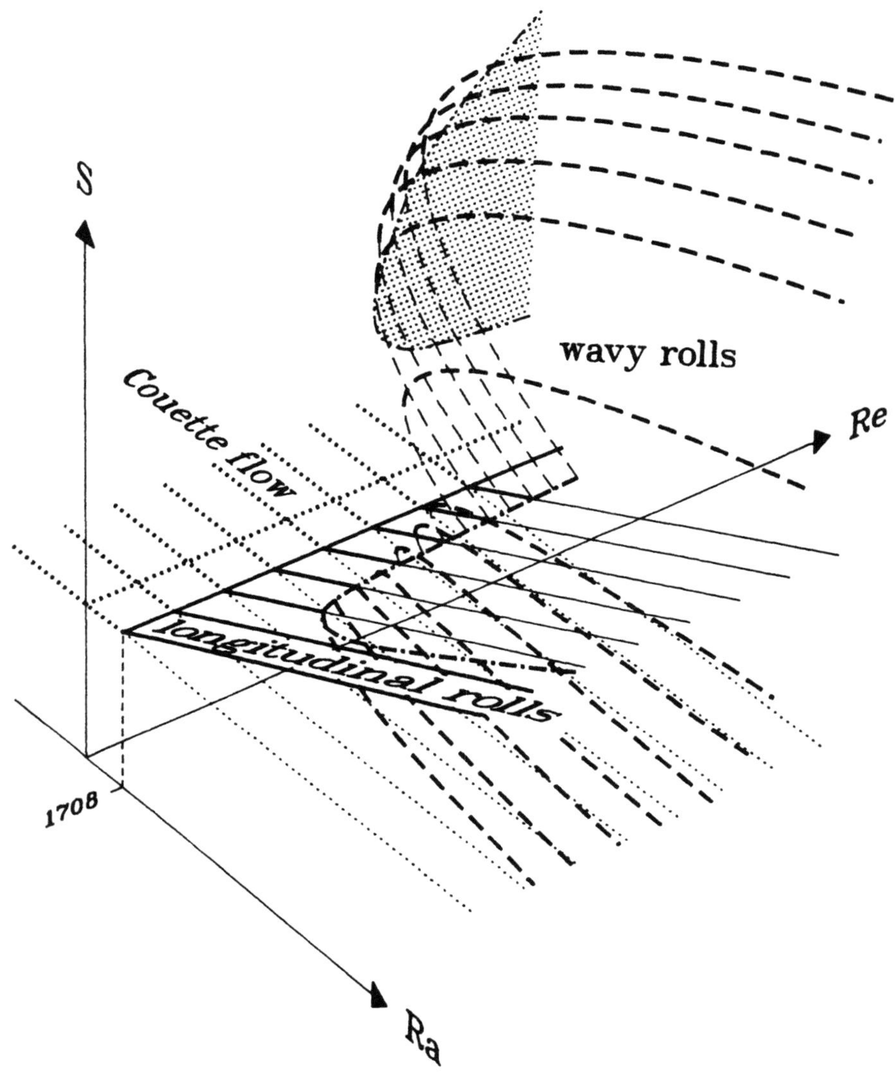

Fig. 4. A qualitative sketch of the bifurcation structure of convection in the presence of a mean shear. The shear Nusselt number is shown as a function of the Rayleigh number Ra and the Reynolds number Re. The dotted plane corresponds to plane Couette flow without convection. The solid lines indicate the bifurcating longitudinal roll solutions while the dashed lines indicate steady wavy roll convection. In the stippled area these solutions intersect the plane Ra = 0.

Rayleigh numbers. A qualitative sketch of the bifurcation structure is shown in figure 4 and computed properties of the wavy rolls are shown in figure 5 for Re = 700. A typical example of the wavy roll structure is shown in figure 6.

A main result of the computations is the three-dimensional solution for the planar Couette problem in the case Ra = 0. This case is known for the absence of any bifurcation from the basic state of pure Couette flow. The expansion of the parameter space through the introduction of the Rayleigh number has allowed the capture of the submanifold of three-dimensional solutions. In a similar way Nagata [13] has reached these solutions earlier by using the rotation about the y-axis as additional parameter and by continuing his three-dimensional solutions to the case of vanishing rotation rate.

A stability analysis [17] carried out along the lines we have discussed in connection with expression (3.2) demonstrates that the steady three-dimensional wavy roll solutions are typically unstable. A complete stability analysis with finite values b and d has not yet been done and the attention has been restricted so far to the case b = d = 0 in which the disturbances do not change the horizontal periodicity interval. The so called vacillation instability which exhibits the same spatial symmetry as the steady wave rolls, but corresponds to an imaginary eigenvalue, appears to predominate. The time depending solutions evolving from this instability exhibit relaxation oscillations during which the flow varies between a nearly steady longitudinal roll state and a wavy roll state resembling closely the corresponding steady solution of figure 6. This process is evident from the plots shown in figure 7.

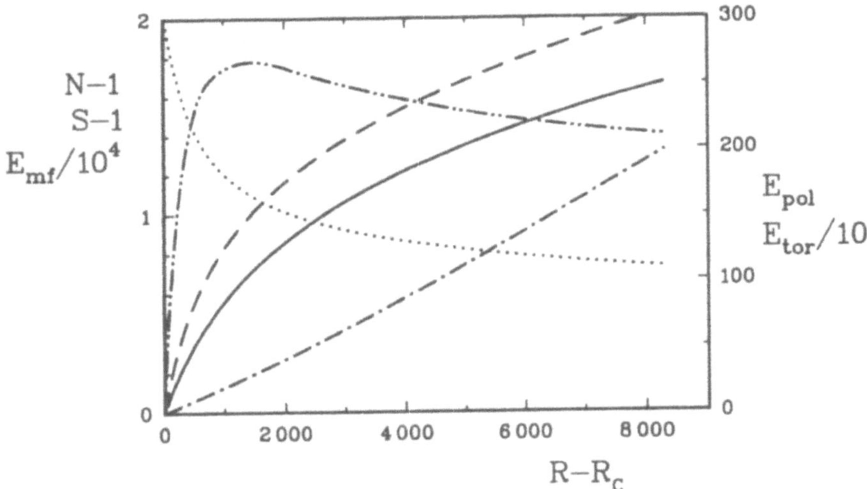

Fig. 5. Properties of wavy convection rolls in the case Re = 700, P = 0.71,
α_x= 1.5, α_y= 3.117 as function of the Rayleigh number. The heat
(solid) and shear (dashed) Nusselt numbers and the kinetic energies of
the poloidal (dash-dotted), toroidal (dash double-dotted) and mean
components of the velocity field have been plotted.

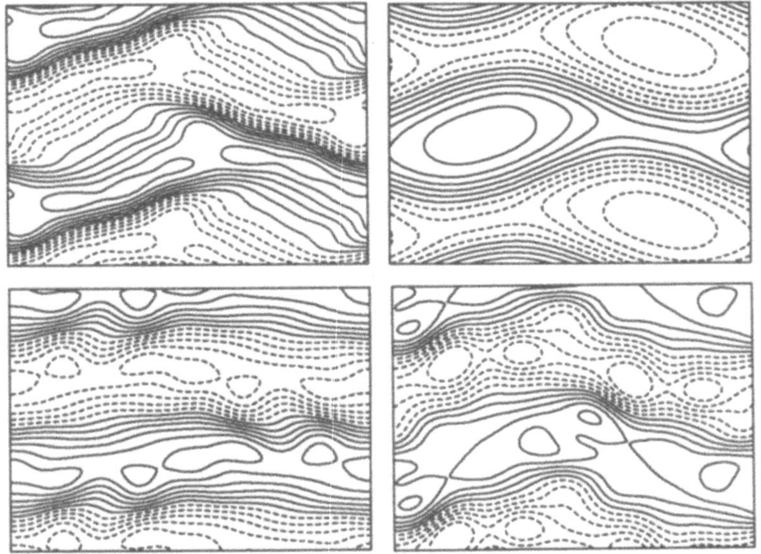

Fig. 6. Lines of constant vertical velocity in the planes z = -0.3 (upper
left) and z = 0 (upper right), stream lines of the toroidal velocity
field, ψ = const., in the plane z = 0 (lower left) and isotherms in
the plane z = 0 lower right in the case P = 0.71, Ra = 3000, Re = 400
α_x = 1.5, α_y = 3.117. Solid (dashed) lines indicate positive
(negative) values except for the solid lines adjacent to the dashed
lines which indicates zero.

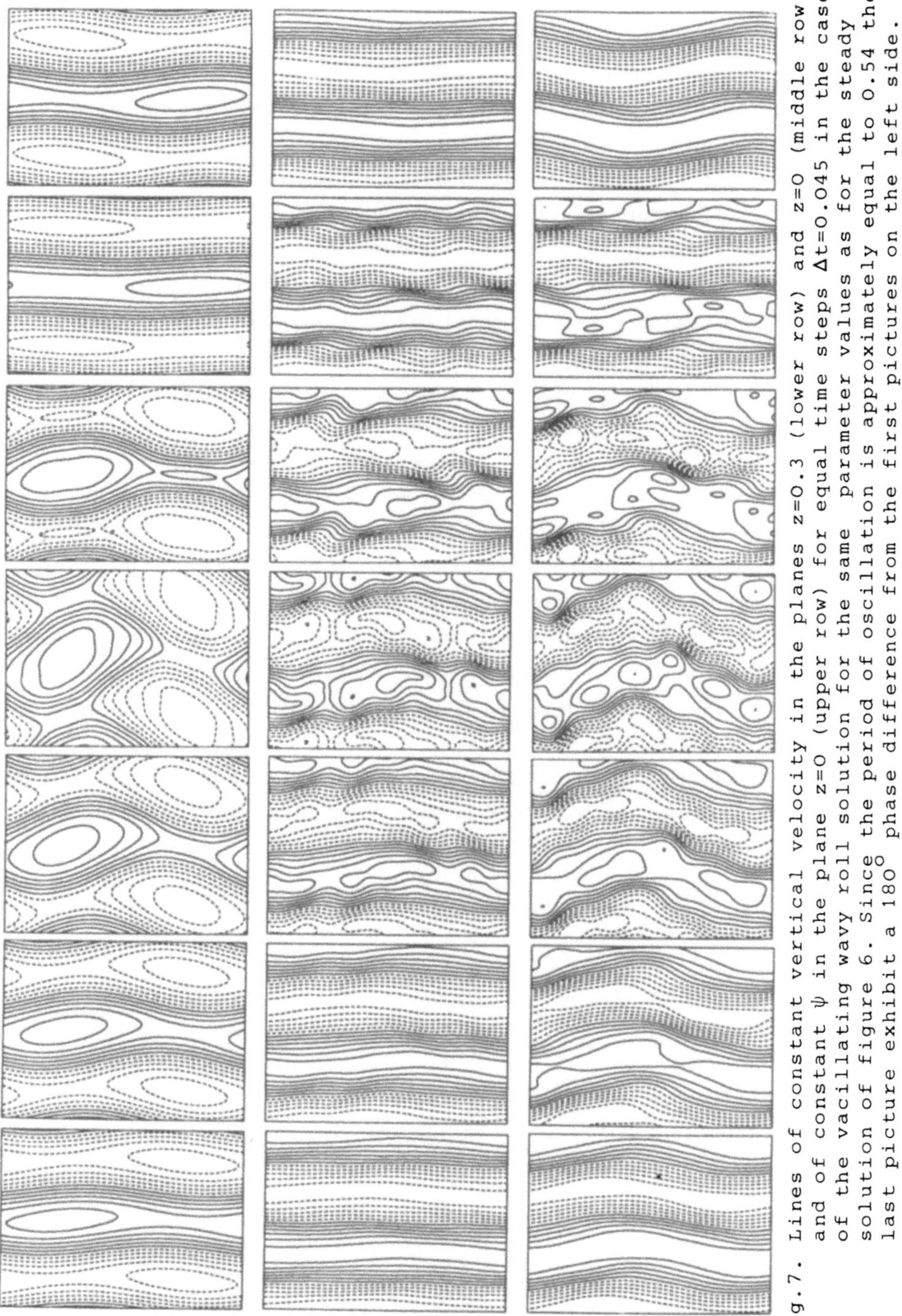

Fig.7. Lines of constant vertical velocity in the planes z=0.3 (lower row) and z=0 (middle row) and of constant ψ in the plane z=0 (upper row) for equal time steps $\Delta t=0.045$ in the case of the vacillating wavy roll solution for the same parameter values as for the steady solution of figure 6. Since the period of oscillation is approximately equal to 0.54 the last picture exhibit a 180° phase difference from the first pictures on the left side.

6. Concluding Remarks

In comparison with the ubiquitous roll structure of secondary solutions, tertiary solutions exhibit a rich variety of spatial patterns. The wavy rolls structure is one of the more commonly realized structures, but pattern such as bimodal convection can also be easily realized in a laboratory experiment. The basin of attraction for stable regular steady tertiary solutions is often relatively small and controlled initial condition may be required to permit their realisation. With increasing control parameter the tertiary solution typically become unstable to time dependent processes. Only in the numerical analysis where rigorous spatial periodicity can be prescribed can limit cycles be observed. Their realisation in experimental settings remains a profound challenge.

Although many of the steady or time-periodic three-dimensional solutions obtained in numerical studies can not be observed in realistic systems, they are important for an understanding for the physical properties of flow systems. In spite of their instability the fixpoints and limit cycles characterize properties of general turbulent states since the trajectories of flow systems in the high dimensional phase space tend to spend more time in the neighborhood of these low-dimensional manifolds than in other areas. Considerable insights can thus be provided by studies of tertiary and quaternary solutions.

The support of the Atmospheric Science Section of the U.S. National Science Foundation for the research reported in this paper is gratefully acknowledged.

References

[1] Busse, F.H., Nonlinear properties of thermal convection, Rep. Prog. Phys. 41, 1929-1967, 1978

[2] DiPrima, R.C., and Swinney,H.L., Instabilities and Transition in Flow Between Concentric Rotating Cylinders pp. 139-180 in "Hydrodynamic Instabilities and the Transition to Turbulence" (H.L. Swinney & J.P. Gollub, eds.) vol. 45, Topics in Applied Physics, Springer, 1981

[3] Busse, F.H., and Bolton, E.W.: Instabilities of convection rolls with stress-free boundaries near threshold. J. Fluid Mech. 146, 115-125, 1984

[4] Bolton, E.W., Busse, F.H., and Clever, R.M.: Oscillatory instabilities of convection rolls at intermediate Prandtl numbers. J.Fluid Mech. 164, 469-485, 1986

[5] Busse, F.H.: On the stability of two-dimensional convection in a layer heated from below. J. Math. Phys. 46, 140-150, 1967

[6] Frick, H., Busse, F.H., and Clever, R.M.: Steady three-dimensional convection at high Prandtl number. J. Fluid Mech. 127, 141-153, 1983

[7] Clever, R.M., and Busse, F.H.: Three-dimensional knot convection in a layer heated from below. J. Fluid Mech. 198, 345-363, 1989

[8] Clever, R.M., and Busse, F.H.: Instabilities of longitudinal rolls in the presence of Poiseuille flow. J. Fluid Mech. 229, 517-529, 1991

[9] Clever, R.M., and Busse, F.H.: Nonlinear oscillatory convection. J. Fluid Mech. 176, 403-417, 1987

[10] Clever, R.M., and Busse, F.H.: Nonlinear oscillatory convection in the presence of vertical magnetic field. J. Fluid Mech. 201, 507-523, 1989

[11] Clever, R.M., and Busse, F.H.: Instabilities of longitudinal convection rolls in an inclined layer. J. Fluid Mech. 81, 107-127, 1977

[12] Clever, R.M., Busse, F.H., and Kelly, R.E.: Instabilities of Longitudinal Convection Rolls in Couette Flow. J. of Appl. Math. and Physics (ZAMP) 28, 771-783, 1977

[13] Nagata, M.: On wavy instabilities of the Taylor-vortex flow between corotating cylinders. J. Fluid Mech. 188, 585-598, 1988

[14] Weisshaar, E., Busse, F.H., and Nagata, M.: Twist vortices and their instabilities in Taylor-Couette systems. J.Fluid Mech. 226, 549-564, 1991

[15] Marcus, P.S.: Simulation of Taylor-Couette flow Part 2; Numerical results for wavy-vortex flow with one travelling wave. J.Fluid Mech. 146, 65-113, 1984

[16] Moser, R.D., Moin, P., and Leonhard, A.: A spectral numerical method for the Navier-Stokes equations with applications to Taylor-Couette flow. J. Comput. Physics 52, 524-544, 1983

[17] Clever, R.M., and Busse, F.H.: Three dimensional convection in a horizontal fluid layer subjected to a constant shear, J. Fluid Mech., 234, 511-527, 1992

International Series of Numerical Mathematics, Vol. 104, © 1992 Birkhäuser Verlag Basel

Hopf-type bifurcations in the presence of linear and nonlinear symmetries

G. Cicogna

Dipartimento di Fisica, Università - p.a Torricelli 2; I - 56126 - PISA, Italy

In this note, we discuss the existence and some properties of different types of bifurcating solutions (in particular, "Hopf-type" bifurcations) to dynamical systems: a common feature of all these solutions will be the close relationship with some nontrivial symmetry property.

In Sect. 1, we focus our attention on the (usually called) "equivariant bifurcation lemma": more precisely, we concentrate on the group-theoretical aspects of this lemma, and give a survey of various hypotheses and situations which ensure the "reduction property", i.e. the possibility of restricting the original dynamical problem to a lower dimensional problem. More attention will be paid to the extension of this method in order to include *nonlinear* symmetries, in the sense of Lie. This opens really new possibilities, as discussed in Sect. 2. One still obtains a "reduction lemma", but now the restricted problem lives on a manifold (instead of a linear subspace), specified by some well defined symmetry requirement. The appeal to (nonlinear) Lie symmetries can also allow us, as discussed in Sect. 3, to state some other results, concerning the existence of nontrivial bifurcating solutions, in terms of some properties of the invariant curves under the action of the symmetry. Some explicit, even if simple, examples are given to illustrate the ideas.

1. The "equivariant reduction lemma" (linear symmetries)

A peculiar property of symmetries in bifurcation (see e.g. [2,8-12,20-23,26,28]) and - more generally - in nonlinear problems is, as well known, that of being able to localize some special directions in the space, where nontrivial solutions can be expected, or - more precisely - to indicate some proper subspace X_0 of the original space, where the problem can be restricted.

Our notations are the following: the dynamical problem is written

$$\dot{u} = f(\lambda, u) \qquad u \in R^n \;, \quad u = u(t) \;, \; \lambda \in R \tag{1}$$

with $f : R \times R^n \to R^n$ smooth (analytic, just for simplicity), and with all standard hypotheses in bifurcation problems. We assume the problem exhibits some symmetry property, described by a group G, i.e.

$$f(\lambda, D(g)u) = D(g) \, f(\lambda, u) \qquad \forall g \in G \tag{2}$$

where D is a linear representation of G acting on the space R^n.

We say that a proper subspace $X_0 \subset R^n$ satisfies the *reduction property* if the problem (1) can be restricted to X_0, i.e. if

$$f|_{R \times X_0} : R \times X_0 \to X_0 \tag{3}$$

This is a list of different group-theoretical situations which allow this reduction.

a) Let the symmetry group G act on the space R^n by means of a (completely) reducible representation D: if in the decomposition of D into irreducible components there appears also the trivial representation D_0 of G (i.e. $D_0(g) = 1, \forall g \in G$) acting as the identity on the subspace X_0, then this subspace X_0 satisfies the reduction property.

a') A (old) result along similar lines [19] (see also [16]): if G is a compact Lie group acting on a manifold M (alternatively, if G acts isometrically under some Riemannian structure of M), and F is a smooth $G-$invariant function defined on M, then a critical point of the restriction $F|_{X_0}$ of F to the set X_0 of fixed points under G, is also a critical point of F.

b) Let now H be a *proper* subgroup of G, and let the representation D of G be decomposed into direct sum of representations T_i of H:

$$D = T_0 \oplus T_1 \oplus \ldots \tag{4}$$

Assume that one of these representations, say T_0, has the following property: none of the symmetric tensor products $T_0^{\otimes p}$ intertwines with the others T_i appearing in (4). Then, the subspace X_0, on which T_0 operates, satisfies the reduction property.

b') If T_0 is the trivial representation (of H), then it obviously satisfies the property stated in b) above, and in this case H is an *isotropy subgroup*. This is precisely the most common situation entering in the usual "equivariant bifurcation lemma" [3,8,11,12,24,26,28], and now X_0 is the *fixed subspace* under H (some interesting special cases of this situation have been examined in full detail [2,6,9-12]).

c) If there is a subspace X_0 and a subgroup H (possibly $H = G$), operating on X_0 through a representation T_0, with the property that for an integer N one has

$$\left(T_0(h)\right)^N = I \qquad \forall h \in H$$

and no other representation of H operating on R^n has the same property (with the same N), then X_0 satisfies the reduction property. This situation is in general unrelated to the previous ones (apart from the trivial case $N = 1$, clearly giving b'), or respectively a) if $H = G$).

d) A nontrivial generalization of the reduction lemma is concerned with the introduction of nonlinear Lie point-symmetries. The next section will be devoted to this topic.

There are many applications of the reduction lemma, especially in the form b'), and in most cases with dim $X_0 = 1, 2$, or odd: we refer e.g. to [2,8-12,24] (and Ref. therein). Some examples of Hopf bifurcations appearing when dim $X_0 = 4$ are presented e.g. in [5], together with a discussion of the role played in such cases by the presence of some symmetry (some typical results: a Z_2 symmetry may be sufficient to allow the existence of a Hopf bifurcation; a problem with explicit $SO(2)$–rotation covariance may possess a rotation-breaking Hopf bifurcation).

2. Reduction lemma (nonlinear Lie symmetries)

In this section we will enlarge the notion of symmetry, to include nonlinear Lie point-symmetries, according to the notion introduced by Lie, and recently reconsidered by many authors [1,17,18,24,25].

A Lie symmetry for a given problem is any continuously differentiable transformation of the independent and dependent variables, such that the solutions $u(t)$ of the problem are mapped into solutions $u'(t')$ of the same problem. The infinitesimal generator η of such a transformation can be written in the general form

$$\eta = \varphi_i(\lambda, u, t)\frac{\partial}{\partial u_i} + \tau(\lambda, u, t)\frac{\partial}{\partial t} \tag{5}$$

where φ_i, τ are functions to be determined (we do not consider here the possibility of including in this scheme also transformations of the parameter λ, i.e. of including an additional term $\nu(\lambda, u, t)(\partial/\partial\lambda)$ in eq. (5); see [7]). It can be shown [1,7,18] that the conditions for the functions φ_i, τ in order that η generates a Lie symmetry for the general dynamical problem, possibly nonautonomous,

$$\dot{u} = f(\lambda, u, t)$$

are given by the following set of *determining equations*

$$\tau\partial_t f_j + f_j\partial_t\tau + f_j f_i\partial_i\tau - \partial_t\varphi_j + \varphi_i\partial_i f_j - f_i\partial_i\varphi_j = 0 \qquad \left(\partial_t \equiv \frac{\partial}{\partial t}\,,\ \partial_i \equiv \frac{\partial}{\partial u_i}\right) \tag{6}$$

We are interested here in autonomous systems (1) and in time-independent Lie point-symmetries, i.e.

$$\eta = \varphi_i(\lambda, u)\frac{\partial}{\partial u_i} \equiv \varphi\partial_u \tag{7}$$

for these, the above determining equations (6) can be written in the following form, in terms of a Lie commutator,

$$[\eta, \eta_f] = 0 \tag{8}$$

where

$$\eta_f \equiv f\partial_u \tag{9}$$

(notice that η_f itself actually generates a Lie symmetry: it generates in fact the time evolution $u(t) \to u(t + \epsilon)$).

The main point is given by the following result [4].

"Nonlinear equivariant bifurcation lemma": Let $\eta = \varphi \partial_u$ be a generator of a Lie symmetry of the given problem (1), and for each fixed λ, let

$$U \equiv U_\eta(\lambda) = \{u \in R^n \mid \varphi_i(\lambda, u) = 0; \quad i = 1, \ldots, n\} \tag{10}$$

Then, U, if not empty, is an invariant manifold under the dynamical flow.

This result comes essentially from (8) and the remark following (9): we can say that U plays here the same role as the "fixed subspace" under a symmetry subgroup in case b') of Sect. 1. Clearly, this lemma allows the reduction of the dynamical problem to the manifold U, i.e. $f|_{R \times U} : R \times U \to U$, just as in the previous section, with the main difference that the reduced problem lies now in some (more or less singular, a priori) manifold, instead of a linear subspace. This leads to various interesting possibilities. For instance, if U is diffeomorphic to the real line, or to R^2, one can expect, generically, a stationary or a Hopf bifurcation, respectively. To illustrate this situation, let us consider the following example.

Example. Let $u \equiv (x, y, z) \in R^3$ and

$$\begin{aligned}
\dot{x} &= X(\lambda, x, y) \\
\dot{y} &= Y(\lambda, x, y) \\
\dot{z} &= Z(\lambda, x, y, z)\big(z - \zeta(\lambda, x, y)\big) + X\zeta_x + Y\zeta_y
\end{aligned} \tag{11}$$

where X, Y, Z, ζ are arbitrary regular functions. A Lie symmetry for this problem is

$$\eta = \big(z - \zeta(\lambda, x, y)\big)\frac{\partial}{\partial z} \tag{12}$$

and the flow-invariant manifold U is given by the surface

$$z = \zeta(\lambda, x, y) \tag{13}$$

On this surface, standard arguments (e.g., Poincaré - Bendixson arguments, or stability-exchange properties [13-15,21]) may very simply ensure the appearance of bifurcating periodic solutions, giving a Hopf-type bifurcation which lies on the surface U (possibly changing with λ: take, for an explicit example, $\zeta = \lambda(x^2 + y^2)$).

Clearly, other situations can be easily imagined, depending on the topological properties of the invariant manifold U. Instead of considering here these possibilities, we present
in the next section some other result based again on the notion of nonlinear Lie symmetry.

3. "Symmetry-invariant curves" and bifurcations

We want to propose now some (preliminary) ideas concerning a different approach
(based, as the previous one, on the introduction of Lie symmetries of the system) to
the problem of finding special (possibly periodic) solutions. Once a Lie symmetry of the
problem (1) is given, let us consider any curve which is left invariant (not necessarily
pointwise) under the action of the symmetry transformations. Such a curve is not left
fixed, in general, under the time evolution, i.e.: symmetry invariance does *not* imply
flow invariance. However, any symmetry-invariant curve has the nice property of being
transformed by the dynamical flow into another symmetry-invariant curve. This may be
symbolically expressed in this way: let η be the symmetry generator, and γ an invariant
curve under η: writing the invariance condition as $\eta\gamma = 0$, and the action of the dynamical
flow as $\gamma \to \eta_f\gamma$, we have (see (8-9))

$$\eta(\eta_f\gamma) = \eta_f\eta\gamma = 0$$

which shows that $\eta_f\gamma$ is another symmetry-invariant curve. This implies that finding a
solution $u(t)$ of the dynamical problem can be viewed as looking for those symmetry-
invariant curves which are left fixed under the action of η_f, or as a "fixed-point problem"
in the space of symmetry-invariant curves γ (cf. also [27]). Now, if such a "fixed" curve
exists, and happens to be closed, then we obtain a periodic solution (at least if this curve
does not contain singular points; otherwise, it corresponds to a set of stationary points if
$f \equiv 0$ along this curve, or to a homoclinic or heteroclinic orbit if $f = 0$ on some point
of the curve). In practice, having parametrized the $(n-1)$-dimensional family of the
symmetry-invariant curves γ by means of a set of $(n-1)$ real parameters c, it can be seen

that the above "fixed-point" condition

$$\eta_f \gamma \equiv \Gamma(\lambda, c) = 0 \tag{14}$$

becomes a set of algebraic (i.e., not differential) conditions involving the parameters c defining the curve, and the bifurcation parameter λ.

Let us clarify this by means of an example.

Example 3: With $u \equiv (x, y, z) \in R^3$, and $r^2 = x^2 + y^2$ and $w = z\, e^{-y}$, consider

$$\begin{aligned}
\dot{x} &= xf(\lambda, r^2, w) - yg(\lambda, r^2, w) \\
\dot{y} &= yf(\lambda, r^2, w) + xg(\lambda, r^2, w) \\
\dot{z} &= yzf(\lambda, r^2, w) + xzg(\lambda, r^2, w) - e^y h(\lambda, r^2, w)
\end{aligned} \tag{15}$$

where f, g, h are regular functions. A Lie symmetry generator for this problem is

$$\eta = x\frac{\partial}{\partial y} - y\frac{\partial}{\partial x} + xz\frac{\partial}{\partial z} \tag{16}$$

Notice that the method presented in Sect. 2 would give in this case simply $U = \{x = y = 0\}$ (i.e. the z axis) as a (trivial) flow-invariant manifold. Instead, the method of symmetry-invariant curves is now more useful: indeed, one has that the symmetry-invariant curves under η describe the 2-dimensional family

$$x^2 + y^2 = c_1, \quad z\, e^{-y} = c_2$$

and condition (14) becomes $f = h = 0$. If, e.g., $f = \lambda - w$, $h = r^2 - w$, then a Hopf-type bifurcation appears along the lines

$$\lambda = z\, e^{-y} = r^2$$

with frequency $\omega = g(\lambda, \lambda, \lambda)$. Notice, incidentally, that in this example the critical subspace (i.e. the space of eigenvectors of the linearized problem, having eigenvalues with real part $= 0$) is the whole R^3. But a fact making "unusual" this situation may be obtained if e.g. $f = h = \lambda - w$: then one has a bifurcation, for each fixed λ, of "nonisolated" periodic cycles, with frequency $\omega = g(\lambda, r^2, \lambda)$, lying on the surface $z = \lambda\, e^y$ (i.e., a continuous family of cycles).

In conclusion, even if the notion of Lie point-symmetry is certainly old, the type of applications above sketched, with their possible improvements and extensions, seems to be new and promising.

References

[1] G.W. Bluman, S. Kumei: "Symmetries and Differential Equations"; Springer, Berlin, 1989

[2] P. Chossat, R. Lauterbach, I. Melbourne: Arch. Rat. Mech. Anal. **113** (1990), 313

[3] G. Cicogna: Lett. Nuovo Cim. **31**, 600 (1981); Boll. Un. Mat. Ital. **3**-A (1984), 131; and J. Phys. A **19** (1986), L368

[4] G. Cicogna: J. Phys. A **23** (1990), L1339

[5] G. Cicogna, G. Gaeta: J.Phys. A **20** (1987), L425; and **21** (1988), L875

[6] G. Cicogna, G. Gaeta: Nuovo Cim. B **102** (1988), 451

[7] G. Cicogna, G. Gaeta: "Lie Point Symmetries in Bifurcation Problems", to be published in Ann. Inst. H. Poincaré; see also J. Phys. A **23** (1990), L799

[8] B. Fiedler: "Global bifurcation of periodic solutions with symmetry"; Springer, Berlin, 1988

[9] M.J. Field, R.W. Richardson: Arch. Rat. Mech. Anal. **106** (1989), 61; and Bull. Amer. Math. Soc. **22** (1) (1990), 79

[10] G. Gaeta: "Bifurcation and symmetry breaking"; Phys. Rep. **189** (1990), 1

[11] M. Golubitsky, I. Stewart: Arch. Rat. Mech. Anal. **87** (1985), 107

[12] M. Golubitsky, I. Stewart and D. Schaeffer: "Singularity and groups in bifurcation theory - vol. II"; Springer, New York (1988)

[13] J. Guckenheimer and P. Holmes: "Nonlinear oscillations, dynamical systems, and bifurcation of vector fields"; Springer, New York, 1983

[14] M. Hirsch, S. Smale: "Differential Equations, Dynamical Systems and Linear Algebra"; Academic Press, New York, 1974

[15] G. Iooss and D.D. Joseph: "Elementary stability and bifurcation theory"; Springer, Berlin, 1981; second edition 1990

[16] L. Michel: C.R. Acad. Sci. Paris A **272** (1971), 433

[17] P.J. Olver: "Applications of Lie groups to differential equations"; Springer, Berlin, 1986

[18] L.V. Ovsjannikov: "Group properties of differential equations"; Novosibirsk, 1962; and "Group analysis of differential equations"; Academic Press, New York, 1982

[19] R.S. Palais: Comm. Math. Phys. **69** (1979), 19

[20] D. Ruelle: Arch. Rat. Mech. Anal. **51** (1973), 136

[21] D.H. Sattinger: "Topics in stability and bifurcation theory"; Springer, Berlin, 1973

[22] D.H. Sattinger: "Group theoretic methods in bifurcation theory"; Springer, Berlin, 1979

[23] D.H. Sattinger: "Branching in the presence of symmetry"; SIAM, Philadelphia, 1983

[24] D.H. Sattinger and O. Weaver: "Lie groups and algebras"; Springer, New York, 1986

[25] H. Stephani: "Differential equations. Their solution using symmetries"; Cambridge Univ. Press 1989

[26] A. Vanderbauwhede: "Local bifurcation and symmetry"; Pitman, Boston, 1982

[27] C.E. Wulfman: "Limit cycles as invariant functions of Lie groups"; J. Phys. A **12** (1979), L73

[28] Proc. of the Warwick Symposium on Singularity theory and applications; Springer, New York, to appear.

On Diffusively Coupled Oscillators

Gerhard Dangelmayr and Michael Kirby [1]

Institut für Informationsverarbeitung
Köstlinstrasse 6, D-7400 Tuebingen, FR Germany

Abstract. Systems of diffusively coupled oscillators are considered. For weak couplings and for oscillators which are close to Hamiltonian systems we derive a canonical system of nonlinear partial differential equations that determines the temporal evolution of fluctuations of the phase and the energy around the uniformly oscillating state. This p.d.e.-system provides a generalization of the generic phase-diffusion equation, but resembles the Kuramoto-Sivashinsky equation. For weakly anharmonic oscillators a globally valid p.d.e. of the Ginzburg-Landau type is derived. Furthermore, we report on some observations made in simulations with oscillators that exhibit tristable behaviour, i.e., besides a stable oscillation also two stable fixed points are present. Here the coupled system shows a sensitive dependence on the diffusion constant and on the initial conditions.

1. Introduction

Many spatially extended systems can be regarded as large assemblies of coupled identical local systems with few degrees of freedom. In particular, reaction diffusion systems are often viewed as arrays of reaction cells [13] which are diffusively coupled to each other. Then, if a local system has the capability of oscillations, we may consider the whole system as a large set of coupled oscillators. Although it is a priori impossible to imagine such local dynamical units in fluid flows, the appearance of organized convection patterns in the form of rolls, hexagons etc. suggests also to consider convection systems as arrays of coupled cells, each possessing the same local dynamics, and to view the observed global pattern as a result of the interaction between the cells. In particular, it is well known that the first instability in binary fluid convection can lead to temporally oscillating rolls so that again the qualitative picture of coupled oscillators emerges.

In this paper we consider a system composed of two-dimensional dynamical "units", with the dynamics of an individual unit governed by

$$\dot{u} = v, \quad \dot{v} = -f(u) + vg(u) \qquad (f(0) = 0). \tag{1}$$

We suppose that (1) admits limit cycle oscillations. Motivated by [8], N such units are coupled diffusively in a one-dimensional chain in the form

$$\dot{u}_n = v_n, \quad \dot{v}_n = -f(u_n) + v_n g(u_n) + \mathrm{D}\Delta v_n, \tag{2}$$

where N ($0 \leq n \leq N - 1$) is a large number and D is the diffusion coefficient. In (1) we use the notation

$$\Delta p_n := p_{n+1} + p_{n-1} - 2p_n \tag{3}$$

---

[1] Permanent address: Department of Mathematics, Colorado State University, Ft. Collins, Co 80523

for any indexed set of variables p_n. For finite N suitable boundary conditions have to be imposed: we choose periodic boundary conditions, i.e., n is taken modulo N. Then (2) is equivariant under a representation of D_N, the symmetry group of cyclic permutations and reflections of a regular N-gon. A theory for the Hopf bifurcation with this symmetry has been developed in [10]. Since, however, this theory restricts the analysis to small amplitude oscillations we will employ another analytical approach that relies on a multiple time scale expansion procedure [15].

By virtue of the diffusive coupling the system (2) admits global states in which all units behave identically. We therefore address first the question of the stability of collective synchroneous oscillations. To make this problem amenable to a perturbation analysis we consider the case of weak couplings and also assume that a single unit is close to a Hamiltonian system. Formally we write

$$g(u) = \epsilon G(u), \quad D = \epsilon D, \tag{4}$$

where $0 < \epsilon \ll 1$ is a small parameter. For the resulting coupled system we will show, in Section 2, that in the continuum limit, i.e., $n/N \to x \in \mathbf{R}$, fluctuations about the uniformly oscillating state are governed by a canonial nonlinear system of partial differential equations for a slowly varying phase and the fluctuating energy. This system of p.d.e.'s resembles the Kuramoto-Sivashinsky equation [14,17], but is more complicated because two variables are coupled. Moreover, it governs generic situations whereas the Kuramoto-Sivashinsky equation is strictly responsible only for phenomena of codimension 1. In fact, the p.d.e.-system derived in Section 2 reduces to the simple, generic phase diffusion equation [2] when the basic units are close to harmonic oscillators.

In Section 3 we specify our model assumptions further and consider the coupling of weakly anharmonic oscillators. The conservative part of (1) is also expanded,

$$f(u) = u + \epsilon f_1(u) \qquad (f_1(0) = df_1(0)/du = 0), \tag{5}$$

i.e., (1) reduces to a simple harmonic oscillator when $\epsilon \to 0$. In this case a multiple time scale expansion (or, equivalently, the averaging method [12]) yields a globally valid system of differential equations for the phases and amplitudes of the individual oscillators which reduces to a partial differential equation of the Ginzburg-Landau type in the continuum limit. This equation allows to determine synchronized states in the form of traveling waves and their stabilities.

Interesting phenomena may occur when the individual units exhibit multistable behaviour. In particular one expects stationary or moving interfaces separating spatial regions with different dynamics, similar to the observations made in recent convection experiments [7]. This kind of spatial multistability has been discussed qualitatively in [8]. We indicate briefly in Section 3 how the dynamics of interfaces can be analyzed for coupled weakly anharmonic oscillators in terms of soliton-type solutions associated with the Ginzburg-Landau equation, but will not pursue this further in this paper. Instead, we report in Section 4 about some observations of simulations with units posessing two stable fixed points and a stable limit cycle. The corresponding coupled system shows a very sensitive dependence on initial conditions and on the diffusion constant. For small

D localized oscillations survive whereas for D above a critical value they die out and the system evolves into a stationary state. The shape of the stationary state as well as the critical value of D depend on the initial conditions.

2. Multiple time scale expansion

In this section we discuss the weak coupling of nearly Hamiltonian systems as described by eqs. (2) - (4). The approach we use is based on the multiple time scale expansion procedure developed in [15]. Let

$$(u, v) = (U(t, E), \dot{U}(t, E))$$ (6)

be a family of periodic orbits of (1) for $\epsilon = 0$, parametrized by the energy E,

$$\frac{1}{2}v^2 + V(u) = E \qquad (V(u) := \int_0^u f(u')du'),$$ (7)

which is constant in the absence of dissipation. Typically, the period

$$T(E) = 2 \int_{u_-}^{u_+} du/\sqrt{2(E - V(u))} \qquad (E = V(u_\pm))$$ (8)

is a smooth function of E, except at isolated values at which saddle connections occur. These are excluded from our analysis. Setting $\omega(E) := 2\pi/T(E)$ we define

$$Q(\varphi, E) := U(\varphi/\omega(E), E), \ P(\varphi, E) := \dot{U}(\varphi/\omega(E), E),$$ (9)

which are 2π-periodic functions of the phase variable φ. The change of variables

$$(u_n, v_n) = (Q(\varphi_n, E_n), P(\varphi_n, E_n))$$ (10)

then leads to the following system of o.d.e.'s for the (φ_n, E_n) :

$$\dot{E}_n/\epsilon = H_n + DP_n\Delta P_n, \quad \dot{\varphi}_n = \omega_n - \epsilon\omega_n(K_n + DR_n\Delta P_n),$$ (11)

where $\omega_n := \omega(E_n)$ and $P_n := P(\varphi_n, E_n)$, $R_n := R(\varphi_n, E_n)$ etc. Here, R, H, K are also 2π-periodic functions of φ, defined by

$$R(\varphi, E) := \frac{\partial}{\partial E}Q(\varphi, E)$$ (12)

$$H(\varphi, E) := P^2(\varphi, E)\, G(Q(\varphi, E))$$ (13)

$$K(\varphi, E) := R(\varphi, E)\, P(\varphi, E)\, G(Q(\varphi, E)).$$ (14)

The presence of possibly complicated resonance surfaces in the space of the E_n makes a general analysis of (11) untractable, because subtle captures in or passages through resonances can appear. We therefore confine the analysis to situations near a $1 : 1 : \cdots : 1$-resonance and also assume that the phases are close to each other, i.e., only small fluctuations around syncroneous oscillations are considered. To the order $O(1)$ this means that all $E_n(t)$ have the same temporal evolution $E_n(t) = E(\tau) + O(\epsilon)$, where $\tau = \epsilon t$ is a slow time. The corresponding frequency is denoted by $\Omega(\tau) := \omega(E(\tau))$. The collective state, in which all oscillators behave identical ($\varphi_n = \varphi_c$, $E_n = E_c$ for all n) is described by

$$E_c(t,\epsilon) = E(\tau) + \epsilon[A(\varphi,\tau) + a(\tau)] + \epsilon^2[B(\varphi,\tau) + b(\tau)] + \ldots \tag{15}$$

$$\varphi_c(t,\epsilon) = \varphi(t,\epsilon) + \epsilon\alpha(\varphi,\tau) + \epsilon^2\beta(\varphi,\tau) + \ldots, \tag{16}$$

where $\varphi(t,\epsilon)$ is a time-like variable which is related to the slow evolution $E(\tau)$ of E_c by

$$\dot\varphi(t,\epsilon) = \Omega(\tau) + \epsilon a_1(\tau) + \epsilon^2 a_2(\tau) + \ldots \tag{17}$$

The $a_j(\tau)$ are at this stage still arbitrary and can be determined during the perturbation expansion. The functions $\alpha(\varphi,\tau), A(\varphi,\tau)$ etc. are required to be 2π-periodic functions of $\varphi(t,\epsilon)$ with zero mean so that the slow drift is captured solely by $a(\tau), b(\tau), a_1(\tau)$ etc. The fluctuations around the collective state (15),(16) are also expanded,

$$E_n(t,\epsilon) - E_c(t,\epsilon) = \epsilon^2[B_n(\varphi,\tau) + e_n(\tau)] + \epsilon^3[C_n(\varphi,\tau) + c_n(\tau)] + \ldots \tag{18}$$

$$\varphi_n(t,\epsilon) - \varphi_c(t,\epsilon) = \epsilon\Phi_n(\tau) + \epsilon^2[\beta_n(\varphi,\tau) + \Phi_{2n}(\tau)] + \ldots, \tag{19}$$

i.e., $O(\epsilon)$-fluctuations in the phase and $O(\epsilon^2)$-fluctuations in the energy are taken into account. Again all φ-dependent functions in (18),(19) are required to be 2π-periodic with zero mean. Making use of the relation

$$\frac{d}{dt} = [\Omega(\tau) + \epsilon a_1(\tau) + \epsilon^2 a_2(\tau) + \ldots]\frac{\partial}{\partial\varphi} + \epsilon\frac{\partial}{\partial\tau}, \tag{20}$$

the ansatz (15) - (19) is now substituted into (11) and expanded in powers of ϵ, yielding a recursive set of equations for the expansion functions. At order ϵ we find

$$\Omega A_\varphi + E_\tau = H(\varphi,\epsilon) \tag{21}$$

$$\Omega\alpha_\varphi + a_1 = (a + A)\omega_E(E) - \Omega K(\varphi,E), \tag{22}$$

where (21) and (22) follow from the equations for E_n and φ_n, respectively, and the subscripts denote partial derivatives. The solvability condition for A is

$$E_\tau = \overline{H}(E), \tag{23}$$

whose critical points $\hat{E}$ $(\overline{H}(\hat{E}) = 0)$ correspond to persistent oscillations of the individual units. Here and in the remainder of this section a bar over a 2π-periodic function denotes its mean value. The solvability condition for α requires that a_1 is equal to the mean value of the r.h.s. of (22). This term depends, however, on a which is not yet determined. To obtain an equation for a we have to consider the 2nd order in the perturbation expansion. From the $O(\epsilon^2)$-term of the first equation of (11) we obtain two equations, one involving B and a and another equation for B_n. The solvability condition for B leads to a linear o.d.e. for a which determines this quantity, and then, from (21), a_1 completely. Although a is not needed for our purpose, we state its governing equation for completeness,

$$a_\tau = \overline{H}_E a + [K,H] + \overline{AH_E}. \tag{24}$$

Here, $[F_1, F_2]$ denotes the covariance of any two periodic functions $F_1(\varphi), F_2(\varphi)$,

$$[F_1, F_2] := \overline{F_1 F_2} - \overline{F}_1 \overline{F}_2. \tag{25}$$

The r.h.s. of the equation

$$\Omega B_{n\varphi} = H_\varphi \Phi_n + DPP_\varphi \Delta\Phi_n \tag{26}$$

for B_n has already a zero mean so that B_n can immediately be written down,

$$B_n = \frac{1}{\Omega}(H - \overline{H})\Phi_n + \frac{D}{2\Omega}(P^2 - \overline{P^2})\Delta\Phi_n. \tag{27}$$

The $O(\epsilon^2)$-term of the second equation of (11) again leads to two equations, one involving n-independent and the other n-dependent quantities. Since the former are not needed we state only the resulting n-dependent equation which takes the form

$$\Omega\beta_{n\varphi} + \Phi_{n\tau} = (B_n + e_n)\omega_E - \Omega K_\varphi \Phi_n - \Omega DRP_\varphi \Delta\Phi_n. \tag{28}$$

The solvability condition for β_n leads to

$$\Phi_{n\tau} = \omega_E e_n - D\Omega\overline{RP_\varphi}\,\Delta\Phi_n. \tag{29}$$

This equation contains $e_n(\tau)$ and must, therefore, be supplemented by a further equation for the temporal evolution of e_n. If ω_E were of order ϵ (as, e.g., for weakly anharmonic oszillators), (29) would constitute a discrete phase diffusion equation and the sign of $\Omega\overline{RP_\varphi}$ would already determine the stability of uniformly oscillating states. Because of its importance we state this term explicitely,

$$\Omega\overline{RP_\varphi}(E) = -\frac{\partial}{\partial E}\{\frac{1}{T(E)}\int_0^{T(E)} V(U(t, E))dt\}. \tag{30}$$

To find an evolution equation for e_n we have to go to third order. The n-dependent part of the $O(\epsilon^3)$-term of the first equation in (11) has the form $C_{n\varphi} + e_{n\tau} = M_n + R_n$, where R_n contains only terms with zero mean. The relevant term M_n is given by the lengthy expression

$$\begin{aligned}
M_n &= (\alpha H_{\varphi\varphi} + AH_{\varphi E})\Phi_n + H_E(B_n + e_n) + H_\varphi \beta_n \\
&\quad + D\{PP_\varphi \Delta\beta_n + PP_E \Delta(B_n + e_n) + \frac{1}{2}PP_{\varphi\varphi}\Delta\Phi_n^2 \\
&\quad + [(\alpha + \Phi_n)P_\varphi^2 + \alpha PP_{\varphi\varphi} + (a + A)(P_E P_\varphi + P_{E\varphi})]\Delta\Phi_n\}.
\end{aligned} \tag{31}$$

The solvability condition for C_n then leads to the desired equation for e_n,

$$e_{n\tau} = \overline{H}_E e_n + D\overline{PP_E}\Delta e_n + Dr\Delta\Phi_n + D^2 q\Delta(\Delta\Phi_n) - \frac{1}{2}D\overline{P_\varphi^2}[(\nabla_+ \Phi_n)^2 + (\nabla_- \Phi_n)^2], \tag{32}$$

where

$$r = \frac{1}{2}[P^2, \frac{\partial}{\partial E}(H/\Omega)] + \overline{RHP_\varphi} \tag{33}$$

$$q = \frac{\Omega}{8}\frac{\partial}{\partial E}\{\Omega^{-2}[P^2, P^2]\} + \frac{1}{2}\overline{RP^2 P_\varphi}, \tag{34}$$

and $\nabla_{\pm}\Phi_n := \Phi_{n\pm1} - \Phi_n$. The three equations (23),(29),(32) determine the temporal evolution of the slowly varying quantities $E(\tau)$, $\Phi_n(\tau)$, $e_n(\tau)$. Of course, (23) is just the averaged equation for a single unit which decouples from the equations for Φ_n, e_n. Generically $E(\tau)$ evolves towards a steady state $\hat{E}$ for which $\overline{H}(\hat{E}) = 0$ and $\overline{H}_E(\hat{E}) < 0$, corresponding to a stable limit cycle of a single unit. Thus we may replace the averaged, E-dependent functions occuring in (29) and (32) by their values evaluated at $E = \hat{E}$, yielding a closed coupled system of equations for the fluctuating variables Φ_n and e_n.

In a continuum approximation, $n/N \to x \in \mathbf{R}$, the quantities $\Delta\Phi_n$, Δe_n, $\Delta(\Delta\Phi_n)$ and $\nabla_{\pm}\Phi_n$ are replaced by Φ_{xx}, e_{xx}, Φ_{xxxx} and $\pm\Phi_x$, respectively. Setting $s := -\overline{H}_E(\hat{E})$ and assuming $s > 0$, we obtain after the rescaling

$$\tau \to \tau/s, \quad e \to -(rq/s\overline{P_\varphi^2})e, \quad \Phi \to -(r/\overline{P_\varphi^2})\Phi, \tag{35}$$

the following canonical form,

$$\Phi_\tau = \rho e + \kappa D\Phi_{xx} \tag{36}$$

$$e_\tau = -e + \nu De_{xx} + D^2\Phi_{xxxx} + \mu D(\Phi_x^2 + \Phi_{xx}), \tag{37}$$

where

$$\rho = q\omega_E/s^2, \quad \kappa = -\Omega\overline{RP_\varphi}/s, \quad \nu = \overline{PP_E}/s, \quad \mu = r/q, \tag{38}$$

and all E-dependent averages are replaced by their values at $\hat{E}$. The appearance of the fourth derivative Φ_{xxxx} and the nonlinearity Φ_x^2 resembles the Kuramoto-Sivashinsky equation [14,17], the main difference being the additional variable e. It governs the fluctuations around a uniformly oscillating state to the order $O(\epsilon)$ in the phase and to $O(\epsilon^2)$ in the energy. The important fact about the system (37),(38) is that it is generic, and not responsible for codimension-1 situations. In contrast, the Kuramoto-Sivashinsky equation governs phase instabilities near a vanishing phase diffusion coefficient which is, of course, a phenomenon of codimension one. Consequently (36),(37) should be viewed as a generalization of the generic phase diffusion equation. We note, however, that the continuum limit may be approached in various ways, depending on the chosen spatial scale. Moreover, rescaling x is equivalent to rescaling the diffusion constant. Thus D may assume any positive value in (36), (37) and needs not to coincide with the original diffusion constant. On the other hand, if we had started with a continuum version of (2) - (4) from the beginning, the diffusion constant would be fixed.

Since already the Kuramoto Sivashinsky-equation exhibits a rich variety of dynamical behaviours, similar and even more complex dynamics are to be expected from the coupled equations (36), (37). In this paper we confine ourselves to an elementary discussion of the stability of the trivial state $\Phi = e = 0$, corresponding to uniform oscillations of the chain of oscillators. Neglecting the nonlinearity Φ_x^2 and making the ansatz

$$(\Phi, e) = (\Phi_0, e_0)\exp(p\tau + ikx/\sqrt{D}), \tag{39}$$

we find that p be an eigenvalue of a 2×2-matrix $M(k^2)$ with

$$\mathrm{Tr}M = -1 - (\kappa + \nu)k^2 \tag{40}$$

$$\det M = k^2[\kappa + \rho\mu + (\kappa\nu - \rho)k^2]. \tag{41}$$

The basic state is stable, if $\mathrm{Tr}\, M < 0$ and $\det M > 0$ for all k^2, i.e., sufficient conditions for stability are

$$\kappa + \nu > 0, \quad \kappa\nu - \rho > 0, \quad \kappa + \rho\mu > 0. \tag{42}$$

If $\kappa + \rho\mu < 0$ and, respectively, $\kappa\nu - \rho < 0$ but for each of these two cases the other two inequalities in (42) still hold, then the basic state is unstable against stationary fluctuations with small and, respectively, large wavenumbers. On the other hand, if $\kappa + \nu < 0$ but the other two inequalities in (42) are still valid, the basic state is unstable against oscillatory fluctuations with large wavelengths. An oscillatory instability of this kind is not present in the Kuramoto Sivashinsky-equation. It is here a consequence of the fact that phase and energy variables are coupled. A more thorough analysis of (36),(37) will be given elsewhere.

We note that a similar pair of coupled p.d.e.'s with fourth order spatial derivatives has been presented in [4]. These p.d.e.'s are responsible for the phase fluctuations of stationary patterns in extended systems in the presence of Galilean invariance, as, for example, in the Boussinesq equations with ideal boundary conditions. In contrast, the coupling of two variables here is a consequence of the fact that the basic unit is close to a conservative system with an energy-dependent frequency. Although the two p.d.e.-systems look similar, there is no direct relation between (36),(37) and the system of [4], apart from both beeing generalizations of the generic phase diffusion equation.

3. Weakly anharmonic oszillators

We now consider weakly anharmonic oscillators (model (2) - (5)), where the potential has the form

$$V(u) = \frac{1}{2}u^2 + \epsilon V_1(u) \quad (d(V_1)(u)/du = f_1(u)). \tag{43}$$

The basic solutions of the conservative part of (1) can be written in the form

$$U(t, B, \epsilon) = B \sin[\omega(B, \epsilon)t] + O(\epsilon), \tag{44}$$

$$\omega = 1 + \epsilon\omega_1(B^2) + O(\epsilon^2), \tag{45}$$

where

$$\omega_1(B^2) = \frac{2}{\pi} \int_0^{\pi/2} \sin^2 t f_{12}(B^2 \sin^2 t) dt. \tag{46}$$

Here, $f_1(u)$ has been split into even and odd parts, $f_1(u) = f_{11}(u^2) + u f_{12}(u^2)$. The amplitude B and the energy E are related via

$$E = B^2[1 + \epsilon\omega_1(B^2)] + O(\epsilon^2), \tag{47}$$

thus it is convenient to choose B instead of E as slowly varying variable. By specifying the general system (11) for $(\dot\varphi_n, \dot E_n)$ to the weakly anharmonic case we obtain, after the transformation $E_n \to B_n$,

$$\dot B_n/\epsilon = B_n \cos^2\varphi_n\, G(B_n \sin\varphi_n) + D \cos\varphi_n\, \Delta(B_n \cos\varphi_n) + O(\epsilon). \tag{48}$$

$$\dot{\varphi}_n = 1 + \epsilon\omega_1(B_n^2) - \epsilon\{G(B_n \sin\varphi_n)\sin\varphi_n \cos\varphi_n + D\frac{\sin\varphi_n}{B_n}\Delta(B_n\cos\varphi_n)\} + O(\epsilon^2). \tag{49}$$

Since, in contrast to the strongly anharmonic case of the preceeding section, the system (48), (49) posesses a pure $1 : 1 : \cdots : 1$-resonance to order 1, we are able to take larger fluctuations around uniform oscillations into account. The following multiple time scale expansion is employed,

$$B_n(t,\epsilon) = R_n(\tau) + \epsilon B_{n1}(t,\tau) + \dots \tag{50}$$

$$\varphi_n(t,\epsilon) = t + \Phi_n(\tau) + \epsilon\varphi_{n1}(t,\tau) + \dots, \tag{51}$$

where the t-dependent functions are required to be 2π-periodic in t. (We do not require that the B_{nj}, φ_{nj} have zero mean because only the equations for (R_n, Φ_n) are relevant.) From the solvability conditions for B_{n1}, φ_{n1} at order ϵ we find immediately the desired system of coupled equations for the slowly varying amplitudes and phases,

$$R_{n\tau} = R_n F(R_n^2) + \frac{D}{2}\{R_{n-1}\cos(\Phi_{n-1} - \Phi_n) + R_{n+1}\cos(\Phi_{n+1} - \Phi_n) - 2R_n\} \tag{52}$$

$$\Phi_{n\tau} = \omega_1(R_n^2) + (D/2R_n)\{R_{n-1}\sin(\Phi_{n-1} - \Phi_n) + R_{n+1}\sin(\Phi_{n+1} - \Phi_n)\}, \tag{53}$$

with

$$F(R^2) = \frac{2}{\pi}\int_0^{\pi/2} \cos^2 t \, G_1(R^2 \sin^2 t)dt, \tag{54}$$

where G is also split into $G(u) = G_1(u^2)+uG_2(u^2)$. We consider again the continuum limit. Replacing $\Phi_{n\pm1} \to \Phi \pm \Phi_x + \Phi_{xx}/2$ and analogeously $A_{n\pm1}$, and expanding the diffusion terms up to the second order in ∂_x yields

$$R\Phi_\tau = R\omega_1(R^2) + D(2R_x\Phi_x + R\Phi_{xx}) \tag{55}$$

$$R_\tau = RF(R^2) + D(R_{xx} - \frac{1}{2}\Phi_x^2), \tag{56}$$

where the diffusion constant has been rescaled to $D \to 2D$, although, in principle, D is arbitrary in this limit. When (R, Φ) are combined to $A = Re^{i\Phi} \in \mathbf{C}$, (55), (56) can be written in the compact form

$$A_\tau = A[F(|A|^2) + i\omega_1(|A|^2)] + DA_{xx}, \tag{57}$$

an equation of the Ginzburg-Landau type.

The equation (57) admits traveling wave solutions of the form

$$\Phi = \omega_1(r^2)\tau + kx, \quad R = r \tag{58}$$

with r and k related to each other by $Dk^2 = F(r^2)$. The stability of these solutions can be analyzed by means of the dispersion relation following from the linearization of (57) around (58). We will not pursue this, but mention two limiting cases. The first is when $r \to 0$ with $k \to k_c \geq 0$. This leads to an amplitude instability of (57) so that a standard (cubic) Ginzburg-Landau equation can be derived to characterize the stability of (57) against plane wave perturbations with wave numbers close to k_c (see, e.g. [13]).

The second case is when $k = 0, r \neq 0$ and $F_{R^2}(r^2) < 0$, corresponding to spatially uniform oscillations which are stable for the individual units. Since the diffusion coefficient D is real and positive, this state is always stable also for the coupled system, in contrast to the strongly unharmonic case discussed in the preceeding section. Indeed, an ansatz with Φ_n and R_n of order ϵ and ϵ^2, respectively, would here directly lead to the phase diffusion equation $\Phi_{n\tau} = D\Delta_n/2$. The same equation also follows from (29), because the term $\omega_E e$, being of order ϵ, must be omitted, and $\Omega\overline{RP_\varphi} = -\frac{1}{2} + O(\epsilon)$.

We conclude this section with some remarks about the possibility of spatial bistability with two different dynamical states separated by stationary or propagating interfaces. If, for example, G is chosen as a quartic polynomial, then $F(R^2)$ takes the form $F = a + bR^2 + cR^4$. For $a, c < 0, b > 0$ and $4ac < b^2$ an individual unit possesses $u = v = 0$ as stable fixed point and a stable/unstable pair of limit cycles, given by the solutions of $F = 0$. For this case discrete averaged equations have been discussed qualitatively in [8]. The continuum limit, eq. (57), then becomes what is called a "subcritical Ginzburg-Landau equation" which has been analyzed in some detail, specifically with respect to the behaviour of interfaces [3,16]. These are identified with soliton-type solutions of (57) and correspond to heteroclinic orbits of a system of o.d.e.'s derived from (57) by a traveling wave ansatz. If $\omega_1 = 0$ one may assume A to be real and apply the results of [3] to characterize the existence and stability of propagating and stationary interfaces. If $\omega_1 \neq 0$ the more refined analysis of [16] for the "complex subcritical Ginzburg-Landau equation" has to be utilized.

4. Simulations of coupled tristable units

One of the most fundamental properties common to many low dimensional, nonlinear physical systems is that of bistability, where the system possesses two stable stationary states. In addition to these, depending on external parameters, one often encounters large stable oscillations surrounding both stable states. Examples for systems exhibiting this kind of tristable behaviour are optically bistable systems [6], SNDC-semiconductors [1], stirred tank chemical reactors and other chemical systems [11]. Since most of these systems are either spatially extended or occur in chains it is tempting to try to understand the dynamics of large arrays of them.

A prototype dynamical system showing various kinds of multistable behaviours is provided by a degenerate version of the Takens-Bogdanov normal form, given by (1) with

$$f(u) = \mu + \nu u + \rho u^2 + u^3, \quad g(u) = \beta - u^2. \tag{59}$$

The dynamics of this system has been analyzed in some detail in [5]. Moreover it is shown in [1] that it acts as a natural organizing center for SNDC-semiconductors. Depending on the parameters, it possesses one or three fixed points with at most two of them stable. Furthermore, there can exist a stable oscillation surrounding all three fixed points. In this paper we confine ourselves to the symmetric case $\mu = \rho = 0$ for which the fixed points of (1) are given by $v = 0$ and, respectively, $u = 0$ (saddle) and $u^2 = -\nu$ (sinks/sources). For $\beta > 0$ there are two cases with tristable behaviour. We focus unto one of these, occuring in the parameter regime $-1.35\beta < \nu < -5\beta/4$. Its phase portrait is shown in Figure 1.

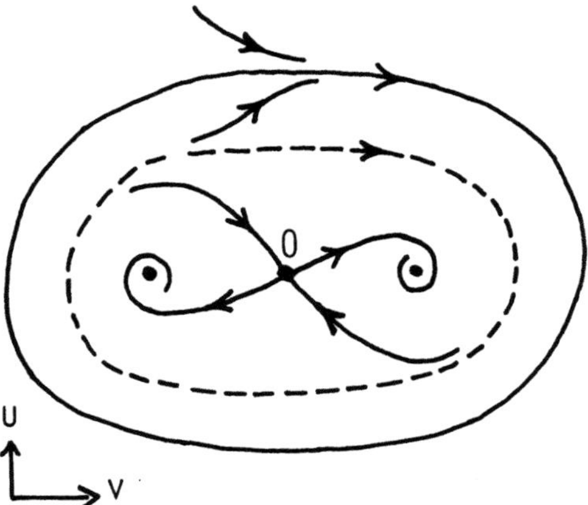

FIGURE 1: TRISTABLE PHASE PORTRAIT

We note that, because the fixed points $v = 0$, $u^2 = -\nu$ are here stable for an individual unit, any global state with $v_n = 0$, $u_n^2 = -\nu$ is a stable equilibrium of the coupled system, no matter what the sign of u_n is. Consequently, (2) has at least 2^N stable equilibria. In a continuum limit of (2), where Δv_n is replaced by v_{xx}, these correspond to infinitely many distinct (not simply related by spatial translations) discontinuous stationary solutions, reflecting the fact that the diffusive part of (2) does not correspondend to a proper elliptic operator in that limit. In principle this could be improved by adding to the $\dot{u}_n$-equation a further small diffusive coupling $\delta\Delta u_n$, or δu_{xx} in the continuum limit, with $\delta \ll D$, but we did not pusue this. As long as N is finite, (2) constitutes a well defined dynamical system and we did not encounter any numerical problems. Moreover, the eigenvalues of the linearization of each of these 2^N equilibria have all negative real parts so that, by continuity, they persist under small perturbations without change in their stability properties.

In our simulations we have coupled 21 symmetric ($\rho = \mu = 0$) units according to (2), choosing $\beta = 1$ and $\nu = -1.26$. The simulations were run with three different initial conditions $IC_{1,2,3}$ for the (u_n, v_n) ($1 \leq n \leq 21$), defined by

   IC_1: $n \neq 11$: $(0,0)$; $n = 11$: $(0,3)$.
   IC_2: $n \leq 5$ and $n \geq 16$: $(-1,0)$; $6 \leq n \leq 15$: $(0,3)$.
   IC_3: $1 \leq n \leq 3$ and $19 \leq n \leq 21$: $(-1,0)$; $4 \leq n \leq 8$ and $14 \leq n \leq 18$: $(0,3)$;
              $9 \leq n \leq 13$: $(1,0)$.

Each of these initial conditions was chosen such that at least one unit would end up in a stable oscillation if no coupling were present. In the case of IC_1 unit 11 would oscillate whereas the other units would strictly remain in the unstable fixed point $(0,0)$, however, small perturbations would throw them into one of the two stable fixed points. For IC_2 the middle part, units 6 - 15, would also be attracted by the limit cycle while the other ones were captured by the left stable fixed point. The third initial condition, IC_3, is slightly more complicated: units 4 - 8 and 14 - 18 would run into oscillations, units 1 - 3 and 19 - 21 would evolve into the left, and the remaining units 9 - 13 would evolve into the right stable fixed point.

Our main purpose was to estimate the critical value D_c such that for $D > D_c$ the oscillations die out, but persist for $D < D_c$. To obtain this, D was successively decreased and for each value of D a die-out time T was calculated as the first value of t for which

$$\sum_n [(u_n^2 - 1)^2 + v_n^2] < 0.1.$$

The critical diffusion D_c is then indicated by a divergence of T as function of D. Figure 2 shows T(D) for the initial condition IC_1, yielding $D_c \approx 0.011$. For large D, T(D) seems to fluctuate around a finite value $T_\infty \approx 70$. The same calculations have been performed for the other two initial conditions, yielding $D_c \approx 0.018$, $T_\infty \approx 90$ for IC_2 and $D_c \approx 0.017$, $T_\infty \approx 85$ for IC_3.

For all three initial conditions the oscillations never penetrated into the whole "space", even for $D < D_c$. For IC_1 and $D < D_c$ the unit 11 continues to oscillate, whereas all other units tend towards the same fixed point, either to the left or to the right one. An example is shown in Figure 3(A). When $D > D_c$ we found for large t either a spatially uniform equilibrium or a state in which $u_n u_{11} = -1.26$ for all $n \neq 11$ as in Figure 3(B). The interesting feature here is the uniform behaviour of the stationary initial regime, because without coupling any combination of left and right fixed points might appear when small ramdom perturbations are added to the initial values.

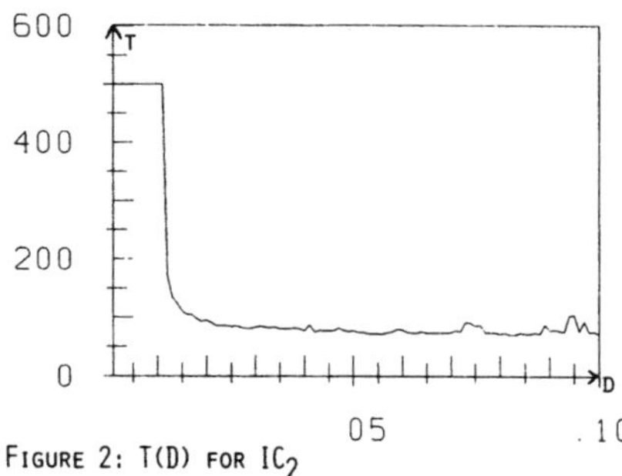

FIGURE 2: T(D) FOR IC_2

For IC_2 and IC_3 and $D < D_c$ the initially oscillating units continue to oscillate synchroneously, whereas the other units quickly evolve towards the fixed point in whose neighbourhood they started. For $D < D_c$ the situation is more complicated than in the case of IC_1. The initial oscillations die out and evolve into equilibrium configurations of various shapes, depending on D. We have not yet made attempts to analyze this dependence carefully. As examples we show in Figure 4 the evolutions towards two different equilibria for IC_2 and in Figure 5 two final states observed for IC_3.

A variety of further kinds of behaviours can be expected for this system. When the initial state is such that all units would oscillate individually, then, for sufficiently large D, the system can evolve into any of the 2^N stable equilibria described before. In contrast, for small D the oscillations will persist, yielding a number of distinct synchronized states. For example, we observed a kind of modulated traveling wave penetrating through a region with uniform oscillations with apparently chaotic behaviour near the boundary. We expect that these kinds of phenomena can be explained on the basis of the system of p.d.e.'s (36), (37). On the other hand, transitions between stationary and oscillatory states are beyond

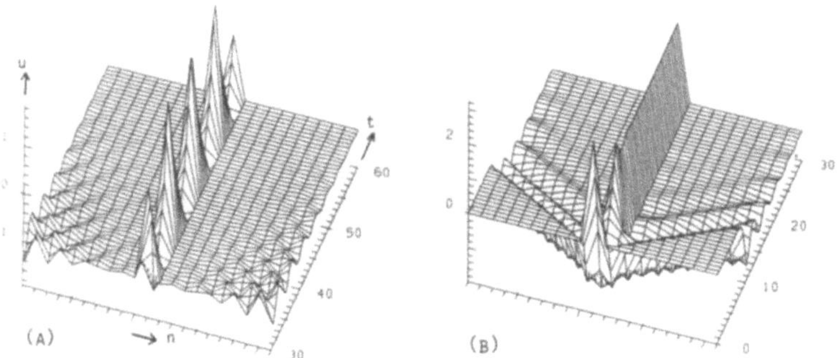

FIGURE 3: TEMPORAL EVOLUTION OF U FOR IC_1. (A) D=0.005. (B) D=0.200.

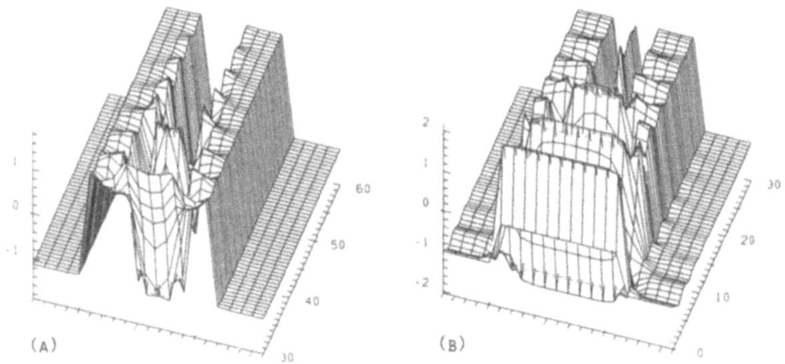

FIGURE 4: SAME AS FIGURE 3 FOR IC_2. (A) D=0.050. (B) D=0.200.

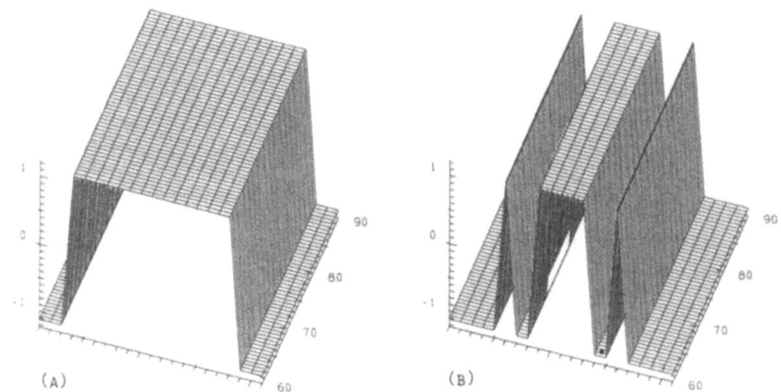

FIGURE 5: FINAL STATES FOR IC_3. (A) D=0.050. (B) D=0.100.

the scope of the applicability of this p.d.e.-system. Also the Ginzburg-Landau approach of section 3 is not adequate for these phenomena because systems exhibiting stationary bistability are far from harmonic oscillators. Thus one has to start directly from the continuum version of (2) ($\Delta v_n \rightarrow v_{xx}$), inserting a traveling wave ansatz $u(x - ct)$ and looking for heteroclinic orbits of the resulting o.d.e.,

$$cDu''' + c^2u'' + c(u^2 - \beta)u' + \nu u + u^3 = 0, \qquad (60)$$

where the prime denotes $d/d\eta$ with $\eta = x - ct$. Techniques like those employed in [9] might then lead to dynamical equations for the positions of interfaces.

Acknowledgement. We are grateful to C. Geiger, W. Güttinger and M. Wegelin for valuable discussions and to B. Dangelmayr for typesetting the manuscript. The work of MK was supported by the Humboldt foundation.

References

1. D. Armbruster and G. Dangelmayr, Physics Letters **A 138** (1989), 46.

2. H. Brandt, in *Nonlinear evolution of spatio-temporal struczures in dissipative continuous systems*, F.H. Busse and L. Kramer (eds.), Plenum Press 1990, p. 206.

3. P. Coullet, L. Gil and D. Repaux, Phys. Rev. Lett. **62** (1989), 2957.

4. P. Coullet and S. Fauve, Phys. Rev. Lett. **55** (1985), 2857.

5. G. Dangelmayr and J. Guckenheimer, Arch. Rat. Mech. Anal. **97** (1987), 321.

6. G. Dangelmayr and M. Wegelin, in *Singularity theory and its applications, part II*, M. Roberts and I. Stewart (eds.), Springer 1991, p. 107.

7. F. Daviaud, P. Berge and M. Dubois, J. Phys. (Paris) **50** (1989), 181.

8. A.-D. Defontaines, Y. Pomeau and B. Rostand, Physica **D 46** (1990), 201.

9. C. Elphick, E. Meron and E.A. Spiegel, Phys. Rev. Lett. **61** (1988), 496.

10. M. Golubitsky and I. Stewart, Contemporary Mathematics **56** (1986), 131.

11. J. Guckenheimer, Physica **D 20** (1986), 1.

12. J. Guckenheimer and P. Holmes, *Nonlinear oscillations, dynamical systems and bifurcations of vector fields*, Springer 1983.

13. Y. Kuramoto, *Chemical oscillations, waves and turbulence*, Springer 1984.

14. Y. Kuramoto and T. Tsuzuki, Progr. Theor. Phys. **55** (1976),356.

15. J.C. Luke, Proc. Roy. Soc. Ser. **A 292** (1966), 403.

16. W. van Saarlos and P.C. Hohenberg, Phys. Rev. Lett. **64** (1990), 749.

17. G.I. Sivashinsky, Acta Astronautica **4** (1977), 1177.

International Series of Numerical Mathematics, Vol. 104, © 1992 Birkhäuser Verlag Basel

Mechanisms of Symmetry Creation

Michael Dellnitz[*]   Martin Golubitsky [†]   Ian Melbourne[‡]

November 1, 1991

1   Introduction

Numerical simulation in [1] indicates that symmetry increasing bifurcations of chaotic attractors occur with great frequency in the dynamics of symmetric mappings. The pictures in [1], [3], [5] also demonstrate that an unexpected kind of pattern formation occurs in symmetric chaotic dynamics where an order based on symmetry is forced on the randomness that is related to chaotic dynamics. Inspection of various numerical simulations in the literature also show that symmetry increasing bifurcations have been observed in ODEs (the *Lorenz equation* in [8]) and in Galerkin approximations of PDEs (the *Ginzburg-Landau equation* in [6]).

In this paper we make a more detailed numerical analysis of how symmetry creation can occur and how it may be related to pattern formation as the term is used in Physics and Engineering. In applications, the fundamental question concerns how the symmetry of an attractor in phase space manifests itself in physical space. The important point to be noted here is that the symmetry of the attractor in phase space is only "on average". One must either iterate the mapping a relatively large number of times, or analogously integrate the differential equation for a relatively long time in order to see that symmetry. It follows that if the symmetry of the attractor is to be seen in physical variables (as a pattern) it must be seen through some averaged quantity.

In Section 2 we indicate by example how averaged quantities can undergo symmetry creation. We consider both the *Brusselator* and the *Ginzburg-Landau equation*. Although these equations each have only a single reflectional symmetry, the types of symmetry increasing they exhibit are quite different. In Section 3 we illustrate these differences by considering the discrete dynamics of odd maps on the line. In this section we also consider how the parameter values where symmetry creation occurs may be computed by methods other than direct simulation. These techniques are based on theoretical results in [2].

---

[*]Department of Mathematics, University of Houston, Houston, TX 77204-3476 and Institut für Angewandte Mathematik der Universität Hamburg, D-2000 Hamburg 13

[†]Department of Mathematics, University of Houston, Houston, TX 77204-3476

[‡]Department of Mathematics, University of Houston, Houston, TX 77204-3476

Generally speaking we think of symmetry creation in systems with finite symmetry as occurring through the "collision" of symmetry related conjugate attractors. As we show in Section 3 this is not always the basis for symmetry creation — but collisions do occur frequently and it is a useful way of thinking. In systems with continuous symmetry there is another method by which symmetry creation can occur — drifting along group orbits. We illustrate this phenomenon in Section 4 by using an example of a mapping on $\mathbf{R}^4$ having O(2) symmetry. We discuss briefly why this type of symmetry creation may be the method by which turbulent wavy vortices turn into turbulent Taylor vortices in the Couette-Taylor experiment.

2   Symmetry creation in PDEs

In this section we document the occurrence of symmetry increasing bifurcations in PDEs by considering two examples: the *Brusselator* and the *Ginzburg-Landau equation*. In each case we detect these bifurcations through the use of appropriately defined time averaged quantities.

(a)   The Brusselator

We consider the following system of reaction-diffusion equations on the interval $[0,1]$

$$
\begin{aligned}
\frac{\partial u}{\partial t} &= \frac{D_1}{\lambda^2}\frac{\partial^2 u}{\partial x^2} + u^2 v - (B+1)u + A \\
\frac{\partial v}{\partial t} &= \frac{D_2}{\lambda^2}\frac{\partial^2 v}{\partial x^2} - u^2 v + Bu.
\end{aligned}
\tag{2.1}
$$

This system is known as the *Brusselator*, in which u, v, A and B represent chemical concentrations and D_1, D_2 are diffusion constants. In this simplified model A and B are assumed to be constant in both space and time while u and v depend on x and t. The parameter λ is a characteristic dimension of the system and we shall treat λ as the bifurcation parameter. We consider (2.1) on the interval $[0,1]$ subject to the Dirichlet boundary conditions

$$
\begin{aligned}
u(0,t) &= u(1,t) = A, \\
v(0,t) &= v(1,t) = B/A.
\end{aligned}
$$

This Dirichlet problem possesses a reflectional symmetry given by

$$
\kappa(u(x,t), v(x,t)) = (u(1-x,t), v(1-x,t)).
$$

Holodniok et. al. [4] have considered this model in detail and found multifrequency motions using numerical simulation. Presuming that chaotic dynamics will occur near multifrequency motion we follow [4] and set

$$
A = 2, \quad B = 5.45, \quad D_1 = 0.008, \quad D_2 = 0.004.
$$

As we mentioned in the introduction the symmetry of a chaotic attractor is expected to be a *symmetry on average*. Therefore it can only be seen by averaging over a long time. But what quantity should be averaged?

The averaged quantity that we use to detect symmetry creation may be interpreted through the following hypothetical story. Suppose that the radical whose concentration is being measured by u is an acid and that u is actually measuring the concentration of this acid along the bottom of a (two dimensional) rectangular container. Suppose further that this acid can eat away or etch the bottom of this container. It would then be reasonable that the rate of etching — at any point along the bottom — would be proportional to the concentration u and that the total amount of the bottom that would be etched is proportional to

$$F_u(x) = \lim_{T \to \infty} \left(\frac{1}{T} \int_0^T u(x, \tau) d\tau \right) . \qquad (2.2)$$

Suppose that the attractor A corresponding to the solution $u(x, t)$ is chaotic and symmetric. We expect that u would sample the whole attractor in phase space rather quickly and that the time average (2.2) would be equal to a (weighted) space average over the attractor in phase space — though the rigorous proof of this point would require an ergodic type theorem to be valid. Presuming this we would expect that (2.2) would be symmetric under $x \to 1 - x$ if the attractor is symmetric. Moreover, generically, we would expect (2.2) to be asymmetric should A be asymmetric.

The graphs of F_u both before and after a symmetry increasing bifurcation are shown in Figure 1. In both cases $T = 20000$ was chosen for the numerical computations. In Figure 2

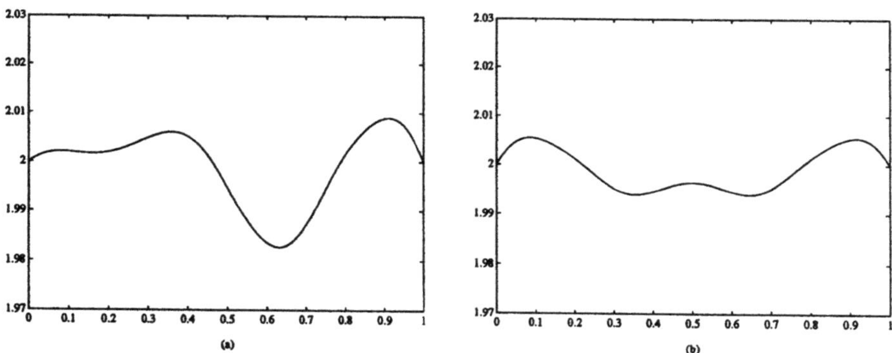

Figure 1: Graphs of F_u for (a) $\lambda = 1.45$, (b) $\lambda = 1.47$.

we indicate why we believe that this symmetry creation is caused by collision of conjugate attractors. If such a collision were to occur, the difference in phase space between the union of A and its conjugate attractor before collision would be approximately equal to A after collision. Applying the presumed ergodic theorem mentioned previously we would expect the average

$$\frac{1}{2}\left(F_u(x) + F_u(1-x)\right) \tag{2.3}$$

to vary continuously. The graph of (2.3) before symmetry creation is given in Figure 2 along with the difference between (2.3) before and after symmetry creation. Note how small this difference is.

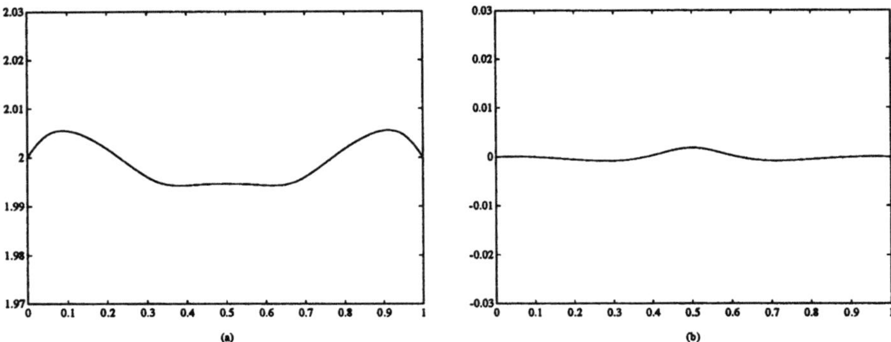

Figure 2: (a) The average of the graphs of F_u for the two conjugate attractors for $\lambda = 1.45$, and (b) the difference between the graph of F_u for $\lambda = 1.47$ and this graph.

(b) The Ginzburg-Landau equation

As a second example of symmetry creation in PDEs we consider the *Ginzburg-Landau equation*:

$$\frac{\partial A}{\partial t} = q^2(i + c_0)\frac{\partial^2 A}{\partial x^2} + \rho A + (i - \rho)A|A|^2. \tag{2.4}$$

The constants q, c_0, ρ are real whereas A is complex. We restrict our attention to Dirichlet boundary conditions:

$$A(0,t) = A(\pi,t) = 0.$$

We regard q as the bifurcation parameter and choose the remaining constants as in [7], namely

$$c_0 = 0.25, \quad \rho = 0.25.$$

In this example we compute the averaged quantity

$$F_A(x) = \lim_{T\to\infty}\left(\frac{1}{T}\int_0^T |A(x,\tau)|^2 d\tau\right) \tag{2.5}$$

for the detection of symmetry increasing bifurcations. The results are shown in Figures 3 and 4. Again, $T = 20000$ was chosen for the numerical computations. In contrast to the symmetry creation in the Brusselator the results in this case indicate that there is no continuous transition from the averaged graphs before the bifurcation to the symmetric graph of F_A afterwards (see Figure 4).

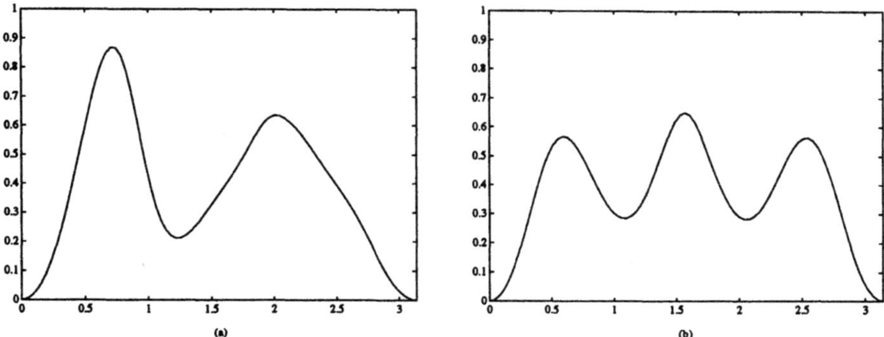

Figure 3: Graphs of F_A for (a) $q = 0.1975$, (b) $q = 0.1925$.

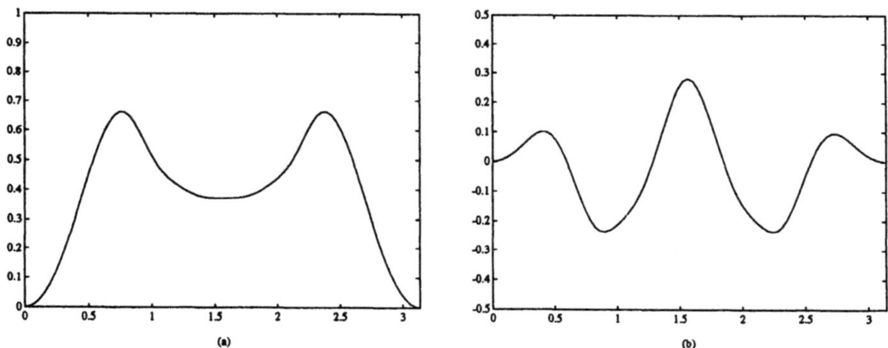

Figure 4: (a) The average of the graphs of F_A for the two conjugate attractors for $q = 0.1975$, and (b) the difference between the graph of F_A for $q = 0.1925$ and this graph.

Similar phenomena corresponding to the *explosion of attractors* have been observed in discrete dynamical systems (see Section 3).

3 Symmetry creation in mappings

In this section, we consider symmetry increasing of attractors for C^1 mappings $f : \mathbf{R}^m \to \mathbf{R}^m$. The technique that we shall describe is documented in [2] and allows us in certain examples to compute the point at which symmetry increasing occurs to any required accuracy. Our aim here is to illustrate by example both the utility and the limitations of the technique.

Given a closed set $S \subset \mathbf{R}^m$, define $\mathcal{P}_S$ to be the set of all points in $\mathbf{R}^m$ that either lie in S or eventually iterate under f to a point in S. The idea is that S can be chosen so that there is a relation between symmetry increasing of attractors and transitions of the set $\mathcal{P}_S$.

Consider first the case of mappings of the line. Suppose that $f_\lambda : \mathbf{R} \to \mathbf{R}$ is a parametrized family of $\mathbf{Z}_2$-equivariant (odd) mappings. In this case S is chosen to be the origin or, more generally, a symmetric periodic orbit. In many examples, as λ is varied two conjugate asymmetric attractors collide at such an orbit to produce a single $\mathbf{Z}_2$-symmetric attractor containing this orbit.

Let λ_c denote the critical parameter value and suppose that the attractor was asymmetric for $\lambda < \lambda_c$ and $\mathbf{Z}_2$-symmetric for $\lambda > \lambda_c$. In [2] we prove that $A \cap \overline{\mathcal{P}_S} = \emptyset$ for $\lambda < \lambda_c$ and $A \subset \overline{\mathcal{P}_S}$ for $\lambda > \lambda_c$. Thus we search for transitions in the set $\mathcal{P}_S$. Determining the value of λ where such a transition occurs is a standard problem in numerical bifurcation theory. A point x is called a *transition point* at λ_c if there is a positive integer m such that

$$f_{\lambda_c}^m(x) \;\in\; S \qquad\qquad (3.6)$$

$$\frac{\partial}{\partial x} f_{\lambda_c}^m(x) \;=\; 0 \qquad\qquad (3.7)$$

The point x is a *transition point of order m* if in addition,

$$f_{\lambda_c}^{m-1}(x) \notin S. \qquad\qquad (3.8)$$

In fact there is a simpler way to find transition points of *minimum* order m_0. That is, at minimum order

$$\frac{\partial}{\partial x} f_{\lambda_c}^{m_0}(x) = 0 \quad \text{iff} \quad \frac{\partial}{\partial x} f_{\lambda_c}(x) = 0.$$

This is easily seen using the chain rule.

As an example, we consider the *cubic logistic map*

$$f_\lambda(x) = \lambda x(1 - x^2)$$

whose bifurcation diagram is shown in Figure 2 of [1]. The lowest order for a transition point is $m = 2$. A simple calculation shows that the only transition points of order 2 are $x = \pm\frac{1}{\sqrt{3}}$ when $\lambda_c = \frac{3\sqrt{3}}{2}$. It is known that this corresponds to the symmetry increasing bifurcation documented in [1]. This is the first of the symmetry increasing bifurcations

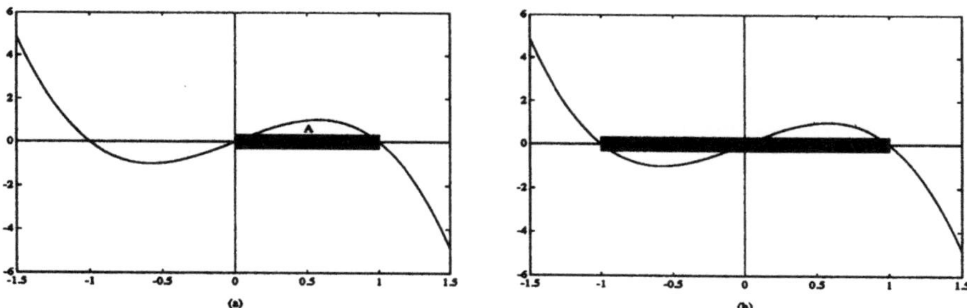

Figure 5: The attractor of the cubic logistic map f_λ for (a) $\lambda = 2.598076$ and (b) $\lambda = 2.598077$.

that occur in the cubic logistic map, see Figure 5. In this figure the graph of f_λ is shown along with the associated attractor in bold.

Further symmetry increasing bifurcations occur after each periodic window in the bifurcation diagram. As an example, the collision of two conjugate attractors at a symmetric period 6 point is shown in Figure 6. Here the corresponding transition point is of order 12. (This can be regarded as a transition point of order 2 for f^6 at a nonsymmetric fixed point.)

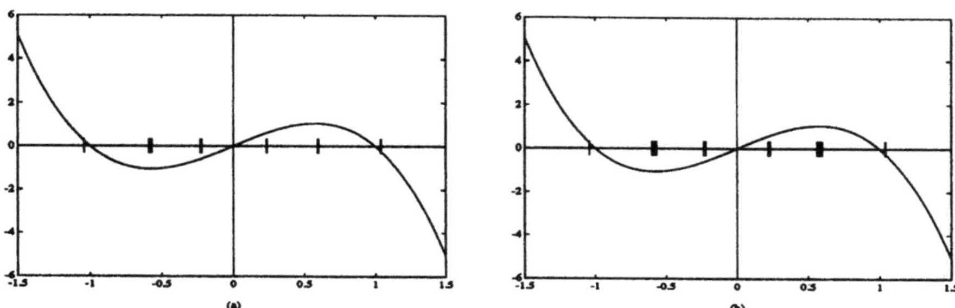

Figure 6: An attractor of the cubic logistic map f_λ for (a) $\lambda = 2.704431$ and (b) $\lambda = 2.704432$.

We end our discussion of one-dimensional mappings by giving an example of a second type of symmetry increasing bifurcation where there is no transition point at criticality. Rather there is a sequence x_k of transition points of order m_k at λ_k where $m_k \to \infty$ and $\lambda_k \to \lambda_c$. This symmetry increasing bifurcation was observed in the family f_λ^2, where f_λ is the cubic logistic map. It differs from the others in that the asymmetric attractors do not continuously approach a symmetric periodic point for $\lambda < \lambda_c$. Rather they explode to a Z_2-symmetric attractor containing 0 when $\lambda \geq \lambda_c$ (see Figure 7). It should be pointed out that this behavior is not related to a hysteresis (observe that the nonsymmetric attractor before bifurcation is covered by the symmetric one after symmetry creation).

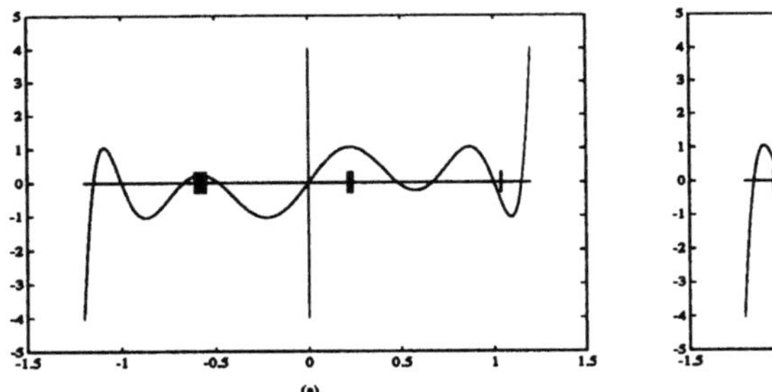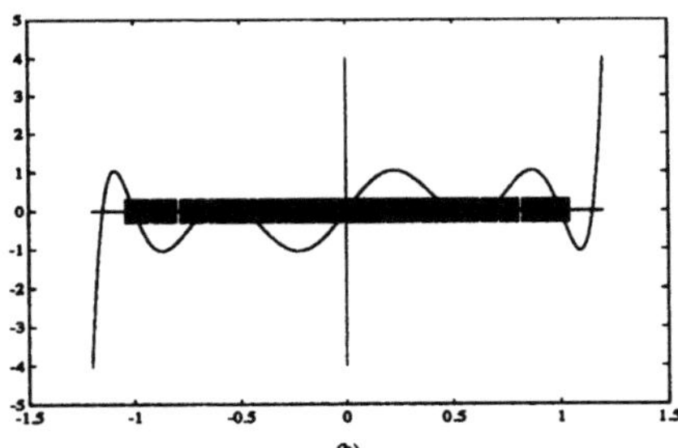

Figure 7: An attractor of f_λ^2 for (a) $\lambda = 2.705639$ and (b) $\lambda = 2.705640$.

Next we consider $\mathbf{D}_3$, the symmetry group of a regular triangle, acting on $\mathbf{R}^2$. The results in [2] suggest that a symmetry increasing bifurcation to an attractor with full $\mathbf{D}_3$-symmetry should be accompanied by a transition in a suitable preimage set $\mathcal{P}_S$, i.e. $A \cap \overline{\mathcal{P}_S} = \emptyset$ before and $A \subset \overline{\mathcal{P}_S}$ after symmetry creation. Here, S is the union of any two axes of symmetry not intersecting the attractor before symmetry creation (two such axis exist by a result in [1]).

As an example we consider the $\mathbf{D}_3$-equivariant map (see [1])

$$f(z,\lambda) = (\alpha u + \beta v + \lambda)z + \gamma \bar{z}^2 \, , \tag{3.9}$$

where $z \in \mathbf{C}$, $\lambda \in \mathbf{R}$ is the bifurcation parameter, $\alpha, \beta, \gamma \in \mathbf{R}$ are fixed constants and

$$u = z\bar{z} \, , \qquad v = \frac{(z^3 + \bar{z}^3)}{2} \, .$$

The group $\mathbf{D}_3$ acts by

$$\kappa z = \bar{z}, \quad \theta z = e^{i\theta} z,$$

where $\theta = \frac{2\pi}{3}$. We use the system

$$\mathrm{Im}(f^m)(z,\lambda) = 0$$
$$\frac{\partial}{\partial z}\mathrm{Im}(f^m)(z,\lambda) = 0 \tag{3.10}$$

in order to compute symmetry increasing bifurcations. Some numerical results are given in Table 1. In analogy to (3.8) "order" means the smallest value of m for which a solution of (3.10) can be found numerically and which corresponds to a symmetry increasing bifurcation. Infinite order in the last line of the table corresponds to a symmetry creation caused by explosion. For all the bifurcations the attractor has Z_2-symmetry before the symmetry increasing bifurcation and D_3-symmetry afterwards.

Order	z	α	β	γ	λ
2	$0.516121 + i\,0.915104$	-1.0	0.0	-0.5	2.269928
2	$0.532793 + i\,0.751622$	1.0	0.0	0.1	-2.371198
3	$-0.214232 + i\,0.288230$	1.8	0.0	1.34164	-1.648899
4	$0.753472 + i\,0.772307$	-1.0	0.1	-0.8	1.519215
11	$0.347859 + i\,0.917966$	-1.1	0.212	0.6	1.891572
24	$0.465201 + i\,0.159120$	1.0	0.0	0.5	-1.798928
∞	—	1.0	0.7	-0.8	≈ -1.98356

Table 1: Data for symmetry-increasing bifurcations

4 Symmetry creation via drifts along group orbits

In the previous sections we have seen two different mechanisms by which symmetry creation can occur in dynamical systems with discrete symmetry: *collisions* of conjugate attractors or *explosions*. In this section we describe another possibility of symmetry creation that can be found in systems with continuous symmetry: the *drifting* of a chaotic attractor along its group orbit.

We consider the following O(2)-equivariant mapping

$$f(z_1, z_2, \lambda) = \begin{pmatrix} (\alpha + \beta_1 u_1 + \gamma_1 v)z_1 + \delta_1 z_2 \\ (\lambda + \beta_2 u_2 + \gamma_2 v)z_2 + \delta_2 z_1 \end{pmatrix},$$

where

$$u_j = |z_j|^2, \quad v = \mathrm{Re}(z_1 \bar{z}_2).$$

Set

$$\alpha = -2.6, \quad \beta_1 = 1.5, \quad \beta_2 = 0.4, \quad \gamma_1 = 0.7, \quad \gamma_2 = 0.5, \quad \delta_1 = -0.5, \quad \delta_2 = 0.3$$

and regard λ as the bifurcation parameter. In Figure 8 the projection of attractors of f onto the z_1-plane is shown for different values of λ. In (a) the attractor is Z_2 symmetric and a small change in λ causes a drift of this attractor along its SO(2) group orbit. The resulting attractor in (b) then has full O(2) symmetry.

In [1] the possibility was pointed out that the transition to turbulent Taylor vortices in the Couette-Taylor experiment may be an example of symmetry increasing bifurcation. We

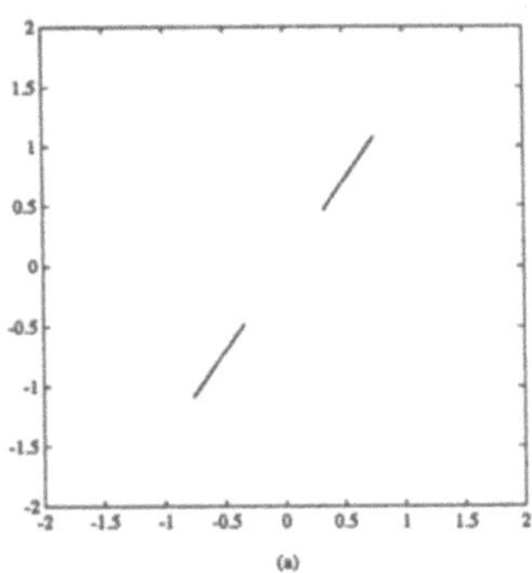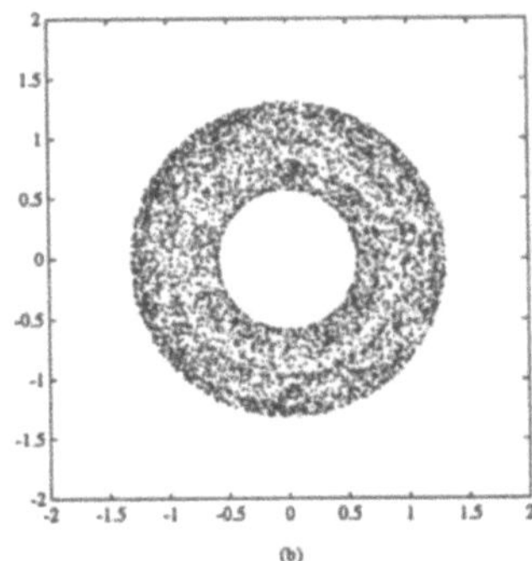

Figure 8: Projection of an attractor of f onto the z_1-plane for (a) $\lambda = -0.477$ and (b) $\lambda = -0.479$.

will explain why we still believe that this may be true by describing in terms of symmetry the drifting along group orbits that might be responsible for this symmetry creation.

In that experiment a fluid is contained between two concentric circular cylinders with the inner one rotating, at speed (or Reynold's number) λ. When λ is small the flow is laminar Couette flow. As λ is increased Couette flow loses stability to Taylor vortices and then to wavy vortices.

In the analysis of this experiment one often assumes periodic boundary conditions in the axial direction which introduces $O(2)$ axial symmetry. The total symmetry group is $O(2) \times SO(2)$ where the $SO(2)$ symmetry comes from the azimuthal geometry of the apparatus. In terms of symmetry the solutions described previously have symmetry types

$$
\begin{array}{ccccc}
\text{Couette flow} & \longrightarrow & \text{Taylor vortices} & \longrightarrow & \text{Wavy vortices} \\
O(2) \times SO(2) & \longrightarrow & Z_2(\kappa) \times SO(2) & \longrightarrow & Z_2(\kappa, \pi)
\end{array}
$$

where κ is a reflection in axial direction and π is a half-period rotation in the azimuthal direction.

As λ is further increased, the flow becomes chaotic and turbulent. However, for large λ, there is a turbulent flow with the pattern of Taylor vortices superimposed. This flow evolves from a turbulent wavy vortex pattern as λ is increased. A transition from turbulent Taylor vortices to homogeneous turbulence takes place at even larger λ.

We believe that the transition from turbulent wavy vortices to turbulent Taylor vortices may be associated with a symmetry increasing bifurcation of a chaotic attractor with $Z_2(\kappa, \pi)$ symmetry forming a chaotic attractor with $Z_2(\kappa) \times SO(2)$ symmetry. Such a change could in principle be generated by drifting along the azimuthal $SO(2)$ group orbits. Much investigation is needed in order to verify such a mechanism.

Acknowledgment

We are grateful to Mike Gorman for a number of helpful discussions. The research of MD was supported in part by the Deutsche Forschungsgemeinschaft and by NSF Grant DMS-9101836. The research of MG and IM was supported in part by NSF Grant DMS-9101836.

References

[1] P. Chossat and M. Golubitsky. Symmetry-increasing bifurcation of chaotic attractors. *Physica D* **32**, 423-436, 1988.

[2] M. Dellnitz, M. Golubitsky and I. Melbourne. The structure of symmetric attractors. University of Houston. Preprint, 1991.

[3] M. Field and M. Golubitsky. Symmetric chaos. *Computers in Physics* Sep/Oct 1990, 470-479, 1990.

[4] M. Holodniok, M. Kubiček and M. Marek. Desintegration of an invariant torus in a reaction-diffusion system. *Prague Institute of Chemical Technology*, Preprint 1989.

[5] G.P.King and I.N.Stewart, Symmetric chaos, in *Nonlinear Equations in the Applied Sciences* (eds. W.F.Ames and C.F.Rogers), Academic Press 1991, 257-315.

[6] J.D. Rodriguez and L. Sirovich. Low-dimensional dynamics for the complex Ginzburg-Landau equation. *Physica D* **43**, 77-86, 1990.

[7] L. Sirovich. Chaotic dynamics of coherent structures. *Physica D* **34**, 126-145, 1989.

[8] C. Sparrow. *The Lorenz Equations*, Springer, 1982.

International Series of Numerical Mathematics, Vol. 104, © 1992 Birkhäuser Verlag Basel

Generic Bifurcations of Pendula

Michael Dellnitz[*]
Department of Mathematics, University of Houston
Houston, TX 77204-3476, USA

Jerrold E. Marsden[†]
Department of Mathematics, University of California at Berkeley
Berkeley, CA 94720, USA

Ian Melbourne[‡]
Department of Mathematics, University of Houston
Houston, TX 77204-3476, USA

Jürgen Scheurle
Institut für Angewandte Mathematik, Universität Hamburg
D-2000 Hamburg 13, Germany

December 9, 1991

1   Introduction

In a parameter dependent Hamiltonian system, an equilibrium might lose its stability via
a socalled *Hamiltonian Krein-Hopf bifurcation* ([1], [12]): Two pairs of purely imaginary
eigenvalues of the linearized system collide $(1-1$ *resonance*) and split off the imaginary
axis into the complex plane. In the following we will refer to this scenario as the *splitting
case*, see Figure 1. It is well known that in one parameter problems without external

[*]Research is supported by the Deutsche Forschungsgemeinschaft and by NSF DMS-9101836. Permanent
address: Institut für Angewandte Mathematik, Universität Hamburg, D-2000 Hamburg 13, Germany
[†]Research partially supported by a Humboldt award and DOE Contract DE-FG03-88ER25064.
[‡]Supported in part by NSF DMS-9101836

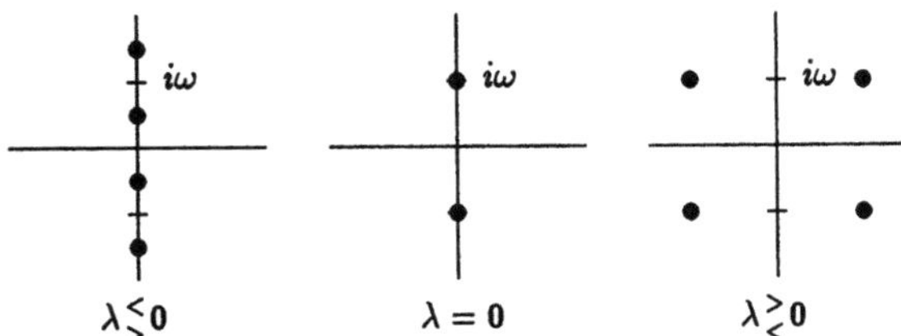

Figure 1: The *splitting* case for the $1-1$ resonance

symmetry this is the only eigenvalue behavior that generically occurs in $1-1$ resonances
([4], [9], [10]).

When there is symmetry present, the situation changes. Under certain circumstances
the eigenvalues might also pass while remaining on the imaginary axis (the *passing* case,
see Figure 2). In this case the linear stability properties of the corresponding equilibrium

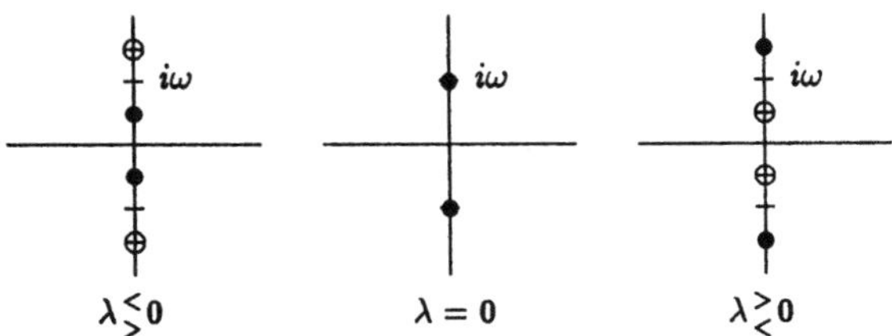

Figure 2: The *passing* case for the $1-1$ resonance

do not change and in this sense the collision is not "dangerous" as in the splitting case.
The question arises naturally whether these essentially different eigenvalue movements can
be characterized so that the occurence of the one or the other in a given system could
in principle be predicted. The answer to this question is given in [3]. There the generic
movement of eigenvalues through a $1-1$ resonance is completely classified by use of group
theory and energetics.

The main purpose of this paper is to show the usefulness of this type of result for
analyzing the dynamical behavior of mechanical systems. We describe briefly the main

result of [3] in Section 2 and consider rotating pendula problems in Section 3.

These examples clearly point out the fact that in specific mechanical systems, both passing and splitting can occur generically. In [3] this behavior is explained in the context of systems with *symplectic* symmetries. The examples suggest there is a corresponding result for systems with antisymplectic symmetries as well. We expect that the techniques of [14] will be useful toward this end.

2 Generic movement of eigenvalues

In this section we briefly describe the main result of [3] for the case of $1-1$ resonances in Hamiltonian systems with symmetry.

Let Z be a symplectic vector space with symplectic form ω. Assume there is a compact Lie group Γ acting symplectically on Z, that is,

$$\omega(\gamma v, \gamma w) = \omega(v, w) \quad \text{for all } \gamma \in \Gamma \text{ and } v, w \in Z.$$

We denote by $\mathrm{sp}_\Gamma(Z)$ the Lie algebra of linear infinitesimally symplectic maps commuting with Γ:

$$B \in \mathrm{sp}_\Gamma(Z) \iff \begin{cases} (i) & \omega(Bv, w) + \omega(v, Bw) = 0 \quad \text{for all } v, w \in Z, \\ (ii) & \gamma B = B\gamma \quad \text{for all } \gamma \in \Gamma. \end{cases}$$

Suppose that $A(\lambda)$ is a one-parameter family in $\mathrm{sp}_\Gamma(Z)$ and that $A(\lambda)$ undergoes a $1-1$ resonance at $\lambda = 0$. After rescaling we may assume that the purely imaginary eigenvalues which are involved are $\pm i$. It is well known (eg [4], [9], [10]) that without symmetry, generically the eigenvalues split off the imaginary axis in a $1-1$ resonance (see Figure 1). When there is symmetry present, this is no longer true: for certain symmetry types the passing case may occur generically (see Figure 2). In [3] it is shown that for symplectic symmetries the generic movement of eigenvalues through a $1-1$ resonance can be completely characterized in terms of group theory and energetics, but by neither of them alone.

To state the corresponding result precisely, it is necessary to recall some terminology from [13]. If U is a symplectic representation of Γ then — by ignoring the symplectic structure of U — we obtain an ordinary representation, which is called the *underlying representation*. A *Γ-irreducible symplectic representation* is a representation that has no proper nonzero Γ-invariant symplectic subspaces. Irreducible symplectic representations are either nonabsolutely irreducible (i.e., are irreducible but some linear map that is not a real multiple of the identity commutes with Γ) or the sum of a pair of isomorphic absolutely irreducible subspaces (see [5]). We now use the fact that the space of linear maps commuting with Γ is isomorphic to $\mathbf{R}$ (the absolutely irreducible case), to $\mathbf{C}$ or to $\mathbf{H}$, the quaternions, see for example [6]. It can be shown (Theorem 2.1 in [13]) that in the real and quaternionic cases the isomorphism type of the irreducible symplectic representation is uniquely determined by that of its underlying representation, whereas in the complex case there are precisely two isomorphism types of irreducible symplectic representations for a given complex irreducible underlying representation. They are said to be *dual* to each other. According to these two different possibilities we will speak of *complex irreducibles of the same type* and *complex duals*.

Theorem 2.1 ([3]) *Let $E_{\pm i}$ be the generalized (real) eigenspace of $A(0)$ belonging to the eigenvalues $\pm i$ and let Q denote the quadratic form induced on $E_{\pm i}$ via*

$$Q(z) = \omega(z, A(0)z) \,. \tag{2.1}$$

Then

$$E_{\pm i} = U_1 \oplus U_2 \,,$$

where, generically, precisely one of the following holds:

(a) U_1 *and* U_2 *are not isomorphic and the eigenvalues pass independently along the imaginary axis. (Q may be indefinite or definite.)*

(b) $U_1 = U_2 = V \oplus V$, *where V is real, or* $U_1 = U_2 = W$, *where W is quaternionic, the eigenvalues split, and Q is indefinite.*

(c) U_1 *and* U_2 *are complex of the same type, the eigenvalues pass and Q is indefinite.*

(d) U_1 *and* U_2 *are complex duals and the eigenvalues pass or split depending on whether Q is definite or indefinite.*

This theorem gives a complete characterization of the generic eigenvalue movement through $1-1$ resonances in Hamiltonian systems with a symplectic symmetry group. The result is summarized in Table 1.

	Eigenspace structure	Induced quadratic form	
		definite	indefinite
(a)	$U_1 \oplus U_2$ nonisomorphic	"independent passing"	
(b)	$V \oplus V \oplus V \oplus V$ real, or $W \oplus W$ quaternionic	not generic	splitting
(c)	$W \oplus W$ complex of the same type	not generic	passing
(d)	$W \oplus W$ complex duals	passing	splitting

Table 1: Generic eigenvalue movement in 1–1-resonances with symplectic symmetry group

Example 2.2 We consider the space $\mathbf{C}^2$, where the symplectic form ω is induced by $J(z_1, z_2) = (z_2, -z_1)$. Let the group S^1 act symplectically on $\mathbf{C}^2$ by

$$\theta(z_1, z_2) = (e^{i\theta} z_1, e^{i\theta} z_2).$$

Then the spaces

$$\mathbf{C}\{z, -iz\}, \quad \mathbf{C}\{z, iz\}$$

are complex duals with respect to this S^1-action: both are irreducible and J is acting as $-i$ on the first and as i on the second subspace. We consider the S^1-invariant quadratic Hamiltonian

$$H(z_1, z_2, \lambda) = \frac{1}{8}|z_1|^2 + \frac{\lambda}{2}\mathrm{Im}(\bar{z}_1 z_2) + 2(1 - \frac{3}{4}\lambda^2)|z_2|^2 ,$$

or in real coordinates,

$$H(q_1, q_2, p_1, p_2, \lambda) = \frac{1}{8}(q_1^2 + q_2^2) + \frac{\lambda}{2}(q_1 p_2 - q_2 p_1) + 2(1 - \frac{3}{4}\lambda^2)(p_1^2 + p_2^2). \qquad (2.2)$$

According to part (d) of Theorem 2.1, we expect to see definite passing or indefinite splitting in 1–1 resonances while varying the parameter λ. A computation of the eigenvalues of $A(\lambda) = JD^2 H(0, 0, \lambda)$ yields

$$\sigma(\lambda) = \frac{i}{2}\left(\lambda \pm \sqrt{1 + 3(1 - \lambda^2)}\right).$$

These together with their complex conjugates are the four eigenvalues for the system induced by (2.2). In fact, definite passing occurs as λ passes through 0 and indefinite splitting occurs as λ passes through $\pm\frac{2}{\sqrt{3}}$ (see also Figure 3). To verify the definiteness properties observe that in this case the induced quadratic form in (2.1) is simply given by H itself.

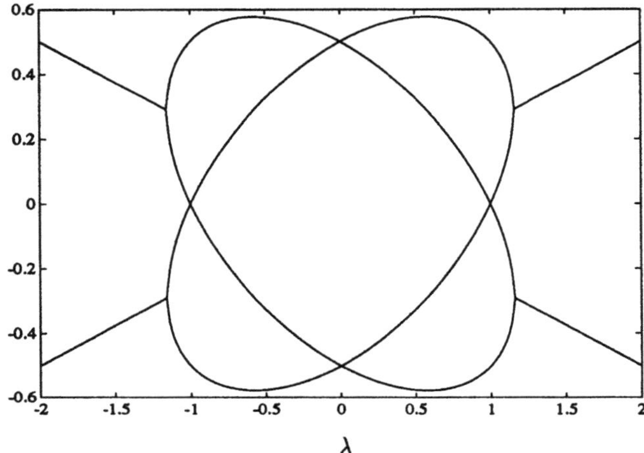

Figure 3: Imaginary parts of all four eigenvalues against λ. The sequence *indefinite splitting* $\rightarrow$ *passing at 0* $\rightarrow$ *definite passing* $\rightarrow$ *passing at 0* $\rightarrow$ *indefinite splitting* can be observed as λ is varied.

3 Generic bifurcation of spinning pendula with symmetry

3.1 The forcibly rotated orthogonal planar double pendulum

As in [2] we consider a rotating orthogonal planar double pendulum. The angular velocity of the rotation is assumed to be Ω_*. The two masses m_1 and m_2 are forced to move in two planes, which are orthogonal to each other. The pendula are assumed to have equal length. We set

$$\Omega^2 = \frac{\Omega_*^2}{g}, \quad m = \frac{m_1}{m_2}.$$

We regard Ω as the bifurcation parameter. After scaling time and making a symplectic change of coordinates (cf [2]) one obtains

$$H_2 = \frac{1}{2(m+1)}(q_1^2 + q_2^2) + \frac{\Omega}{\sqrt{m+1}}(q_1 p_2 - q_2 p_1) + \frac{1}{2}(m+1)\left(1 - \frac{m}{m+1}\Omega^2\right)(p_1^2 + p_2^2)$$

as the quadratic part of the Hamiltonian H describing the behavior of this system. But this Hamiltonian is exactly of the form as the one in (2.2) (set $m = 3$) and, moreover, the underlying symplectic structures are the same. Therefore Example 2.2 shows that both indefinite splitting and definite passing occur in this mechanical system.

Remark 3.1 The S^1 symmetry of H_2 is not a symmetry of the full nonlinear mechanical system which is described in coordinates inside the rotating frame. In fact, this system only possesses a (nonsymplectic) $Z_2 \times Z_2$ symmetry. Thus, although H_2 clearly has the S^1 symmetry and the results apply, its origin as a mechanical symmetry is not so clear.

First, the S^1 symmetry is only a symmetry at quadratic level which can easily be seen by looking at the higher order terms of H (cf [2]). Although it is not yet completely clear why this symmetry is present, there are the following facts which seem to play an important role:

- the nonsymplectic $Z_2 \times Z_2$ symmetry is generated by one symplectic and one anti-symplectic involution and forces some quadratic terms of the Hamiltonian to vanish.

- the underlying mechanical structure as well as the requirement that there exists a $1-1$ resonance in the problem forces additional restrictions on the quadratic terms of the Hamiltonian.

It will be part of subsequent work to clarify the situation. In particular, this example and the next one suggest it would be useful to extend the above theorem to include antisymplectic symmetries as well.

3.2 The double spherical pendulum

Consider the double spherical pendulum, as shown in Figure 4. The relative equilibria and their stabilities are found in [11]. For angular momentum $\mu \neq 0$ the relative equilibria do

118   M. Dellnitz et al.

not have any obvious symmetry properties (except that they are invariant under reflection in a uniformly rotating plane). Consistent with the generic (nonsymmetric) theory, one only sees eigenvalue splitting.

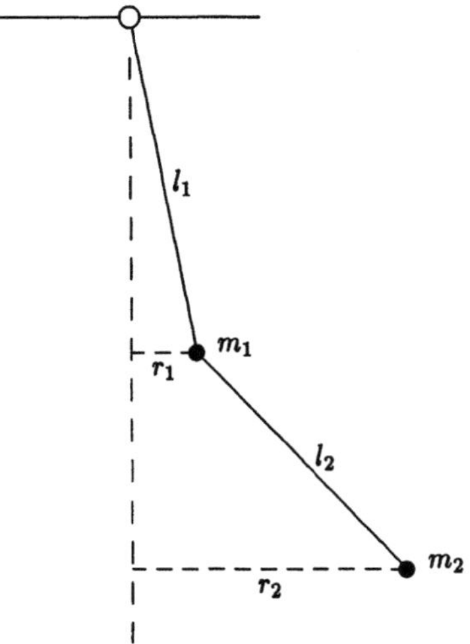

Figure 4: The double spherical pendulum

To get more interesting behavior, we look at the straight down state with $\mu = 0$. This state is however, singular in the sense that the overall S^1 action is not free there and, correspondingly, the set $\mu = 0$ is not a smooth manifold, but has a conic singularity.

To get around this difficulty, we regularize the system near this singular state. This is done as follows. Using ideas of Lagrangian reduction, one finds that the linearized dynamics at a relative equilibrium with angular momentum $\mu = $ constant (about the vertical axis) is given by a certain Lagrangian whose quadratic terms we shall denote

$$\mathcal{L}_2(\mu, \delta r_1, \delta r_2, \delta \varphi, \delta \dot{r}_1, \delta \dot{r}_2, \delta \dot{\varphi})$$

where r_1 and r_2 are the distances of the two pendula from the vertical axis and φ is the angle between the two vertical planes through the symmetry axis and the two masses. Variation of these variables are denoted $\delta r_1, \delta r_2$ and $\delta \varphi$.

The regularized Lagrangian at the straight down state is given at quadratic order by

$$\mathcal{L}_2^0(s_1, s_2, \theta, \dot{s}_1, \dot{s}_2, \dot{\theta}) = \lim_{\mu \to 0} \left[\frac{1}{|\mu|} \mathcal{L}_2(\mu, \sqrt{|\mu|} s_1, \sqrt{|\mu|} s_2, \theta, \sqrt{|\mu|} \dot{s}_1, \sqrt{|\mu|} \dot{s}_2, \dot{\theta}) \right].$$

This regularization procedure is akin to those used in celestial mechanics and corresponds to blowing up of singularities in algebraic geometry.

The regular Lagrangian that results from this procedure in this example is given by the following expression

$$
\begin{aligned}
2\mathcal{L}_2^0 \;=\;& m\dot{s}_1^2 + 2r\dot{s}_1\dot{s}_2 + r^2\dot{s}_2^2 + (m-1)g(g+1)\dot{\theta}^2 \\
& + \frac{Q}{r}\left(\frac{3m\beta - 4g^2(m-1)}{\beta} + m \right) s_1^2 \\
& + 2Q\left(\frac{3\beta + 4g(m-1)}{\beta} \right) s_1 s_2 \\
& + \frac{Q}{r}\left(\frac{3\beta - 4(m-1)}{\beta} + r \right) s_2^2 + \frac{\alpha}{r}\theta^2 \\
& + \frac{1}{\sqrt{r}}\frac{2g^2(m-1)}{\beta}(\dot{\theta}s_1 - \dot{s}_1\theta) \\
& - \sqrt{r}\frac{2g(m-1)}{\beta}(\dot{\theta}s_2 - \dot{s}_2\theta)
\end{aligned}
$$

where $r = l_2/l_1$ is the ratio of the lengths,

$$m = (m_1 + m_2)/m_2, \quad g = \frac{1}{2}\left[m(r-1) - \sqrt{m^2(r-1)^2 + 4rm} \right],$$

$$\beta = g^2 + 2g + m, \text{ and } Q = g/(1+g).$$

Notice that $\mathcal{L}_0^2$ still has two free parameters r and m. If one wishes, one can easily get the corresponding Hamiltonian via Legendre transform. The Lagrangian is invariant under the transformation

$$
\begin{aligned}
s_1 &\mapsto s_1, & s_2 &\mapsto s_2, & \theta &\mapsto -\theta, \\
\dot{s}_1 &\mapsto -\dot{s}_1, & \dot{s}_2 &\mapsto -\dot{s}_2, & \dot{\theta} &\mapsto \dot{\theta},
\end{aligned}
$$

which yields an antisymplectic involution on the Hamiltonian side. This symmetry appears to be crucial to what follows and is a reason for suggesting a generalization of the results of Sec. 2 to include antisymplectic transformations. In addition, this may help put the previous example (Sec. 3.1) into a better context.

The equations of motion

$$\frac{d}{dt}\frac{\partial \mathcal{L}_2^0}{\partial \dot{q}^i} - \frac{\partial \mathcal{L}_2^0}{\partial q^i} = 0$$

for $(q^i) = (s_1, s_2, \theta)$ have the form

$$M\ddot{q} + S\dot{q} + \Lambda q = 0$$

for 3×3 matrices M, S and Λ. The characteristic polynomial of the system is defined by

$$\det[\lambda^2 M + \lambda S + \Lambda] = 0$$

and is a polynomial $p(x)$ in $x = \lambda^2$. It is explicitly given as follows. Defining r, m, g and Q as above, further define the following quantities

$$
\begin{aligned}
G &= 2\frac{m-1}{m+g(g+2)} & A &= \left(1 + \tfrac{g}{m}\right)^2 Q^2 \\[4pt]
f &= \frac{A-r^2}{A-g^2} & R &= r^2 - g^2 f \\[4pt]
L &= 1 - f & a &= QL^{\frac{3}{2}}(3m - 2g^2 G) + m\sqrt{R} \\[4pt]
b &= QL\sqrt{R}(3 + 2gG) & c &= QR\sqrt{L}(3 - 2G) + r^2\sqrt{L} \\[4pt]
s &= \sqrt{LR} & d &= s + g^2 f \\[4pt]
e &= mr^2 - d^2 & B &= mc + ar^2 - 2bd \\[4pt]
F &= ac - b^2 & U &= (m-1)(g+1)R^{\frac{3}{2}}eL \\[6pt]
V &= eRL + (m-1)(g+1)RB\sqrt{L} + RLG^2 g(mR + g(2ds + r^2 gL)) \\[4pt]
W &= sB + (m-1)(g+1)F\sqrt{R} + sG^2 g(aR + g(2bs + cgL)).
\end{aligned}
$$

Then a straightforward, but lengthy calculation shows that

$$p(x) = Ux^3 + Vx^2 + Wx + F.$$

Using this expression for p we compute numerically $1-1$ resonances for the straight down state in the double spherical pendulum. The results are shown in Figure 5. Note especially that both the splitting and the passing cases occur generically.

The phenomena seen in this example of the straight down state of the double spherical pendulum can be expected to be ubiquitous in mechanical systems with symmetry at symmetric solutions. For example, we hope that techniques like this will be useful for rotating elastic and fluid masses in three dimensions (see [7] and [8]).

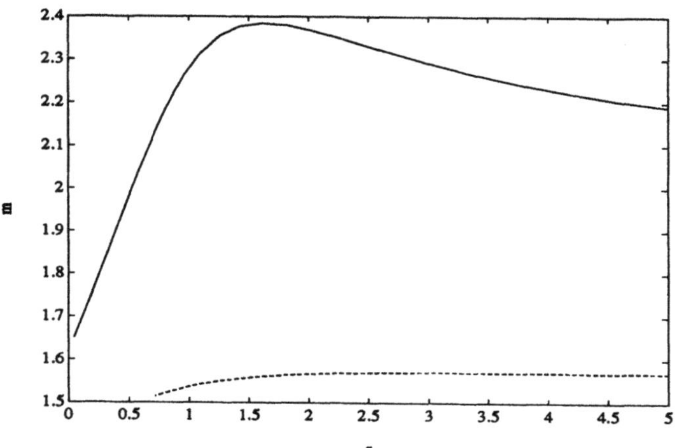

Figure 5: Curves of splitting (solid) and passing cases (dashed) in the spherical double pendulum

References

[1] R. Abraham and J. Marsden [1978] *Foundations of Mechanics*, 2nd ed., Addison-Wesley, New York.

[2] T.J. Bridges [1990] Branching of periodic solutions near a collision of eigenvalues of opposite signature. *Math. Proc. Camb. Phil. Soc.* 108, 575-601.

[3] M. Dellnitz, I. Melbourne and J.E. Marsden [1991] Generic bifurcation of Hamiltonian vector fields with symmetry. University of Houston (preprint).

[4] D.M. Galin [1982] Versal deformations of linear Hamiltonian systems. *AMS Transl.* 2 118, 1-12. (1975 Trudy Sem. Petrovsk. 1 63-74).

[5] M. Golubitsky and I. Stewart [1987] Generic bifurcation of Hamiltonian systems with symmetry. *Physica* 24D, 391-405.

[6] M. Golubitsky, I. Stewart and D. Schaeffer [1988] *Singularities and Groups in Bifurcation Theory*. Vol. 2, Springer.

[7] D.R. Lewis [1989] Nonlinear stability of a rotating planar liquid drop. *Arch. Rat. Mech. Anal.* **106**, 287-333.

[8] D.R. Lewis and J.C. Simo [1990] Nonlinear stability of rotating pseudo-rigid bodies. *Proc. Roy. Soc. Lon.* A **427**, 281-319.

[9] R.S. MacKay [1986] Stability of equilibria of Hamiltonian systems. In *Nonlinear Phenomena and Chaos*, edited by S. Sarkar, 254-270.

[10] R.S. MacKay and P.G. Saffman [1986] Stability of water waves. *Proc. Roy. Soc. Lond.* A **406**, 115-125.

[11] J.E. Marsden and J. Scheurle [1991] Lagrangian reduction and the double spherical pendulum (preprint).

[12] J.C. van der Meer [1985] *The Hamiltonian Hopf Bifurcation*. Lecture Notes in Mathematics **1160**.

[13] J. Montaldi, M. Roberts and I. Stewart [1988] Periodic solutions near equilibria of symmetric Hamiltonian systems. *Phil. Trans. R. Soc.* **325**, 237-293.

[14] Y.H. Wan [1989] Versal deformations of infinitesimally symplectic transformations with involutions. State University of New York at Buffalo (preprint).

[15] Y.H. Wan [1989] Codimension two bifurcations of symmetric cycles in Hamiltonian systems with an antisymplectic involution. State University of New York at Buffalo (preprint).

International Series of Numerical Mathematics, Vol. 104, © 1992 Birkhäuser Verlag Basel

SYMMETRY ASPECTS OF 3-PERIODIC MINIMAL SURFACES

Werner Fischer & Elke Koch

Institut für Mineralogie, Petrologie und Kristallographie (und Wissenschaftliches Zentrum für Materialwissenschaften), Hans-Meerwein-Straße, D-3550 Marburg

Abstract - Symmetry properties of 3-periodic minimal surfaces subdividing R^3 into two congruent regions are discussed. The relation between the order of a flat point and its site symmetry is established. Explicit formulae are given for the calculation of the genus of such a surface depending on the kind of surface patches that build up the surface. Making use of 2-fold axes that have to be embedded in a surface with given symmetry new families of minimal balance surfaces have been derived. Two examples of bifurcations related to minimal surfaces are mentioned.

1. Symmetry and derivation

A **minimal surface** in 3-dimensional space R^3 is defined as a surface with mean curvature zero at each of its points, i.e. the two extreme values of curvature (main curvatures) are equal in magnitude but opposite in sign for each point of the surface. Thus all points of a minimal surface are saddle points.

In crystallography, especially those minimal surfaces have attracted attention that are periodic in three independent directions and, therefore, may be related to crystal structures. In this connection mainly those surfaces that are free of self-intersections seem to be of interest. Such a surface subdivides R^3 into two regions or labyrinths such that each labyrinth is connected but not simply connected. If the two labyrinths are congruent the intersection-free, 3-periodic minimal surface is called a **minimal balance surface** [1].

The symmetry of a minimal balance surface is best characterized by a pair of space groups G-H: G describes the full symmetry of the (non-oriented) surface, and H is that subgroup of G with index 2 which consists of all symmetry operations that do not interchange the two sides of the surface and the two labyrinths. Obviously,

the pairs G-H correspond uniquely to the proper black-white space
groups [2].

Let s be a symmetry operation of G that does not belong to H. Then
s interchanges the two sides of each minimal balance surface with
symmetry G-H, and all fixed points of s must lie on the surface.
This property, however, is inconsistent with the absence of self-
intersections for minimal balance surfaces if s is a 3-, 4- or 6-
fold rotation, a reflection or a 6-fold rotoinversion. As a consequence, certain space-group pairs G-H are incompatible with minimal balance surfaces. A detailed examination of the 1156 types of
group-subgroup pairs with index 2 shows that - for the reasons described above - only 547 of them are not incompatible with minimal
balance surfaces.

For these 547 types of space-group pairs all 2-fold rotation axes
and all (roto)inversion centres $\bar{1}$, $\bar{3}$ and $\bar{4}$ have been tabulated
that must be located on each minimal balance surface with that
symmetry [3]. This knowledge gives an aid for the derivation of
new families of minimal balance surfaces. Especially useful are 2-
fold rotation axes which exist for 352 out of the 547 types. Considering only the sets of all 2-fold axes belonging to G but not
to H, 52 different configurations of straight lines on minimal
balance surfaces result. In 18 of these cases all 2-fold axes are
3-dimensionally connected, in 12 cases they form infinite sets of
parallel plane nets. Each of these two situations is favourable
for the derivation of minimal balance surfaces, because the 2-fold
axes on the one hand may define a Plateau problem (resulting in a
surface patch) and on the other hand may be used to generate the
entire surface from one surface patch by means of Schwarz's reflection principle [4]:

(1) In a 3-dimensionally connected set of 2-fold axes skew poly-
    gons are formed that may be spanned by **disc-like surface
    patches.** If the original skew polygon has been adequately cho-
    sen the resulting infinite surface is free of self-intersec-
    tions, i.e. it is a minimal balance surface. An adequately
    chosen skew polygon has to fulfill the following conditions:
    (i) All its vertex angles must be chosen as small as possible;
    in particular, no angles larger than 90° are allowed. (ii) The

skew polygon must not be penetrated by a further 2-fold axis
belonging to the same set.

The 18 configurations of 3-dimensionally connected 2-fold axes
give rise to 15 families of minimal balance surfaces that may
be generated from disc-like spanned skew polygons [1,3,5].
Eight of these families had not been known before.

(2) The 12 configurations of 2-fold axes that disintegrate into
parallel plane nets are compatible with different kinds of
surface patches:

(i) If all plane nets are congruent and if at least half the
polygon centres for a pair of adjacent nets lie directly above
each other, **catenoid-like surface patches** may be spanned between neighbouring polygons from adjacent nets. Such catenoids
give rise to seven families of minimal balance surfaces [3,5],
one of which had not been described before.

(ii) If plane nets of two different kinds are stacked alter-
nately upon each other surface patches may be spanned that
have been called **branched catenoids**. A branched catenoid is
bounded by a convex polygon at one end and by a concave polygon with one point of self-contact at its other end. The convex polygon stems from one of the more wide-meshed nets,
whereas the concave polygon is formed by two, three or four
polygons with a common vertex of an adjacent close-meshed net.
Branched catenoids refer to three new families of minimal balance surfaces [6]. Recently a rigorous proof for the existence
of such surface patches has been published [7].

(iii) Congruent parallel plane nets stacked directly upon each
other allow surface patches that have been called **multiple
catenoids**. A multiple catenoid may be imagined as resulting
from fusion of two, three, four or six neighbouring catenoids.
It is bounded by two congruent concave polygons with one point
of self-contact each. Multiple catenoids give rise to eight
new families of minimal balance surfaces [8,9].

(iv) Configurations of 2-fold axes that disintegrate into parallel plane quadrangular nets are compatible with 1-dimension-
ally infinite surface patches, called **infinite strips**. Such an

infinite strip is bounded by two infinite (zigzag or meander) lines. The strips may be regarded as resulting from fusion of an infinite row of neighbouring catenoids. In most cases infinite strips constructed in this way produce minimal surfaces that may be also built up from finite surface patches as described above. In two cases, however, minimal surfaces of new families are formed [10].

(v) For configurations of 2-fold axes that disintegrate into congruent plane parallel nets stacked directly upon each other the catenoid-like surface patches [cf. (i)] may be replaced by more complicated ones, called **catenoids with spout-like attachments.** For this, spouts are attached to the "faces" of the catenoids resulting in surface patches with two, three or four additional ends that are not bounded by straight lines. Spouts of neighbouring catenoids are united to handles or to three-armed or four-armed handles, respectively. Six families of minimal balance surfaces correspond to such surface patches [11]. One of these families has been derived before by another method [5].

In addition, two families of balance surfaces without positive Gaussian curvature have been derived which contain skew 2-fold axes in three independent directions and which also seem to be minimal surfaces [1,12].

Two families of minimal balance surfaces without 2-fold axes are known so far, the gyroid surfaces [5] and orthorhombically distorted P surfaces [9]. Both have been included into further discussions [12,13,14].

2. Genus and Euler characteristic

A non-periodic surface in R^3 is said to be of **genus** g if it may topologically be deformed to a sphere with g handles. For 3-periodic minimal surfaces a modified definition must be used [5] counting only the number of handles per primitive unit cell. In other terms, the 3-periodic surface is embedded in a (flat) 3-torus T^3 to get rid of all translations, and then the conventional definition of the genus may be applied. This procedure corresponds to identifying the opposite faces of a primitive unit cell.

The genus of a 3-periodic minimal surface without self-intersec-
tion may be calculated in different ways, two of which will be
discussed in the following:

(1) Labyrinth graph: Each of the two labyrinths may be represented
by a graph that is entirely located within its labyrinth; each
branch of a labyrinth contains an edge of its graph; each circuit
of a labyrinth graph encircles at least one edge of the other
graph [5].

Any of the two labyrinth graphs may be used to represent topological properties of the surface. As each circuit of the graph corresponds to a handle of the surface the number of circuits per primitive unit cell may be counted to get the genus of the surface. In
case of a minimal balance surface the genus has to refer to a
primitive unit cell of the subgroup H. There exist two different
possibilities to derive the genus with the aid of labyrinth
graphs:

(i) In modification of a procedure proposed before [15], a con-
    nected subgraph containing no translationally equivalent ver-
    tices may be separated from a given labyrinth graph. Then the
    genus of the surface may be calculated as
    $$g = \frac{p}{2} + q,$$
    where p is the number of edges connecting the finite subgraph
    to the rest of the infinite labyrinth graph, and q is the num-
    ber of edges that has to be omitted to make the subgraph sim-
    ply connected. As p equals at least 6 the genus of a 3-period-
    ic minimal surface without self-intersection is at least 3.

(ii) Keeping in mind the embedding of the minimal surface in the
     torus T^3 a more crystallographic formula for the genus may be
     derived. g equals the difference between the number e of edges
     in the embedded labyrinth graph and the number e_s of edges in
     any simply connected subgraph with the same number v of verti-
     ces. With
     $$e = \frac{1}{2} \sum_i m_i e_i \qquad \text{and} \qquad e_s = v - 1 = \sum_i m_i - 1$$
     it follows:
     $$g = e - e_s = 1 + \sum_i m_i \left(\frac{e_i}{2} - 1\right).$$

Here m_i means the multiplicity of the i-th kind of vertices referred to a primitive unit cell of H, and e_i is the number of edges meeting in this vertex. The summation runs over all kinds of symmetrically equivalent vertices of the labyrinth graph.

Details on the labyrinth graphs and the genera of the known minimal balance surfaces have been published [12].

(2) Euler characteristic: An intersection-free surface in R^3 may also be characterized by a number χ, its **Euler characteristic.** χ is related to g by

$$g = 1 - \frac{\chi}{2} .$$

The Euler characteristic of an intersection-free surface may be derived in a simple way by defining a tiling on the surface, i.e. by subdividing the surface into tiles (disc-like surface patches). For such an arbitrary tiling the equation

$$\chi = f - e + v$$

holds, where f, e and v are the numbers of tiles (faces), edges and vertices, respectively, in the tiling. For a 3-periodic surface the tiling must be compatible with the translations of the surface, and the tiles, edges and vertices have to be counted per primitive unit cell of H [12].

For a minimal balance surface generated from disc-like surface patches that span skew polygons of 2-fold axes these surface patches may be used as tiles. Then χ may be calculated as

$$\chi = f(1-\frac{e_P}{2}) + \sum_i v_i ,$$

where e_P is the number of edges of such a skew polygon, f is the number of skew polygons and v_i the multiplicity for the i-th kind of symmetrically equivalent vertices. f and v_i are both referred to a primitive unit cell of H.

If a minimal balance surface consists of catenoid-like surface patches spanned between parallel plane nets of 2-fold axes its Euler characteristic is given by

$$\chi = v_N - e_N ,$$

where v_N and e_N refer to the plane nets of 2-fold axes. v_N means the number of vertices, e_N the number of edges counted for all nets of 2-fold axes and per primitive unit cell of H. Making use of the relation $f_N - e_N + v_N = 0$ for nets, the genus may be calculated from χ as

$$g = k + 1$$

where $k = f_N/2$ gives the number of catenoids per primitive unit cell of H. The same formula holds for minimal balance surfaces generated from infinite strips spanned between plane nets of 2-fold axes. Then k must be understood as the number of original catenoids (per primitive unit cell of H) that have been united to infinite rows.

Similar formulae have been derived for the other three kinds of minimal balance surfaces spanned between parallel plane nets of 2fold axes. Here k means the number of surface patches per primitive unit cell of H:

If a minimal balance surface is made up from multiple catenoids

$$g = km + 1$$

holds. m gives the number of catenoids that must be united to form one multiple catenoid.

The genus of a minimal balance surface built up from branched catenoids is given by

$$g = \frac{k(1+b)}{2} + 1.$$

b is the number of branches at one of the ends of a branched catenoid.

If a minimal balance surface may be generated from catenoids with s spouts attached, its genus may be calculated as

$$g = ks + 1.$$

3. Flat points

For each point of a minimal surface the defining condition

$$K_1 + K_2 = 0$$

must be fulfilled, where K_1 and K_2 are the main curvatures in that point. Normally $K_1 = -K_2$ differs from zero, i.e. the point is an ordinary saddle point. For exceptional points, however,

$$K_1 = K_2 = 0$$

may be fulfilled. Such points are called flat points of the surface. In contrast to ordinary saddle points, the surrounding of a flat point shows n valleys separated by n ridges (n>2). The simplest example is the "monkey saddle" with n=3.

For any point on an intersection-free minimal surface its degree of flatness may be characterized by an integer number β, called its order. The order of a point P_0 with normal vector n_0 can be derived as follows: A second point P with normal vector n is moved on the surface around P_0. If P_0 is an ordinary point, n rotates once around n_0 during one revolution of P. If, however, P_0 is a flat point, n rotates more than once (e.g. p times) around n_0 per revolution of P. Then the order β of P_0 is defined as

$$\beta = p - 1.$$

Accordingly, a normal point has order $\beta=0$, and the order of a flat point may equal any positive integer. For 3-periodic minimal surfaces flat-point orders up to $\beta=4$ have been observed so far. The number n of valleys (or ridges) in the surrounding of a flat point is

$$n = \beta + 2.$$

Each order of a (flat) point corresponds to a maximal site symmetry of such a point. This symmetry is $\bar{4}m2$ for $\beta=0$, $\bar{3}m$ for $\beta=1$, $\bar{8}m2$ for $\beta=2$, $\bar{5}m$ for $\beta=3$ and $\bar{12}m2$ for $\beta=4$. Therefore, most site symmetries of points on a minimal surface enforce the existence of a flat point. Conversely, only points with site symmetry $\bar{4}$, 222, 2mm, 2, m or 1 can be non-flat points of a minimal surface.

There exists a relation between the genus of an intersection-free minimal surface and the order of its flat points [15,16]:

$$g = 1 + \frac{1}{4} \sum_i \beta_i .$$

The sum runs over all flat points within a primitive unit cell of H. This formula can be used in different ways:

(1) If all flat points with their orders are known the genus of a minimal surface may be calculated.

(2) If the symmetry of a minimal surface and its genus are known the relation between flat-point symmetry and flat-point order in combination with the above formula may be used to derive a complete list of flat points [13].

4. Bifurcation properties of minimal surfaces

A minimal surface (or one of its surface patches) may be considered as the solution $f(x,y)$ of the partial differential equation
$$(1+f_y{}^2)f_{xx} - 2f_x f_y f_{xy} + (1+f_x{}^2)f_{yy} = 0$$
with the restriction
$$1 + f_x{}^2 + f_y{}^2 > 0$$
and under certain boundary conditions e.g. due to the 2-fold axes
embedded in the surface. If these boundary conditions contain free
parameters, their variation may give rise to bifurcations [17,18,
19 §396], i.e. the occurrence of additional solutions above or
below some critical value(s). Two different kinds of examples
shall be mentioned here:

(1) Deformation of disc-like surface patches: The surface patch
for a tD surface generated by symmetry $P\bar{4}n2-I\bar{4}2d$ is bounded by a
skew octagon with equal outer angles $\alpha=\pi/2$ [3]. It is made up by 8
of the 12 edges of a square prism, i.e. by the 4 edges parallel to
the prism axis (length c) and by two opposite edges from each the
basic and the top square (length a). This frame by itself has symmetry $\bar{4}2m$ (D_{2d}) which is also the symmetry of the spanned minimal-
surface patch. As the total curvature of the boundary curve is
$8\alpha=4\pi$, the solution of the Plateau problem is unique. It maintains
this property if the ratio c/a is used as parameter. The situation
is different, however, if α is varied with the edge lengths kept
fixed. Then the symmetry of the octagon is not changed. If α is
enlarged by moving each two edges from the same square face of the
prism toward each other, the total curvature exceeds the critical
value of 4π that guarantees the uniqueness [20]. For c/a=1 it has
been shown before that a bifurcation takes place at some unknown
value $\alpha_B>\pi/2$ [17,19 §396]. Soap-film experiments using wire frames
with different values of c/a and varying α gave the following results: Bifurcation is observed if c/a is large enough. Then only
one surface with symmetry $\bar{4}2m$ is formed in the range between the
minimal value $\alpha_{min}=\pi/4$ (flat octagon) and the bifurcation value
$\alpha_B>\pi/2$. For $\alpha>\alpha_B$ this surface becomes physically unstable and either one of two new stable surfaces with symmetry mm2 (C_{2v}) occurs. These two surfaces can be mapped onto each other by any symmetry operation out of the coset of the subgroup mm2 (index 2) of
the symmetry group $\bar{4}2m$ of the frame. The soap film can easily be

made to jump from one of these positions to the other (a frame
with a=4cm, c=6cm, α=100° results in a nice model). The value of
α_B increases with decreasing c/a. If c/a is too small (approximate
limit 0.5) no bifurcation could be observed. It has to be noted
that only for α=π/2 the surface patch can be expanded into an in
tersection-free 3-periodic surface.

(2) Change of the distance between parallel nets of 2-fold axes:
As an example the case of nets with equilateral triangles (edge
length a) stacked directly onto each other (distance c) will be
discussed. Here, the trivial solution is to span each triangle
disc-like; only this solution is compatible with all values of the
parameter p=a/c. Catenoids (H surfaces [4]), multiple catenoids
(MC1 surfaces [8,9]) and catenoids with spout-like attachments
(C(H) surfaces [5,11]) can only be spanned between triangles of
adjacent nets if p exceeds p_c , p_M and p_s , respectively. Soap-film
experiments indicate that $p_M < p_c$ and $p_s < p_c$. It was not possible,
however, to decide whether p_M or p_s is the smallest bifurcation
value. Whereas the trivial solution has the non-crystallographic
symmetry of a set of parallel equidistant planes with continuous
translations in two independent directions, the H and C(H) surfaces show the symmetry P6$_3$/mmc-P$\bar{6}$m2. The symmetry of an MC1 surface is further decreased (index 3) to P6$_3$/mcm-P$\bar{6}$2m.

References

[1] W. Fischer & E. Koch, Z. Kristallogr. **179**, 31 - 52 (1987).

[2] A. L. Mackay & J. Klinowski, Comp. & Maths. with Appls. **12B**, 803 - 824 (1986).

[3] E. Koch & W. Fischer, Z. Kristallogr. **183**, 129 - 152 (1988).

[4] H. A. Schwarz, Gesammelte Abhandlungen, Band 1, Springer-Verlag, Berlin (1890).

[5] A. H. Schoen, NASA Technical Note D-5541 (1970).

[6] W. Fischer & E. Koch, Acta Crystallogr. **A45**, 166 - 169 (1989).

[7] J. C. C. Nitsche, Coll. Phys., Tome **51**, Colloque **C7**, 265 - 271 (1990).

[8] E. Koch & W. Fischer, Acta Crystallogr. **A45**, 169 - 174 (1989).

[9] H. Karcher, Manuscripta Math. **64**, 291 - 357 (1989).

[10] W. Fischer & E. Koch, Acta Crystallogr. **A45**, 485 - 490 (1989).

[11] E. Koch & W. Fischer, Acta Crystallogr. **A45**, 558 - 563 (1989).

[12] W. Fischer & E. Koch, Acta Crystallogr. **A45**, 726 - 732 (1989).

[13] E. Koch & W. Fischer, Acta Crystallogr. **A46**, 33 - 40 (1990).

[14] W. Fischer & E. Koch, Coll. Phys., Tome **51**, Colloque **C7**, 131 - 147 (1990).

[15] S. T. Hyde, Z. Kristallogr. **187**, 165 - 185 (1989).

[16] H. Hopf, Springer Lecture Notes in Mathematics No. 1000 (1983).

[17] J. C. C. Nitsche, Arch. Rat. Mech. Anal. **30**, 1 - 11 (1968).

[18] J. C. C. Nitsche, Ann. Acad. Sci. Fenn. AI **2**, 361 - 373 (1976).

[19] J. C. C. Nitsche, Lectures on minimal surfaces, Vol. 1, Cambridge Univ. Press, Cambridge (1989).

[20] J. C. C. Nitsche, Arch. Rat. Mech. Anal. **52**, 319 - 329 (1973).

International Series of Numerical Mathematics, Vol. 104, © 1992 Birkhäuser Verlag Basel

Hopf bifurcation at non-semisimple eigenvalues :

a singularity theory approach

J.E.Furter
Mathematics Institute
University of Warwick
Coventry CV4 7AL

Abstract

We consider the problem of Hopf bifurcation in an ODE when the linearisation at an equilibrium has a pair of non-semisimple purely imaginary eigenvalues with geometric multiplicity one and algebraic multiplicity k. In [8] Vanderbauwhede derived an equivariant bifurcation equation via a Lyapounov-Schmidt reduction. In this work we give a systematic way to study those equations, including the discussion of the effects of small perturbations, via an ad hoc version of singularity theory. We give more detailed results on the 1:1 resonance and the generic 1:1:1 resonance, although we have not studied the stability of the bifurcating solutions.

1   Introduction

Consider the following autonomous ODE on $\mathbf{R}^n$

$$\dot{y} = f(y, \tilde{\alpha}) \tag{1}$$

where f is smooth enough and $\tilde{\alpha} \in \mathbf{R}^m$ are parameters. We assume that $f(0,0) = 0$. We want to study the bifurcation of small amplitude periodic solutions from the trivial steady state $(0,0)$ ("Hopf bifurcation") and the effects of the perturbations induced by $\tilde{\alpha}$ ("universal unfolding") under the following conditions for the linearisation of (1).

Denote $f_y(0, \tilde{\alpha})$ by $A(\tilde{\alpha})$, we assume that $A_o \overset{\text{def}}{=} A(0)$ has eigenvalues $\pm i$ of *geometric* multiplicity one and *algebraic* multiplicity $1 \leq k \leq \frac{1}{2} n$, and, that A_o has no other resonant eigenvalues on the imaginary axis (i.e. of the form $\pm j\,i, j = 0, 2, 3 \ldots$). Hence, note that the linearized system of (1) has a natural period of 2π, and that, without limiting the generality, we can assume that $f(0, \tilde{\alpha}) \equiv 0$ in a neighbourhood of $(0,0)$.

Those conditions will be assumed throughout this work. They were fundamental in Vanderbauwhede [8], where a clear account was given on the way of describing the problem (with one parameter) by a reduced bifurcation equation obtained via a Lyapounov-Schmidt reduction. In particular Vanderbauwhede unified the results and methods of Ashkenazi-Chow [1], $k = \frac{1}{2} n$, and Caprino-Maffei-Negrini [2], $k = 2$ and $4 \leq n$. In the generic situation there are either 0 or 2 bifurcating branches when k is even or one branch when k is odd. More recently vanGils-Krupa-Langford [5] studied extensively the case when $k = 2$ using normal form theory and a three parameters unfolding.

We use singularity theory to systematically study the qualitative behaviour of the bifurcation equation. Qualitative means modulo changes of coordinates respecting the structure of the solution set. Note that the changes of coordinates will mix the period and amplitude of the solutions, although the amplitude is only scaled at the lowest order. In practice we often have a distinguished parameter, either directly from the model (or experiment) or at least from the drawing of 1-D bifurcation diagrams. In that situation, it is a natural requirement to preserve the behaviour of 1-D slices of the zero-set of (1). Hence, we also consider the case when the parameter space is split via $(\lambda, \alpha) \in \mathbf{R} \times \mathbf{R}^{m-1}$ where λ is the main bifurcation parameter and α now represents the perturbation ("unfolding") parameters, that is, they are considered as fixed with respect to variations in λ. The method have the advantage of being straightforward to implement independently of the dimension n of the space we work with, depending as it stands on the Taylor serie expansion of f, with the drawback of not keeping much of the rest of the dynamics of (1). We have not investigated the stability properties of the bifurcating solutions.

In Section 2, we follow the derivation of Vanderbauwhede to fix notations and obtain a $\mathbf{Z}_2$-invariant bifurcation equation on $\mathbf{R}^2$ with $\tilde{\alpha}$ as parameters. In Section 3, the necessary singularity theory is developed to study that kind of bifurcation equation and its unfoldings ("perturbations"). We concentrate on the distinguished parameter case as the results for the "plain" situation are easily obtained from them. We use a simple modification of a Golubitsky-Schaeffer [6] [7] type of theory (contact-equivalence with one distinguished parameter). It belongs to the family of singularity theories analysed by Damon [3] where the fundamental theorems are proved. In Section 4, we first classify all problems without distinguished parameter with at most two degeneracy conditions. Even if that classification can be done independently of k, the codimension of those problems increases fast with k, of the order of $2k$. Then follows the classification of the equations with distinguished parameter up to codimension at least 4. To help structuring those results, we use the "path formulation" ([6]). In Section 5, we analyse some diagrams for the 1:1 resonance. As expected, in the generic situation there are at most k solutions for (1) with that number being attained for some values of the parameters. We finish on the generic situation for the 1:1:1 resonance.

I would like to thank SERC for support through a research grant.

2 Bifurcation equation

We follow the derivation of the bifurcation equation of Vanderbauwhede [8]. Our main concern here is to fix the notations and obtain a priori information on the structure of the bifurcation equation.

Let $y(t)$ be a T-periodic solution of (1) and define $\tau = \frac{2\pi}{T} - 1$, then $x(t) \stackrel{\text{def}}{=} y((1+\tau)^{-1}t)$ is a 2π-periodic solution of the equation

$$(1 + \tau)\dot{x} = f(x, \tilde{\alpha}). \qquad (2)$$

Since we are interested in small amplitude periodic solutions of (1), T must be near the period of the linearized equation, that is 2π. Hence, we will look for small amplitude 2π-periodic solutions of (2) with τ near 0.

To fix a functional set-up for (2), consider the Banach subspaces $X \subset Y$ of $C^o(\mathbf{R}, \mathbf{R}^n)$ consisting of the C^1, resp. C^o, 2π-periodic functions equipped with their usual sup-norms.

If we define $F : X \times \mathbf{R} \times \mathbf{R}^m \to Y$ by $F(x,\tau,\tilde{\alpha})(t) = -(1+\tau)\dot{x}(t) + f(x(t),\tilde{\alpha})$, our problem (2) is then to solve the following equation, for $(x,\tau,\tilde{\alpha})$ near $(0,0,0)$ in $X \times \mathbf{R} \times \mathbf{R}^m$

$$F(x,\tau,\tilde{\alpha}) = 0. \tag{3}$$

The operator $L(x)(t) = -\dot{x}(t) + A_o x(t)$ is the linearisation of F at the origin. By hypothesis there exists a basis $\{\xi_j\}_{j=1}^k \subset \mathbf{C}^n$ of $\ker(A_o - iI)^k$ such that $(A_o - iI)\xi_1 = 0$, $(A_o - iI)\xi_j = \xi_{j-1}, j = 2,\ldots,k$. Similarly $\ker(A_o^t + iI)^k$ has a basis $\{\xi_j^*\}_{j=1}^k \subset \mathbf{C}^n$ such that $(A_o^t + iI)\xi_k^* = 0$, $(A_o^t + iI)\xi_j^* = \xi_{j+1}^*, j = 1,\ldots,k-1$. Denote by $(.,.)$ the usual scalar product on $\mathbf{C}^n$, we normalize the previous bases such that $(\xi_l^*,\xi_j) = 2\delta_{jl}$, $1 \le j,l \le k$. Define also a scalar product on the complexification Y^c of Y by $<x,y> = \frac{1}{2\pi}\int_0^{2\pi} (x(t),y(t))\, dt$. For $1 \le j \le k$, define $\zeta_j(t) = e^{it}\xi_j$, $\zeta_j^*(t) = e^{it}\xi_j^*$ and the projectors $P,Q : Y \to Y$ by $P(x) = \Re(<\zeta_1^*,x> \zeta_1)$, $Q(x) = \Re(<\zeta_k^*,x> \zeta_k)$. A simple calculation shows that $\ker L = \operatorname{im} P$ and that $\operatorname{im} L = \ker Q$. Splitting $x = v + w \in X$, using $v = P(x)$, we are ready to apply the Lyapounov-Schmidt reduction to (3). We get the system

$$(I - Q)\, F(v + w,\tau,\tilde{\alpha}) = 0 \tag{4}$$
$$Q\, F(v + w,\tau,\tilde{\alpha}) = 0. \tag{5}$$

Near the origin, (4) can be implicitly solved for w to get $w = \bar{w}(v,\tau,\tilde{\alpha})$, replacing into (5), we obtain the following bifurcation equation on $\ker L$

$$\tilde{G}(v,\tau,\tilde{\alpha}) = Q\, F(v + \bar{w}(v,\tau,\tilde{\alpha}),\tau,\tilde{\alpha}).$$

To get hold on $\tilde{G}$ we can introduce some coordinates. The maps $\chi : \mathbf{C} \to \operatorname{im} P$, defined by $\chi(z) = \Re(z\zeta_1)$, and $\chi^* : \mathbf{C} \to \operatorname{im} Q$, $\chi^*(z) = \Re(z\zeta_k^*)$, are linear isomorphisms between two dimensional real vector spaces. Using those isomorphisms we obtain the bifurcation equation (with $w(z,\tau,\tilde{\alpha}) = \chi(z) + \bar{w}(\chi(z),\tau,\tilde{\alpha}))$:

$$G(z,\tau,\tilde{\alpha}) = <\zeta_k^*, F(w(z,\tau,\tilde{\alpha}),\tau,\tilde{\alpha})> . \tag{6}$$

Our ODE being autonomous it is equivariant with respect to the phase-shifts $t \mapsto t + \phi$, $\forall \phi \in S^1$. Via our choice of coordinates the equivariance translates as the usual $SO(2)$-action, $e^{i\phi}$, on $\mathbf{C}$. In particular there exists $H : \mathbf{C} \times \mathbf{R} \times \mathbf{R}^m \to \mathbf{C}$ such that $G(z,\tau,\tilde{\alpha}) = \tilde{H}(|z|^2,\tau,\tilde{\alpha})\,z$. Hence, the nontrivial solutions of (6) are in 1-1 correspondence with the solutions of the $\mathbf{Z}_2$-equivariant equation $\tilde{H}(\rho^2,\tau,\tilde{\alpha}) = 0$, where $\rho = |z|$.

The next step is to get some insight on the structure of $\tilde{H}$. After some calculations (expanding on Vanderbauwhede [8]), we find that $\tilde{H}(0,\tau,0) = -i^k\tau^k$ and that $\tilde{H}_{\tilde{\alpha}}(0,\tau,0) = \frac{1}{2}\sum_{j=1}^k i^{j-1}\tau^{j-1}(\xi_k^*,A_{\tilde{\alpha}}(0)\xi_j) \in \mathbf{C}$. For a treatment independent of k, we find it more convenient to use the bifurcation equation $H(\rho^2,\tau,\tilde{\alpha}) \stackrel{\text{def}}{=} (-i)^k\tilde{H}(\rho^2,\tau,\tilde{\alpha})$ instead of $\tilde{H}$. And so, there are two functions $\tilde{H}_1$ and $\tilde{H}_2$ such that

$$H(\rho,\tau,\tilde{\alpha}) = -\tau^k + (-i)^k\,\tilde{H}_1(\tau,\tilde{\alpha}) + (-i)^k\,\tilde{H}_2(\rho^2,\tau,\tilde{\alpha}), \tag{7}$$

where $\tilde{H}_1(\tau,0) \equiv 0$. The particularity of the structure of H is that if $\tilde{H}_1$ and $\tilde{H}_2$ take values a priori in $\mathbf{C}$, $H(0,\tau,0)$ is always *real*, whatever f is. Denote by $h(\rho,\tau)$ the "unperturbed" problem, that is $H(\rho,\tau,0)$. Identifying the real and imaginary parts of h as the two components of a real map, it means that we are dealing with functions of the type

$(\hat{p}(\rho^2, \tau), \hat{q}(\rho^2, \tau))$ where $\hat{q}(0, \tau) \equiv 0$. We are going to preserve that "triangular" structure as well as the Z_2-invariance when developing the part of the singularity theory dealing with the recognition and classification of the bifurcation problems.

As we have just seen, in the unfolding H, the coefficients of powers of τ and $\tilde{\alpha}$ can assume *complex* values. This means that the part of the singularity theory dealing with the unfolding theory should have no restriction other than the Z_2-invariance. This explains why the generic situation has codimension $2k - 1$. In the presence of a distinguished parameter λ, we assume that $\tilde{\alpha}$ is split like (λ, α). Note that, as far as the Taylor expansion of H is concerned, $\tilde{\alpha}$ and λ (and also α) have the same role, that is, we can use the same formulas for all derivatives with respect to the parameters.

3   Singularity theory

As the special role of λ is a complication, we are going to give the singularity theory we need in that case. The adaptation to the situation without distinguished parameter will be straightforward. We suppose the reader acquainted with the mechanics of singularity theory applied to bifurcation problems as developed by Golubitsky-Schaeffer [6],[7]. Let $\mathcal{E}$ denote the ring of Z_2-invariant germs of functions $r : (\mathbf{R}^{2+1}, 0) \to \mathbf{R}$ satisfying $r(-\rho, \tau, \lambda) = r(\rho, \tau, \lambda)$. We work in the $\mathcal{E}$-module $\vec{\mathcal{E}}$ of germs of maps (p, q) where p and q belong to $\mathcal{E}$. The maps h satisfying (7) belong to $\vec{\mathcal{E}}_1$, the $\mathcal{E}$-submodule of $\vec{\mathcal{E}}$ generated by $X_1 = (1, 0)$, $X_2 = (0, \lambda)$ and $X_3 = (0, u)$. Moreover, $\mathcal{E}_\lambda$ denotes the ring of real functions in λ, and m, m_λ the maximal ideals of $\mathcal{E}$, $\mathcal{E}_\lambda$. We denote by a $^\circ$ superscript the evaluation of the functions or derivatives at the origin.

We define a group $\mathcal{K}$ of changes of coordinates on $\vec{\mathcal{E}}_1$ from the set of triples $\mathcal{K} = \{(M, X, \Lambda)\}$ where $M \in \mathcal{M}$, the $\mathcal{E}$-module of matrices generated by $\begin{pmatrix} 1 & 0 \\ 0 & 0 \end{pmatrix}$, $\begin{pmatrix} 0 & 1 \\ 0 & 0 \end{pmatrix}$, $\begin{pmatrix} 0 & 0 \\ 0 & 1 \end{pmatrix}$, $\begin{pmatrix} 0 & 0 \\ \lambda & 0 \end{pmatrix}$ and $\begin{pmatrix} 0 & 0 \\ u & 0 \end{pmatrix}$, where $X \in \mathcal{X}$, the set of λ-parametrized changes of coordinates on $\mathbf{R}^2$ given by $\mathcal{X} = \{ (r, \rho s) \mid r, s \in \mathcal{E}, \ r^\circ = 0, \ r_\tau^\circ \cdot s^\circ \neq 0 \}$ and $\Lambda \in \mathcal{E}_\lambda$ with $\Lambda^\circ = 0$, $\Lambda_\lambda^\circ > 0$. Under $\mathcal{K}$, only the λ-direction of the bifurcating branches is preserved. $\mathcal{K}$ has a structure of group through the classical action on $\vec{\mathcal{E}}_1$: $(M, X, \Lambda)((p, q)) = M(p \circ (X, \Lambda), q \circ (X, \Lambda))$, $\forall (M, X, \Lambda) \in \mathcal{K}$. Note that, as we are not considering the stability properties of the solutions, we do not fix any particular sign properties for the (M, X)-part of the changes of coordinates in $\mathcal{K}$.

We will need the following algebraic objects. The *extended* tangent space of $(p, q) \in \vec{\mathcal{E}}_1$ is the vector space

$$T(p, q) = \mathcal{E}\{(p, 0), (q, 0), (0, p), (0, q), (up_u, uq_u), (p_\tau, q_\tau)\} + \mathcal{E}_\lambda\{(p_\lambda, q_\lambda)\} \ .$$

The *unipotent* tangent space $UT(p, q)$ is the subspace of $\vec{\mathcal{E}}_1 \cap T(p, q)$ given by

$$m\{(p, 0), (q, 0), (0, q), (up_u, uq_u)\} + \mathcal{E}\{(0, \lambda p), (0, up)\} + m^2\{(p_\tau, q_\tau)\} + m_\lambda^2\{(p_\lambda, q_\lambda)\} \ .$$

We can then define the *codimension* of $(p, q) \in \vec{\mathcal{E}}_1$ as the real dimension of the vector space $\vec{\mathcal{E}}/T(p, q)$, denoted by $\mathrm{codim}(p, q)$.

The asymmetry between the use of $\vec{\mathcal{E}}$ and $\vec{\mathcal{E}}_1$ is explained by the asymmetry between the structures of h and H. The natural space for the perturbation is $\vec{\mathcal{E}}$. Let us briefly describe the unfolding problem for a $(p, q) \in \vec{\mathcal{E}}_1$. An *unfolding* H of h with m parameters is a germ of a map $(\mathbf{R}^{2+1+m}, 0) \to \mathbf{R}^2$ such that $H(u, \tau, \lambda, 0) = h(u, \tau, \lambda)$. We denote by $\vec{\mathcal{E}}(m)$ the space of all unfoldings with m parameters of all elements $h \in \vec{\mathcal{E}}_1$. The group $\mathcal{K}$ extends naturally to a group of changes of coordinates on $\vec{\mathcal{E}}(m)$, denoted by $\mathcal{K}(m)$, by considering the quadruples (M, X, Λ, Φ) acting on $\vec{\mathcal{E}}(m)$ where (M, X, Λ) is a m parameter unfolding of an element in $\mathcal{K}$ and $\Phi : (\mathbf{R}^m, 0) \to (\mathbf{R}^m, 0)$ is a diffeomorphism. Contact equivalence actually induces another fundamental relation on the unfoldings. We can compare two unfoldings of the same $h \in \vec{\mathcal{E}}_1$ with different numbers of parameters. Let $H \in \vec{\mathcal{E}}(m)$ and $G \in \vec{\mathcal{E}}(l)$ be two such unfoldings of $h \in \vec{\mathcal{E}}_1$. We say that G *maps into* H if there is $(M, X, \Lambda, \Phi) \in \mathcal{K}(l)$, unfolding of the identity in $\mathcal{K}$, and $\Psi : (\mathbf{R}^l, 0) \to (\mathbf{R}^m, 0)$ such that $G = (M, X, \Lambda, \Psi \circ \Phi) \cdot H$. The important property here is versality. H is said to be a *versal* unfolding of h if any other unfolding G of h maps into H. In particular, the versal unfoldings with the minimal number of parameters, say r, are called *universal*. The universal unfoldings of h are all equivalent in the sense that they belong to the same $\mathcal{K}(r)$-orbit.

We can now state the fundamental properties of germs of finite codimension.

Theorem 1 *Let $(p, q) \in \vec{\mathcal{E}}_1$ be of finite codimension, then*

(1) (p, q) *is finitely determined, that is, is $\mathcal{K}$-equivalent to a polynomial, say $(\hat{p}, \hat{q})$, called a normal form. The coefficients of the normal form are polynomial combinations of a finite number of coefficients of the Taylor series expansion of (p, q).*

(2) If $\mathrm{codim}(p, q) = 0$ *then* (p, q) *is its own universal unfolding. Otherwise, let* $\{(p_j, q_j)\}_j$, $1 \le j \le \mathrm{codim}(p, q)$, *be a base of* $\vec{\mathcal{E}}/T(p, q)$, *then* $(p, q) + \sum_{j=1}^{\mathrm{codim}(p,q)} \alpha_j (p_j, q_j)$ *is a universal unfolding of* (p, q). *Note that, modulo a change of coordinates, we can instead use the normal form* $(\hat{p}, \hat{q})$ *and a base* $\{(\hat{p}_i, \hat{q}_i)\}$ *of* $\vec{\mathcal{E}}/T(\hat{p}, \hat{q})$.

(3) An unfolding (P, Q) *of* (p, q) *with m-parameters, $\alpha \in \mathbf{R}^m$, is versal iff* $\{(P_{\alpha_j}, Q_{\alpha_j})\}_j$, $1 \le j \le m$, *projects into a base of* $\vec{\mathcal{E}}/T(p, q)$.

The last useful quantity is the *module of higher order terms* defined by

$$\mathcal{P}(p, q) = \{ (r, s) \in \vec{\mathcal{E}} \mid (\tilde{p} + r, \tilde{q} + s) \sim_{\mathcal{K}} (p, q), \ \forall (\tilde{p}, \tilde{q}) \sim_{\mathcal{K}} (p, q) \}.$$

In other words, the elements of $\mathcal{P}(p, q)$ are those we can discard from the Taylor expansion of *any* germs in the $\mathcal{K}$-orbit of (p, q). The remaining ("intermediary") terms are to be tackled with explicit changes of coordinates.

To estimate $\mathcal{P}(p, q)$, we need the following. *Intrinsic ideals* are $\mathcal{K}$-invariant ideals and, in particular, it is an exercise to show that finite sums and products of $\{u, \tau\}$, $\{u\}$ and $\{\lambda\}$ are intrinsic ideals. *Intrinsic subspaces* are $\mathcal{K}$-invariant subspaces of $\vec{\mathcal{E}}_1$. In our case they are of the form $I \oplus J$, where I, J are intrinsic ideals and satisfy $\{u, \lambda\}I \subset J \subset I$ (We denote by $\{u, \lambda\} \subset \mathcal{E}$ the ideal generated by u and λ). From an adaptation of the results of Bruce-du Plessis-Wall [4] : $\mathcal{P}(p, q) \supset \mathrm{Intr}\, \mathcal{U}T(p, q)$, where $\mathrm{Intr}\, \mathcal{N}$ denotes the maximal intrinsic subspace of an $\mathcal{E}$-submodule $\mathcal{N}$.

¿From the results of this section, note that the singularity theory for the problems without distinguished parameter follows exactly the same line, in particular Theorem 1 holds provided we use the correct algebraic quantities. Those different algebraic quantities are simply obtained by discarding the dependence on λ from the one we have defined.

4 Classification of the bifurcation equations

When $k = 1$ our theory, as expected, reduces to the study of degenerate Hopf bifurcation of Golubitsky-Langford (see [6]). By an explicit change of coordinates $h \sim_{\mathcal{K}} (\tau, \hat{q}(u, \lambda))$. A simple calculation shows that $\operatorname{codim} h = \operatorname{codim}_{\mathbf{Z}_2} \hat{q}$ with $\bar{\mathcal{E}}/T(h)$ isomorphic to $\mathcal{E}_{\mathbf{Z}_2}/T_{\mathbf{Z}_2}(\hat{q})$, where the Z_2-subscript denotes the quantities coming from the Z_2-equivariant theory applied to $\hat{q}(x^2, \lambda)\, x$ (see [6]). Moreover, an explicit characterization of $\mathcal{P}(h)$ from $\mathcal{P}_{\mathbf{Z}_2}(\hat{q})$ is related to the terms we need in the previous explicit change of coordinates.

Now assume that $k \geq 2$. First, we classify the diagrams without distinguished parameter up to two degeneracy conditions. We find four types, we denote by the roman numbers I to IV, and keep track of k as a parameter.

- If $q_u^o \neq 0$, H is $\mathcal{K}(2k-1)$-equivalent to

$$\mathrm{I}(k) : \quad (-\tau^k, u) + \Big(\sum_{j=0}^{k-2} \tilde{\alpha}_j \tau^j, \sum_{j=0}^{k-1} \tilde{\beta}_j \tau^j \Big), \tag{8}$$

- If $q_u^o = 0$ but $p_u^o \cdot q_{u\tau}^o \neq 0$, H is $\mathcal{K}(2k)$-equivalent to

$$\mathrm{II}(k) : \quad (-\tau^k + \epsilon_1 u, u\tau) + \Big(\sum_{j=0}^{k-2} \tilde{\alpha}_j \tau^j, \sum_{j=0}^{k-1} \tilde{\beta}_j \tau^j + \tilde{\beta}_k u \Big),$$

- If $q_u^o \cdot p_u^o = 0$ but $q_{u\tau}^o \cdot A \neq 0$, H is $\mathcal{K}(2k+1)$-equivalent to

$$\mathrm{III}(k) : \quad (-\tau^k + \epsilon_2 u^2, u\tau) + \Big(\sum_{j=0}^{k-2} \tilde{\alpha}_j \tau^j + \tilde{\alpha}_{k-1} u, \sum_{j=0}^{k-1} \tilde{\beta}_j \tau^j + \tilde{\beta}_k u \Big),$$

- If $q_u^o \cdot q_{u\tau}^o = 0$ but $p_u^o \cdot B \neq 0$, H is $\mathcal{K}(2k+1)$-equivalent to

$$\mathrm{IV}(k) : \quad (-\tau^k + \epsilon_1 u, u\tau^2) + \Big(\sum_{j=0}^{k-2} \tilde{\alpha}_j \tau^j, \sum_{j=0}^{k-1} \tilde{\beta}_j \tau^j + \tilde{\beta}_k u + \tilde{\beta}_{k+1} u\tau \Big),$$

where $\epsilon_1 = \operatorname{sign} (p_u^o)^{k+1}$, $\epsilon_2 = \operatorname{sign} A^k$, $B = (p_u^o)^2 q_{u\tau\tau}^o + (\text{only when } k = 2)\, p_u^o q_{uu}^o$,

and $A = (q_{u\tau}^o)^2 p_{uu}^o - p_{u\tau}^o q_{u\tau}^o q_{uu}^o - (\text{only when } k = 2)\, \frac{1}{2}(q_{uu}^o)^2$.

The algebraic data for the higher order terms are

$$\mathcal{P}(\mathrm{I}(k)) \supset (m\{u\} + m^{k+1}, m\{u\}) \quad , \quad \mathcal{P}(\mathrm{II}(k)) \supset (m\{u\} + m^{k+1}, \{u\}^2 + m^2\{u\})$$
$$\mathcal{P}(\mathrm{III}(k)) \supset (m^2\{u\} + m^{k+1}, m^2\{u\}) \quad , \quad \mathcal{P}(\mathrm{IV}(k)) \supset (m\{u\} + m^{k+1}, m\{u\}^2 + m^3\{u\}).$$

Theorem 2 *There are at most k solutions to the equation $I(k) = 0$ with k being reached for some values of the unfolding parameters. The same conclusion holds for the other cases with the bound k being replaced by $k+1$ for $II(k)$ and by $k+2$ for $III(k)$ and $IV(k)$.*

The proof of Theorem 2 is an exercise. We are going now to classify the problems with a distinguished parameter as "paths" in the unfolding parameter spaces of I ... IV. Let $h(x, \lambda)$ be a bifurcation diagram with one distinguished parameter, of universal unfolding $H(x, \lambda, \alpha)$. We can consider h as a one parameter unfolding of $h_o \stackrel{\text{def}}{=} h(x, 0)$, of universal unfolding $H_o(x, \tilde{\alpha})$. And so, h maps into H_o, that is, h is $\mathcal{K}(1)$-equivalent to $H_o(x, \phi(\lambda))$. Note that ϕ is a curve in the space of the $\tilde{\alpha}$'s. We call ϕ the *path* associated with h. A similar conclusions can be reached for the unfolding H, giving rise to an unfolding $\Phi(\lambda, \alpha)$ of ϕ. In particular, note that the number of solutions as λ (and α) varies is determined by the number of solutions of H_o as $\tilde{\alpha}$ varies that is, we can use Theorem 2 to get a priori bounds. This time the classification is more complex and so we concentrate on getting all the (k)-families with the codimension of at least one member smaller or equal to 4. There are, furthermore, some isolated problems of codimension 4 when $k = 2$. We add letters to the roman numbers to distinguish between different paths in the same unfolding parameter space. Note that we have split the conditions of the recognition problem between those related to h_o and those related to the path ϕ and the $\{\delta_i\}_{i=1}^{12}$ are ± 1 coefficients.

- $\mathbf{q_{uu}^o} \neq \mathbf{0}$ Let $(\tilde{\alpha}_0 \ldots \tilde{\alpha}_{k-2}; \tilde{\beta}_0 \ldots \tilde{\beta}_{k-1})$ be the unfolding parameters of $\mathrm{I}(k)$.

If $(p_u^o q_\lambda^o - p_\lambda^o q_u^o) \cdot q_\lambda^o \neq 0$, H is $\mathcal{K}(2k - 2)$-equivalent to

$$\mathbf{Ia}(k): \quad (\delta_2 \lambda + \alpha_0, \ldots \alpha_{k-2}; \delta_1 \lambda, \ldots \beta_{k-1}).$$

If $(p_u^o q_\lambda^o - p_\lambda^o q_u^o) = 0$ but $q_\lambda^o \cdot C \neq 0$, H is $\mathcal{K}(2k - 1)$-equivalent to

$$\mathbf{Ib}(2) : \quad (\delta_3 \lambda^2 + \alpha_1 \lambda + \alpha_0; \delta_1 \lambda, \beta_1),$$
$$(k \geq 3) \quad \mathbf{Ib}(k) : \quad (\alpha_{k-1} \lambda + \alpha_0, \delta_3 \lambda + \alpha_1, \ldots \alpha_{k-2}; \delta_1 \lambda, \ldots \beta_{k-1}).$$

If $q_\lambda^o = 0$ but $p_\lambda^o \cdot q_{\tau\lambda}^o \neq 0$, H is $\mathcal{K}(2k - 1)$-equivalent to

$$\mathbf{Ic}(k): \quad (\delta_4 \lambda + \alpha_0, \ldots \alpha_{k-2}; \beta_1 \lambda + \beta_0, \lambda, \ldots \beta_{k-1}).$$

If $q_\lambda^o = q_{\tau\lambda}^o = 0$ but $p_\lambda^o \cdot \delta_5 \neq 0$ and $k = 2$, H is $\mathcal{K}(4)$-equivalent to

$$\mathbf{Id}(2): \quad (\delta_4 \lambda + \alpha_0; \delta_5 \lambda^2 + \beta_2 \lambda + \beta_0, \beta_1 \lambda).$$

If $(p_u^o q_\lambda^o - p_\lambda^o q_u^o) = C = 0$ but $q_\lambda^o \cdot \delta_6 \neq 0$ and $k = 2$, H is $\mathcal{K}(4)$-equivalent to

$$\mathbf{Ie}(2): \quad (\delta_6 \lambda^3 + \alpha_2 \lambda^2 + \alpha_1 \lambda + \alpha_0; \delta_1 \lambda, \beta_1).$$

If $p_\lambda^o = q_\lambda^o = 0$ but $(a^2 - \delta_7) \cdot q_{\tau\lambda}^o \neq 0$ (a is modal) and $k = 2$, H is $\mathcal{K}(4)$-equivalent to

$$\mathbf{If}(2): \quad (\delta_7 \lambda^2 + \alpha_1 \lambda + \alpha_0; a\lambda^2 + \beta_1 \lambda + \beta_0, \lambda).$$

- $\mathbf{q_{uu}^o} = \mathbf{0}$, $\mathbf{p_u^o} \cdot \mathbf{q_{u\tau}^o} \neq \mathbf{0}$, $(\tilde{\alpha}_0 \ldots \tilde{\alpha}_{k-2}; \tilde{\beta}_0 \ldots \tilde{\beta}_k)$ the unfolding parameters of $\mathrm{II}(k)$.
If $q_\lambda^o \neq 0$, H is $\mathcal{K}(2k - 1)$-equivalent to

$$\mathbf{IIa}(k): \quad (\alpha_0, \ldots \alpha_{k-2}; \delta_8 \lambda, \ldots \beta_k).$$

If $q_\lambda^o = 0$ but $a \cdot (a - \epsilon_1 \delta_9) \cdot p_\lambda^o \cdot q_{\tau\lambda}^o \cdot \delta_{10} \neq 0$ and $k = 2$, H is $\mathcal{K}(4)$-equivalent to

$$\mathbf{IIb}(2): \quad (\delta_9 \lambda + \alpha_0; \delta_{10} \lambda^2 + \beta_0, a\lambda + \beta_1, \beta_2).$$

where a is a modal parameter.

The quantities we need explicitly for next section are $\delta_1 = \text{sign}\, q_u^o q_\lambda^o$, $\delta_2 = \text{sign}\,\{q_u^o(p_\lambda^o q_u^o - q_\lambda^o)\}^{k+1}$, $\delta_3 = \text{sign}\, C$, $\delta_4 = \text{sign}\, p_\lambda^o(q_u^o q_{\tau\lambda}^o)^k$, $\delta_8 = \text{sign}\,(p_u^o q_\lambda^o q_{u\tau}^o)^k$ and

$$
\begin{aligned}
C \;=\; \Big[& 2({q_u^o}^2 p_{\lambda\lambda}^o - 2q_u^o q_\lambda^o p_{u\lambda}^o + {q_\lambda^o}^2 p_{uu}^o) - 2\tfrac{p_\lambda^o}{q_\lambda^o}({q_u^o}^2 q_{\lambda\lambda}^o - 2q_u^o q_\lambda^o q_{u\lambda}^o + {q_\lambda^o}^2 q_{uu}^o) + \\
& \{(q_u^o q_{\tau\lambda}^o - q_\lambda^o p_{u\tau}^o) - \tfrac{p_\lambda^o}{q_\lambda^o}(q_u^o q_{\tau\lambda}^o - q_\lambda^o q_{u\tau}^o)\}^2 \Big] .
\end{aligned}
$$

The algebraic data for the higher order terms are

$$
\mathcal{P}(\text{Ia}(k)) \supset (m\{u,\lambda\} + m^{k+1}, m\{u,\lambda\}),\; \mathcal{P}(\text{Ib}(k)) \supset (\{u,\lambda\}^3 + m^{k+1}, m^2\{u,\lambda\})
$$
$$
\mathcal{P}(\text{Ic}(k)) \supset (m\{u,\lambda\} + m^{k+1}, m\{u\} + \{\lambda\}^2 + m^2\{\lambda\})
$$
$$
\mathcal{P}(\text{IIa}(k)) \supset (m\{u,\lambda\} + m^{k+1}, m\{\lambda\} + \{u\}^2 + m^2\{u\}) .
$$

- $q_u^o = p_u^o = 0$, $q_{u\tau}^o \cdot A \neq 0$, $(\tilde{\alpha}_0, \tilde{\alpha}_1 ; \tilde{\beta}_0 \ldots \tilde{\beta}_2)$ the unfolding parameters of III(2).

If $a \cdot (1 + 4\epsilon_2 a^2) \cdot q_\lambda^o \neq 0$ (a is modal), H is $\mathcal{K}(4)$-equivalent to

$$
\text{IIIa}(2): \quad (\delta_{11}\lambda, \alpha_1 ; a\,\lambda + \beta_0, \beta_1, \beta_2) .
$$

- $q_u^o = q_{u\tau}^o = 0$, $p_u^o \cdot B \neq 0$, $(\tilde{\alpha}_0 ; \tilde{\beta}_0 \ldots \tilde{\beta}_3)$ the unfolding parameters of IV(2).

If $q_\lambda^o \neq 0$, H is $\mathcal{K}(4)$-equivalent to

$$
\text{IVa}(2): \quad (\alpha_0 ; \delta_{12}\lambda, \beta_1, \beta_2, \beta_3) .
$$

We can now sum-up the classification of the problems of low codimension.

Theorem 3 *Let $k \geq 2$. Without a distinguished parameter, the problems of codimension 3 ("generic") up to 5 are :*
<u>*codim=3*</u> *: I(2)*   <u>*codim=4*</u> *: II(2)*   <u>*codim=5*</u> *: I(3), III(2), IV(2).*
With a distinguished parameter, the problems of codimension 2 up to 4 are :
<u>*codim=2*</u> *: Ia(2)*   <u>*codim=3*</u> *: Ib(2), Ic(2),IIa(2)*
<u>*codim=4*</u> *: Ia(3), Id(2), Ie(2), If(2), IIb(2), IIIa(2), IVa(2)*

5 Analysis of the bifurcation equations

5.1 1:1 resonance

Here $k = 2$. The generic bifurcation without a distinguished parameter is of codimension 3, given by (8). Important information is given by the (semi-algebraic) transition varieties, surfaces in the parameter space delimiting the changes under perturbation in the behaviour of the solution set of the bifurcation equation. Recall that in the present case only the *number* of solutions for fixed values of the parameters is preserved under $\mathcal{K}$. For II(2) we have two transition varieties : $\mathcal{F} = \{\, (\tilde{\alpha}, \tilde{\beta}_0, \tilde{\beta}_1) \mid \tilde{\alpha} = 0,\; \tilde{\beta}_0 < 0 \,\}$ and $\mathcal{B}_0 = \{\, (\tilde{\alpha}, \tilde{\beta}_0, \tilde{\beta}_1) \mid \tilde{\beta}_0^2 = \tilde{\alpha}\,\tilde{\beta}_1^2 \,\}$, respectively representing the collapse of two solutions in a fold and the birth (or death) of a solution through the u-component becoming negative. This is represented in Figure 1. Note that we have one solution only in a quadratic wedge, explaining why the generic one parameter bifurcation is to 0 or 2 branches.

To illustrate the classification, we follow [5] and consider a model equation where the Taylor series expansion of H is rather straightforward to compute. In complex notations,

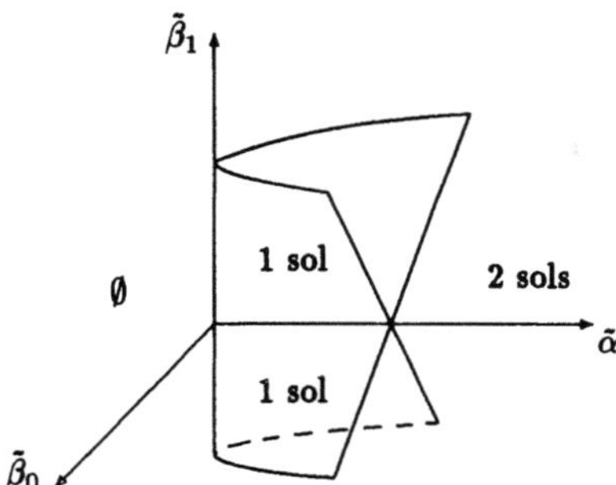

Figure 1: Transition varieties for I(2)

take $z_1 = x_1 + i\,x_2$ and $z_2 = x_3 + i\,x_4$. In the centre manifold, the unfolded (3 parameters) **Birkhoff normal form for the 1:1 resonance is the following :**

$$
\begin{aligned}
f_1(z_1, z_2, \nu, \mu) &= (i+\nu)\,z_1 + z_2 + o(N, |z_1|^2 + |z_2|^2)\\
f_2(z_1, z_2, \nu, \mu) &= \mu\,z_1 + (i+\nu)\,z_2 + z_1\,g_1(|z_1|^2, \mathcal{I}m(\bar{z}_1 z_2), \nu, \mu)\\
&\quad + z_2\,g_2(|z_1|^2, \mathcal{I}m(\bar{z}_1 z_2), \nu, \mu) + o(N, |z_1|^2 + |z_2|^2)\,.
\end{aligned}
$$

where $\nu \in (\mathbf{R}, 0)$, $\mu \in (\mathbf{C}, 0)$ **are the unfolding parameters,** N **is any natural number as big as we need and** g_i **is** $o(1)$, $i = 1, 2$. **Let** $\bar{a} + i\,\bar{b} = D_1 g_1(0, \ldots, 0)$, $\bar{c} + i\,\bar{d} = D_2 g_1(0, \ldots, 0)$, $\bar{e} + i\,\bar{f} = D_1 g_2(0, \ldots, 0)$ **and** $\bar{g} + i\,\bar{h} = \frac{1}{2} D_1^2 g_1(0, \ldots, 0)$. **After some calculations, we find that the Taylor series expansion of** H **starts as :**

$$
\left(-\tau^2 - \mu_1 + \bar{a}\,u + (\bar{f} - \bar{c})\,u\tau + \ldots,\ \mu_2 - \bar{b}\,u - \nu\,\tau + (\bar{d} + \bar{e})\,u\tau + \ldots \right).
$$

We can verify that if $\bar{b} \neq 0$ then H is $\mathcal{K}(3)$-equivalent to I(2), fully unfolded by (ν, μ). The rest of the classification follows in a similar way. If $\bar{b} = 0$ but $\bar{a} \cdot (\bar{d} + \bar{e}) \neq 0$ then H is $\mathcal{K}(4)$-equivalent to II(2) and is fully unfolded by $(\bar{b}, \nu, \mu)$, and so on.

With a distinguished parameter the bifurcation diagrams are represented by paths through the space $(\tilde{\alpha}, \tilde{\beta})$. And so, the transition varieties in that situation correspond to the values of the *parameters of the paths* when those paths change of position with respect to $\mathcal{F}$ and $\mathcal{B}_0$ (non-transversal intersections). As a simple example we can look at the generic case Ia(2). For fixed values of the unfolding parameters, the paths lie in horizontal planes $\tilde{\beta}_1$ =constant. The transition varieties are $\{\ (\alpha, \beta) \mid \alpha = 0\ \}$ and $\{\ (\alpha, \beta) \mid \alpha = -\frac{1}{4}\beta^2\ \}$. The bifurcation diagrams are described in Figure 2.

There are three diagrams of codimension 3.

1. Ib(2) correspond to parabolic paths in horizontal planes. There are three transition varieties : $\mathcal{H}_0$ (hysteresis at $u = 0$) of equation $\alpha_0 = 0$, $\mathcal{B}_0$ (bifurcation at $u = 0$), $\alpha_0 = -\frac{1}{4}\beta_1^2\alpha_1^2$, and $\mathcal{B}$ (bifurcation when $u \neq 0$), $\alpha_0 = \frac{1}{4}\delta_3\alpha_1^2$ with $\delta_1\delta_3\alpha_1 > 0$. When $\delta_3 < 0$, the bifurcation diagrams are like those found in Figure 7.4b in [6] (p. 278), as when $\delta_3 > 0$ the $(\|(u, \tau)\|, \lambda)$-diagrams are like those in Figure 7.4 (p. 277, [6]) (Note that the (u, τ)-projections are like in Figure 7.3 (p. 276, [6]).

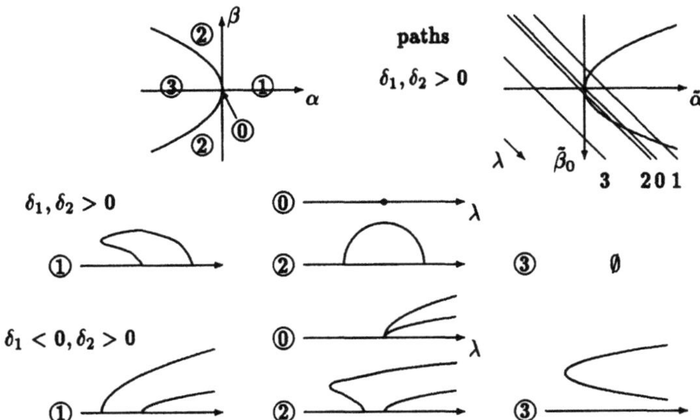

Figure 2: Transition varieties and bifurcation diagrams for Ia(2)

2. The paths in case Ic(2) are lines in 3D. The transition varieties are better described as topological cones. Let $\alpha_0 = a_0\beta_1^2$ and $\beta_0 = b_0\beta_1^3$, then $\mathcal{B}$ is $b_0 = \delta_4 a_0$ and $\mathcal{B}_0 \cup \mathcal{H}_0$ is $b_0 = -\frac{2}{27}[\delta_4(1-9a_0) \pm (1+3a_0)^{3/2}]$ delimiting 6 regions of diagrams like in Figure 8.3 (p. 287, [6]).

3. The last case IIa(2) has also its transition varieties with a conical structure, $\alpha_0 = a_0\beta_2^2$, $\beta_1 = b_2\beta_2^2$. They are given by a parabola $\mathcal{B}_0$, $(2a_0+\epsilon_1 b_2)^2 = 4a_0$, $\mathcal{H}_0$, $a_0 \cdot b_2 = 0$, and $\mathcal{H}$ (hysteresis with $u \neq 0$), $a_0 = \epsilon_1 b_2 - \frac{1}{3}$ with $b_2 < \frac{4}{9}\epsilon 1$, delimiting 13 regions.

For our model problem it is natural to look for bifurcation diagrams with respect to one of the parameter, keeping the two others fixed. In that way, we will actually cover the cases we just described.

If we choose $\lambda = \mu_2$, then provided $\bar{a} \cdot \bar{b} \neq 0$, we have the generic situation Ia(2). When $\bar{a} = 0$, provided $\bar{b} \cdot C \neq 0$, with $C = -2\bar{g} + \bar{b}^2 + \frac{1}{2}\bar{b}(\bar{d}+\bar{e}) + (\bar{c}-\bar{f})^2$, we have Ib(2) with $(\bar{a}, \nu, \mu_1)$ providing a complete unfolding. When $\bar{b} = 0$, if $\bar{a} \cdot (\bar{d}+\bar{e}) \neq 0$ then H is equivalent to IIa(2).

If we choose $\lambda = \mu_1$, then provided $\bar{b} \neq 0$, we have the case Ic(2). If we choose $\lambda = \nu$ we get at best codimension 4.

5.2   1:1:1 resonance

The generic 1:1:1 resonance I(3) is of codimension 5. Even so, it is possible to describe the transition varieties in an efficient way. Some remarks first on the structure of the solution set. Obviously there are at most 3 solutions for any values of the parameters $(\tilde{\alpha}, \tilde{\beta})$. We can be even more precise. If $\beta_1^2 - \beta_0\beta_2 < 0$ there are either no solutions when $\beta_2 > 0$ or, when $\beta_2 < 0$, as many as the number of τ solving the first component equation that is, 1 or 3 unless we are at a fold point. It is when $\beta_1^2 - \beta_0\beta_2 > 0$ that the situation is more complicated. It is possible to describe the transition varieties as $\tilde{\beta}$-parametrized curves in the $\tilde{\alpha}$-space. Explicitely, $\mathcal{F}$ is contained in the cusp $\alpha_0^2 = \frac{4}{27}\alpha_1^3$, where we have to take care of τ giving a positive u component. This can be done checking with the second transition variety $\mathcal{B}_0$. When $\beta_1^2 - 4\beta_0\beta_2 > 0$, $\mathcal{B}_0$ is non-empty, given by the two lines

$$2\beta_2\alpha_0 - (\beta_1 \pm \sqrt{\beta_1^2 - 4\beta_0\beta_2})\,\alpha_1 = \beta_2^{-2}[3\beta_0\beta_2 - \beta_1^3 \mp (\beta_1^2 - 4\beta_0\beta_2)^{\frac{3}{2}}],$$

when $\beta_2 \neq 0$. When $\beta_2 = 0$ $\mathcal{B}_0$ is given by $\beta_1^3\alpha_0 - \beta_0\beta_1^2\alpha_1 + \beta_0^3 = 0$.

References

[1] M.Ashkenazi, S.N.Chow. *A Hopf's bifurcation theorem for nonsimple eigenvalues*, Adv.Appl.Math. 1 (1980), 360-372.

[2] S.Caprino, C.Maffei, P.Negrini. *Hopf bifurcation at 1:1 resonance*, Nonlin.Anal., Th.Meth. & Appl. 8 (1984), 1011-1032.

[3] J.Damon. *The unfolding and determinacy theorems for subgroups of $\mathcal{A}$ and $\mathcal{K}$*, Memoirs AMS 306, Providence (1988).

[4] J.Bruce, A.duPlessis, C.Wall. *Determinacy and unipotency*, Inv.Math. 88 (1987), 521-534.

[5] S.vanGils, M.Krupa, W.Langford. *Hopf bifurcation with non-semisimple 1:1 resonance*, Nonlinearity 3 (1990), 825-850.

[6] M.Golubitsky, D.G.Schaeffer. *Singularities and groups in bifurcation theory, Vol.1*, Springer-Verlag, New York (1985).

[7] M.Golubitsky, I.N.Stewart, D.G.Schaeffer. *Singularities and groups in bifurcation theory, Vol.2*, Springer-Verlag, New York (1988).

[8] A.Vanderbauwhede.
Hopf bifurcation at non-semisimple eigenvalues, AMS.Contemporary Mathematics. 56 (1986), 343-353.

International Series of Numerical Mathematics, Vol. 104, © 1992 Birkhäuser Verlag Basel

On Trigonometric Collocation in Hopf Bifurcation

by

Eckart W. Gekeler

Abstract. Trigonometric collocation methods are used to approximate simple Hopf bifurcation problems, $\omega y' + f(y,\mu) = 0$, with 2π-periodic solutions. The discrete system is solved by a constructive Ljapunov-Schmidt reduction originally due to Keller and Langford. In this method no component of the solution is fixed in advance therefore codes for the Fast Fourier Transform can be applied in the iteration which is of the form

$$F_z(z_0,0)(z_{new} - z) = F(z,\varepsilon), \quad |\varepsilon| \text{ small.}$$

This way yields a fast algorithm with low storage requirements for the matrix $F_z(z_0,0)$.

1. Introduction

In the present contribution we deal with the numerical computation of 2π-periodic solutions of the autonomous dynamical system

(1) $M(y,\omega,\mu)(x) := \omega y'(x) + f(y(x),\mu) = 0,$

in a neighborhood of a simple Hopf bifurcation point $(y_0,\mu_0) \in D \subset \mathbb{R}^P \times \mathbb{R}$ under the assumption that $f: D \to \mathbb{R}^P$ is three times continuously differentiable in the open domain D.

Since the work of E. Hopf [15] there exists a large literature on the theory of Hopf bifurcation and the detection of Hopf bifurcation points including the study of symmetry properties so that we mention here only contributions of Dellnitz [4], Fiedler [9], Golubitsky et al. [12, 13], Roose and de Dier [18], and the references there. As concerns the numerical solution of the above problem, we name the work of Doedel et al. [7, 8] in which general collocation methods are applied, and Weber [21].

In the present paper, we use a discretization which is very similar to a Fourier expansion method proposed recently by Dellnitz [6,7]. However, the discrete bifurcation problem is solved by an iterative method developed by Keller and Langford in [16] for solving general bifurcation problems near a single or multiple bifurcation point. This method is a modification of the Ljapunov-Schmidt reduction for numerical computation. Double-periodic nonlinear elliptic eigenvalue problems are treated by this way in [10] and Langford applies it in [17] for solving bifurcation problems in ordinary differential systems with rather general boundary conditions including Hopf bifurcation but without using trigonometric collocation.

No component of the unknown solution is fixed in this iterative Ljapunov-Schmidt reduction but the phase shift itself hence all components are treated equally and codes for the Fast Fourier Transform (FFT) can be applied here. As well-known, the full transformation matrix between the ordinate values and the Fourier coefficients and vice versa is not computed explicitly in FFT. Therefore, also the storage requirement is reduced considerably besides the computational amount of work. Together with stable codes for the computation of trigonometric sums (cf. e.g. [19, vol. 3]) trigonometric interpolation polynomials can be computed with very high precision and therefore the even spacing of the nodes is not a serious drawback but has on the other side the advantage that with one solution y also all phase-shifted functions $y(\cdot + \gamma)$ are direct solutions of the numerical approximation. Moreover, the bifurcation equations of the discretized problem have much the same form as those of the analytical problem and even the Hopf bifurcation points have the same frequency ω_0.

We follow here the paper of Langford [17] as concerns the goal of solving Hopf bifurcation problems by the Ljapunov-Schmidt method. But, using trigonometric collocation, the large linear system arising in every iteration step is solved after p FFT's for the Fourier coefficients. Then p back FFT's yield the ordinate values again. Accordingly, a QR-decomposition is necessary once for a 'skeleton' matrix with about $p^2 n$ instead of $(pn)^2$ entries where n is the node number of the collocation. After this decomposition, the method needs a matrix-vector multiplication with the matrix Q^T and the solution of a linear triangular system of the same skeleton form in every step of iteration.

With a trivial modification, the method can also be used to compute small periodic solutions of Hamiltonian system under the situation of the Ljapunov Center Theorem ; cf. e.g. [1, Th. 26.26; 3]. But it is a local procedure which works in that neighborhood of the bifurcation point in which the Ljapunov-Schmidt reduction of the discretized problem is valid. Beyond this region continuation methods must be applied; cf. e.g. [2].

2. Trigonometric Collocation and Hopf Bifurcation

We omit the parameter μ temporarily and recall first some well-known facts on trigonometric interpolation, see e.g. [19, § 2.6; 20, § 2.3]. Let

$$x_k = k\varrho, \quad k = 0,\ldots,n-1, \quad \varrho = 2\pi/n, \quad n = 2m,$$

be n evenly spaced collocation abszissae, where n is always tacitly assumed to be a power of two, and denote the real p-dimensional vector of ordinate values $y(x_k)$ by y_k.

Then we start out from the complex vector-valued phase polynomial

$$(2) \quad p(x;y) = \Sigma_{j=1-m}^{m-1} y_j^* e^{ijx} + \tfrac{1}{2} y_m^* (e^{imx} + e^{-imx}), \quad y_{-j}^* = \overline{y_j^*},$$

which is uniquely defined by

$$(3) \quad y_j^* = \frac{1}{n}\Sigma_{k=0}^{n-1} y_k\, e^{-ikx_j} = \frac{1}{n}\Sigma_{k=0}^{n-1} y_k\, e^{-ijk\varrho}, \quad j = 1-m,...,m,$$

and has the interpolation property,

$$(4) \quad y_k = \Sigma_{j=1-m}^{m} y_j^*\, e^{ijx_k} = \Sigma_{j=1-m}^{m} y_j^*\, e^{ijk\varrho}, \quad k = 0,...,n-1.$$

Because all node vectors y_k are real, the polynomial itself is also real-valued and

hence its derivative is also real-valued but we obtain

$$(5) \quad p'(x_k;y) = i\,\Sigma_{j=1-m}^{m-1} j y_j^*\, e^{ijk\varrho}.$$

Some caution is however necessary here since most FFT codes use the phase

polynomial $q(x;y) = \Sigma_{j=0}^{n-1} y_j^*\, e^{ijx}$ which has the above interpolation property for real

values, too, because $y_{n-j}^* = y_{-j}^*$, $j = 1,...,m$, but which is not real for real data.

Let us now assume that (1) has a solution y with absolutely continuous derivative.

Then we can write y as a complex Fourier series, $y(x) = \Sigma_{j=-\infty}^{\infty} y_j^* e^{ijx}$, $y_{-j}^* = \overline{y_j^*}$

and, in the same way, $(f \circ y)(x) = \Sigma_{j=-\infty}^{\infty} (f \circ y)_j^*\, e^{ijx}$, where

$$(6) \quad (f \circ y)_j^* = \frac{1}{2\pi}\int_0^{2\pi} (f \circ y)(x)e^{-ijx}dx.$$

Thus we can write instead of (1)

$$(7) \quad M(y,\omega)(x) = \Sigma_{j=-\infty}^{\infty} [(i\omega j y_j^* + (f \circ y)_j^*]e^{ijx} = 0$$

and a self-suggesting discretization consists in summing up here over a finite index set
only, say $j = 1-m,...,m$. The result is the well-known Ritz method with respect to
an expansion in the eigenfunctions of the differential operator d/dx. So, if
$u(x) = \Sigma_{j=1-m}^{m} u_j^* e^{ijx}$, $u = [u_\nu]_{\nu=1}^{p}$, denotes the approximation to the exakt

solution y, then the system for the unknown Fourier coefficients u_j^* reads because (5)

$$(8) \quad i\omega j u_j^* + f(u)_j^* = 0, \quad j = 1-m,...,m-1, \quad f(u)_m^* = 0;$$

cf. also Dellnitz [5, 6]. For the numerical integration of 2π-periodic smooth functions
over the periodicity interval, the trapezoidal sum is particularly well suited because
the error decreases exponentially with the step length. Therefore, regarding $u(0) = u(x_0) = u(x_n) = u(2\pi)$, we replace (6) by

$$(9) \quad (f \circ u)_j^* \doteq \frac{2\pi}{n}\frac{1}{2\pi}\Sigma_{k=0}^{n-1} (f \circ u)(x_k))e^{-ijk\varrho}, \quad j = 1-m,...,m,$$

with a slight abuse of notation. Let

$$F_\nu(U) = [(f_\nu \circ u)(x_0),...,(f_\nu \circ u)(x_{n-1})]^T, \quad \nu = 1,...,p,$$

$$F(U) = [F_1(U),...,F_p(U)]^T, \quad F_\nu(U)^* = [(f_\nu \circ u)_{1-m}^*,...,(f_\nu \circ u)_m^*]^T,$$

$$F(U)^* = [F_1(U)^*,...,F_p(U)^*]^T,$$

and let the same notations hold for the identity $f \circ u = u$. Further, let

$$D = i[1-m,...,m-1,0] \quad \text{and} \quad C = [e^{ijk\varrho}]_{j=0,\,k=1-m}^{n-1\ \ m}$$

a diagonal matrix and the complex Fourier transformation matrix, respectively, and let $P \times Q = [p_{jk}Q]$ be the tensor product of two matrices $P = [p_{jk}]$ and Q. Then e.g. the formulas of discrete Fourier transformation, (3) and (4), can be written for u as

$$U_\nu = CU_\nu^*, \quad U_\nu^* = C^H U_\nu/n, \quad C^H C = nI,$$

$$(10) \quad U = (I \times C)U^*, \quad U^* = (I \times C^H)U/n,$$

$(C^H = \overline{C}^T)$ and (8) becomes

$$(11) \quad \omega(I \times D)U^* + F(U)^* = 0$$

with (9) having the form $F(U)^* = (I \times C^H)F(U)/n$. If we observe (10) and multiply (11) from left by $I \times C$ then we obtain the finite-dimensional system

$$(12) \quad \omega(I \times \Phi)U + F(U) = 0$$

which now replaces the analytical problem (1). The matrix

$$\Phi = [\varphi_{jk}]_{j,k=0}^{n-1} := C D C^H/n$$

replacing the operator d/dx is real and skew-symmetric with n different entries,

$$\varphi_{jk} = -\frac{1}{m}\Sigma_{\ell=1}^{m-1} \ell \sin(\ell(j-k)\varrho), \quad j,k = 0,\dots,n-1.$$

If $f(u) = A u + g(u)$ with a (p,p)-matrix A then the system (11) for the Fourier coefficients has the form $\omega[(I \times D) + (A \times I)]U^* + G(U)^* = 0$. Because of $= u_{-j}^* =$

$\overline{u}_j^*$, $j = 0,\dots,m$, it suffices to solve numerically instead of (11) the equivalent complex system $\omega(I \times D_+)U_+^* + F(U)_+^* = 0$ where $D_+ = i[0,1,\dots,m-1,0]$ is a diagonal matrix and all components with negative indices are dropped, i.e. e.g.,

$$F_\nu(U)_+^* = [(f_\nu \circ u)_0^*,\dots,(f_\nu \circ u)_m^*]^T, \quad F(U)_+^* = [F_1(U)_+^*,\dots,F_p(U)_+^*]^T.$$

Note however that the computation of $F(U)_+^*$ needs the large vector U uniquely determined by U_+.

We now consider simple Hopf bifurcation in the system (1) and assume without loss of generality that $(y, \mu) = (0, 0)$ is the bifurcation point. Then we can write instead of (1)

$$(13) \quad \omega y' + Ay + \mu By + h(y,\mu) = 0$$

where $A = f_y(0,0)$ and $B = f_{y,\mu}(0,0)$.

(14) **Assumption.** (i) The matrix $A + \mu B$ has near $\mu = 0$ two simple conjugate complex eigenvalues, $\sigma(\mu)$ and $\overline{\sigma(\mu)}$, with $\mathrm{Re}\,\sigma(0) = 0$, $\omega_0 := \mathrm{Im}\,\sigma(0) > 0$, $(\mathrm{Re}\,\sigma)'(0) \neq 0$.

(ii) The spectrum $\mathrm{Sp}(A)$ of A satisfies $i\mathbb{R} \cap \{\mathrm{Sp}(A)\} = \{i\omega_0, -i\omega_0\}$.

(iii) $h(y,\mu) = O((\|y\| + |\mu|)^2)$, $h(0,\mu) = 0$, and $h(y,0) \neq 0$ for $(0,0) \neq (y,0) \in D$.

The condition $h(y,0) \neq 0$ for $y \neq 0$ was not regarded in [17] but has to be imposed in order that the constructive Ljapunov-Schmidt procedure does remove from the bifurcation point at all. It can however be reached always by a suitable re-scaling of the system.

The following lemma concerning the kernels of the operators

$$L = \omega_0 \frac{d}{dx} + A \quad \text{and} \quad L^* = -\omega_0 \frac{d}{dx} + A^T \quad \text{(adjoint operator to L)}.$$

is proved in [12, §§ 8.2, 8.3]. The introduction of the shift parameter γ is a trivial modification. (Note however that $\tilde{u}_2(\cdot + \gamma + \pi/2) = \tilde{u}_1(\cdot + \gamma)$.)

(15) **Lemma.** Let $\gamma \in \mathbb{R}$ be fixed.

(i) There exist two vectors $c, d \in \mathbb{C}^p$ such that
$$Ac = -i\omega_0 c, \quad \|c\|^2 = 2, \quad A^T d = i\omega_0 d, \quad d^H c = 2, \quad d^T c = 0.$$

(ii) $\tilde{u}_1(\cdot + \gamma): x \to Re(ce^{i(x+\gamma)})$ and $\tilde{u}_2(\cdot + \gamma): x \to Im(ce^{i(x+\gamma)})$
are a real basis of of Ker(L).

(iii) $\tilde{v}_1(\cdot + \gamma): x \to Re(de^{i(x+\gamma)})$ and $\tilde{v}_2(\cdot + \gamma): x \to Im(de^{i(x+\gamma)})$
are a real basis of Ker(L*).

(iv) $\quad <\tilde{v}_j(\cdot + \gamma), \tilde{v}_k(\cdot + \gamma)> = \|d\|^2 \delta_{jk}/2, \quad <\tilde{v}_j(\cdot + \gamma), \tilde{u}_k(\cdot + \gamma)> = \delta_{jk},$
$$<\tilde{u}_j(\cdot + \gamma), \tilde{u}_k(\cdot + \gamma)> = \delta_{jk}, \quad j,k = 1,2,$$

where δ_{jk} denotes the Kronecker symbol and $<f,g> = \frac{1}{2\pi} \int_0^{2\pi} f(x)^H g(x) dx$.

(v) $\quad \tilde{u}_1'(\cdot + \gamma) = -\tilde{u}_2(\cdot + \gamma), \quad \tilde{u}_2'(\cdot + \gamma) = \tilde{u}_1(\cdot + \gamma)$
$$\tilde{v}_1'(\cdot + \gamma) = -\tilde{v}_2(\cdot + \gamma), \quad \tilde{v}_2'(\cdot + \gamma) = \tilde{v}_1(\cdot + \gamma).$$

(vi) $\quad Re\, d^H Bc = 2\,(Re\,\sigma)'(0), \quad Im\, d^H Bc = 2(Im\,\sigma)'(0).$

Let $|\alpha + i\beta| = 1$ and $\alpha + i\beta = e^{-i\gamma}$ then we find that
$$u_1(x) := \alpha\tilde{u}_1(x) + \beta\tilde{u}_2(x) = \tilde{u}_1(x + \gamma), \quad u_2(x) := -\beta\tilde{u}_1(x) + \alpha\tilde{u}_2(x) = \tilde{u}_2(x + \gamma),$$
$$v_1(x) := (\alpha\tilde{v}_1(x) + \beta\tilde{v}_2(x)) = \tilde{v}_1(x + \gamma), \quad v_2(x) := (-\beta\tilde{v}_1(x) + \alpha\tilde{v}_2(x)) = \tilde{v}_2(x + \gamma).$$
Hence the weights α and β control the the phase shift γ.

(16) **Corollary.** Let $\vartheta_{j1} = <v_j, Au_1>$, and $\vartheta_{j2} = <v_j, Bu_1>$, $j = 1,2$. Then
$$\vartheta_{11} = 0, \quad \vartheta_{21} = \omega_0, \quad \vartheta_{12} = (Re\,\sigma)'(0), \quad \vartheta_{22} = (Im\,\sigma)'(0).$$

Proof. Direct verification by means of Lemma (15); cf. also [11].

By this result, the matrix $\Theta = [\vartheta_{jk}]$ is regular with $\det(\Theta) = -\omega_0(Re\,\sigma)'(0)$

152   E. Gekeler

independently of the choice of the phase angle γ.

Without loss of generality we now compute periodic solutions $\mathbf{y}$ of (13) satisfying $<\mathbf{y}, \mathbf{u}_2> = 0$. Then we can set

$$(17) \quad \mathbf{y} = \varepsilon\mathbf{u}_1 + \mathbf{w}, \quad <\mathbf{w}, \mathbf{u}_1> = <\mathbf{w}, \mathbf{u}_2> = 0,$$

and the system (13) can be written in the form

$$\mathbf{y}' + \lambda_0(1 + \tau)[A\mathbf{y} + \mu B\mathbf{y} + \mathbf{h}(\mathbf{y},\mu)] = 0$$

where $\omega^{-1} = \lambda = \lambda_0(1 + \tau)$, $\omega_0^{-1} = \lambda_0$. Regarding (17), we obtain

$$(18) \quad \tilde{L}\mathbf{w} := \mathbf{w}' + \lambda_0 A\mathbf{w} = -\lambda_0[\tau A\mathbf{y} + (1 + \tau)[\mu B\mathbf{y} + \mathbf{h}(\mathbf{y},\mu)]] =: -\lambda_0\mathbf{g}(\mathbf{y},\tau,\mu).$$

Under condition (17), this system has a unique solution $\mathbf{w}$ if the right side is contained in the range of the operator $\tilde{L}$. Thus, as the range is the orthogonal complement of the kernel of the adjoint operator, we obtain the compatibility conditions or bifurcation equations $<\mathbf{v}_i, \mathbf{g}(\mathbf{y},\tau,\mu)> = 0$, $i = 1,2$, which can be written as,

$$(19) \quad \Theta\,\eta = \mathbf{r} \quad (\text{or } \omega_0\tau = r_2 \text{ in the case where } \mu = 0),$$

with $\Theta = [\vartheta_{jk}]$, $\eta = (\tau,\mu)^T$, $\mathbf{r} = (r_1, r_2)^T$, and

$$r_i = -<\mathbf{v}_i, \tau A\mathbf{w} + \mu B\mathbf{w} + \tau\mu B\mathbf{y} + (1 + \tau)\mathbf{h}(\mathbf{y},\mu)>/\varepsilon.$$

The iteration scheme defined by repeatedly solving at first (19) and then (18) represents the constructive Ljapunov–Schmidt procedure proposed by Langford [17] in a slightly different form.

(20) **Theorem.** Let Ass. (14) be fulfilled. Let $0 \leq \gamma < 2\pi$ be fixed and let $0 \neq \varepsilon$ be sufficiently small. Then, starting from $(\mathbf{w}^{(0)},\tau^{(0)},\mu^{(0)}) = (0,0,0)$, the iterative solution of (19) and (18) generates a sequence $(\mathbf{w}^{(i)},\tau^{(i)},\mu^{(i)})$ converging to $(\mathbf{w},\tau,\mu) \neq (0,0,0)$ such that $(\mathbf{y},\tau,\mu)$ is a solution of (13) with 2π-periodic $\mathbf{y} = \varepsilon\mathbf{u}_1 + \mathbf{w}$.

The proof follows after a suitable normalization from the Contraction Mapping Theorem in the same way as in [16, 17].

For solving the system (13) numerically in a neighborhood of the bifurcation point we use her the approximation suggested by (12),

$$(21) \quad \omega(I \times \Phi)Y + (A \times I)Y + \mu(B \times I)Y + H(Y,\mu) = 0.$$

Because $A\mathbf{c} = -i\omega_0\mathbf{c}$ and $\Phi\mathbf{q} = i\mathbf{q}$ where $\mathbf{q} = [e^{i(k\varrho+\gamma)}]_{k=0}^{n-1}$, $\varrho = 2\pi/n$, we have

$$(22) \quad i[\omega_0(I \times \Phi) + (A \times I)](\mathbf{c} \times \mathbf{q}) = 0.$$

On the other side, because $A^T\mathbf{d} = i\omega_0\mathbf{d}$, we obtain

$$(23) \quad i[-\omega_0(I \times \Phi) + (A^T \times I)](\mathbf{d} \times \mathbf{q}) = 0.$$

But, as the matrix in (22) is real, its kernel is spanned by

$$U_1 = Re(c \times q) \text{ and } U_2 = Im(c \times q),$$

and the kernel of the transposed matrix in (23) is spanned by

$$V_1 = Re(d \times q) \text{ and } V_2 = Im(d \times q).$$

(Recall that $\Phi^T = -\Phi$.) So all results remain valid if every 2π-periodic p-valued function $u = (u_1,...,u_p)^T$ is replaced by the vector of its discrete approximation $U = [[u_i(x_k)]_{k=0}^{n-1}]_{i=1}^{p}$ with the node abszissae $x_k = k\varrho$ of the trigonometric collocation, and if the scalar product $<\cdot,\cdot>$ is replaced by scalar product in $\mathbb{R}^{p \cdot n}$.

In every step of the method we have to compute

$$Y_{new} = \varepsilon U_1 + W_{new}$$

by solving at first the bifurcation equations

$$(24) \quad \tau_{new} V_i^T(A \times I)U_1 + \mu_{new} V_i^T(B \times I)U_1$$
$$= - V_i^T[\tau(A \times I)W + \mu(B \times I)W + \tau\mu(B \times I)Y + (1 + \tau)H(Y,\mu)]/\varepsilon,$$

$i = 1,2,$ and then the linear system

$$(25a) \qquad [(I \times \Phi) + \lambda_0(A \times I)]W_{new} = - \lambda_0 G,$$

$$(25b) \qquad U_1^T W_{new} = 0, \ U_2^T W_{new} = 0,$$

where

$$G = (A \times I)(\tau_{new}\varepsilon U_1 + \tau W) + (B \times I)(\mu_{new}\varepsilon U_1 + \mu W)$$
$$+ \tau\mu(B \times I)(\varepsilon U_1 + W) + (1 + \tau)H(\varepsilon U_1 + W,\mu).$$

However, (25b) is equivalent to

$$(25c) \qquad (c \times q)^H W_{new} = 0, \ (\overline{c} \times \overline{q})^H W_{new} = 0$$

and it is the crucial step of the algorithm that instead of (25a) and (25b) the corresponding system for the Fourier coefficients is solved numerically, namely

$$(26a) \qquad [(I \times D) + \lambda_0(A \times I)]W^* = G^*,$$

$$(26b) \qquad (c \times q)^H(I \times C)W_{new}^* = [(I \times C^H)(c \times q)]^H W_{new}^* = 0,$$

$$(26c) \qquad [(I \times C^H)(\overline{c} \times \overline{q})]^H W_{new}^* = 0.$$

If $e_j = [\delta_{jk}]_{k=0}^{n-1}$ is the j-th column of the identity matrix then $C^H q = e_{m+1}$ hence e.g. (27b) reads $2m(c \times e_{m+1})^H W_{new}^* = 0.$

Splitting W in the same way as above, it suffices to solve instead of (26) the complex linear system of (almost) half dimension

$$(27a) \qquad [(I \times D_+) + \lambda_0(A \times I_+)]W_{new,+}^* = G_+^*,$$

$$(27b) \qquad (c \times e)^H W_{new,+}^* = 0, \ e = [\delta_{1k}]_{k=0}^{m}, \ \delta_{ik} \text{ Kronecker symbol,}$$

whose $(p(m+2),p(m+1))$-matrix has maximum rank $p(m+1)$ by construction. The square-bracketed term in (27a) is a block matrix whose block elements are diagonal matrices therefore this matrix has $p^2(m+1)$ entries only instead of $(2mp)^2$. A complex QR-algorithm can be modified easily in order to take advantage of this special form. After its application, the last row $(d \times e)^H$, being of 'skeleton' form, is to be eliminated by $m + 1$ further steps of complex Householder reduction. For further details see [11].

3. Examples

(1) Van der Pol Equation (cf. e.g. Hairer et al. [14]).
$$\omega\dot{x} - y = 0, \quad \omega\dot{y} + x + \mu(x^2 - 1)y = 0, \quad \mu > 0.$$

Multiplication with $\sqrt{\mu}$ yields the normalized equation ($u = \sqrt{\mu}x$, $v = \sqrt{\mu}y$)
$$\omega\dot{u} - v = 0, \quad \omega\dot{v} + u + (u^2 - \mu)v = 0, \quad \omega_0 = 1.$$

(2) Small Brusselator (cf. Hairer et al. [14]).
$$\omega\dot{x} + (\beta + 1)x - x^2y - \alpha = 0, \quad \omega\dot{y} - \beta x + x^2y = 0.$$
Critical points are $(x,y) = (\alpha, \beta/\alpha)$; Hopf bifurcation arises for $\beta = \alpha^2 + 1$. Substitution of $\beta = \alpha^2 + 1 + \mu$ yields singular points in dependence of μ
$$(x,y) = (\alpha, (\alpha^2 + 1 + \mu)/\alpha).$$
Further substitution of
$$(x,y) = (\alpha + u, \ [(\alpha^2 + 1 + \mu)/\alpha] + v)$$
yields
$$\omega\dot{u} - (\mu + \alpha^2)u - \alpha^2v - (\alpha + \mu + \alpha^{-1})u^2 - 2\alpha uv - u^2v = 0,$$
$$\omega\dot{v} + (\alpha^2 + 1 + \mu)u + \alpha^2v + (\alpha + \mu + \alpha^{-1})u^2 + 2\alpha uv + u^2v = 0.$$
$$A = \begin{bmatrix} -\alpha^2 & -\alpha^2 \\ \alpha^2+1 & \alpha^2 \end{bmatrix}, \quad B = \begin{bmatrix} -1 & 0 \\ 1 & 0 \end{bmatrix}, \quad \omega_0 = \alpha = 1,$$
$$c = \left(-\frac{\alpha(\alpha + i)\gamma}{1 + \alpha^2}, \ \gamma\right)^T, \quad \gamma = \left[\frac{2(\alpha^2 + 1)}{2\alpha^2 + 1}\right]^{1/2},$$
$$d = \left(-\frac{i(1 + \alpha^2)}{\alpha\gamma}, \ \frac{1 - i\alpha}{\gamma}\right)^T.$$

(3) Feedback-Inhibition Oscillator (Langford [17]).
$$y' + Ay + f(y,\mu) = 0,$$
$$A = \begin{bmatrix} 1 & 0 & 2 \\ -2 & 1 & 0 \\ 0 & -2 & 1 \end{bmatrix}, \quad f(y,\mu) = \begin{bmatrix} g(y_3,\mu) \\ -g(y_1,\mu) \\ -g(y_2,\mu) \end{bmatrix}, \quad \omega_0 = \sqrt{3},$$
$$g(y,\mu) = \frac{1}{2}\frac{(1 + 2y)^{\mu+4} - 1}{(1 + 2y)^{\mu+4} + 1} - 2 \quad y = \frac{1}{2}\mu y - 2y^2 - 8y^3 + \dots \ .$$
Therefore we obtain
$$y' + Ay + \mu By + h(y,\mu) = 0,$$

$$B = \frac{1}{2} \begin{bmatrix} 0 & 0 & 1 \\ -1 & 0 & 0 \\ 0 & -1 & 0 \end{bmatrix}, \quad h(y,\mu) = \begin{bmatrix} g(y_3,\mu) - \mu y_3/2 \\ -g(y_1,\mu) + \mu y_1/2 \\ -g(y_2,\mu) + \mu y_2/2 \end{bmatrix}.$$

$$c = (i\sqrt{3} - 1),\ i\sqrt{3} + 1,\ 2)^T/\sqrt{6}, \quad d = c.$$

(4) Periodical Hamiltonian System.

$$\omega\dot{x} - y = 0, \quad \omega\dot{y} + x + x^2 = 0. \quad \omega_0 = 1.$$

In the following table some results are given for $n = 64$ node points. The chosen ε's are large values for which the method is still convergent. The right eigenvector $U_1 \in \mathbf{R}^{pn}$ is normed such that $\|U_1\| := (U_1^T U_1)^{1/2}/n = 1$ and the data are rounded to three digits (rel. Res. $= $ Res$/\|Y\|$).

Table.

Example	Its.	ε	μ	$T = 2\pi/\omega$	rel. Residuum.
(1)	10	1.5	0.571	6.410	6E-6
(2)	20	0.5	0.130	6.335	3E-7
(3)	20	0.3	2.093	3.517	7E-4
(3)	40	0.3	2.105	3.516	7E-6
(4)	20	0.5	0	7.502	3E-6.

(The value of μ in Ex. (3) corresponds to the value $\mu + 4$ in the paper of Langford [17].)

Mathematisches Institut A, Universität Stuttgart,

Postfach 80 11 40, D-7000 Stuttgart 80, Germany

References
[1] Amann, H.: Gewöhnliche Differentialgleichungen. Berlin, New York: De Gruyter 1983.
[2] Allgower, E.L., and Georg, K.: Numerical Continuation Methods. Heidelberg, New York: Springer 1990.
[3] Chow, Sh. N., and Hale, J.K.: Methods of Bifurcation Theory. Berlin, Heidelberg, New York: Springer 1982.
[4] Dellnitz, M.: Hopf-Verzweigung in Systemen mit Symmetrie und deren Numerische Behandlung. Hamburg: Verlag an der Lottbek 1988.
[5] Dellnitz, M.: A computational method and path following for periodic solutions with symmetry. NATO ASI series C313, 1990.
[6] Dellnitz, M.: Computational bifurcation of periodic solutions in systems with symmetry. Research Report, Houston, Texas. To appear in I.M.A. J. Numer. Anal. 1991.
[7] Doedel, E.: AUTO: Software for continuation and bifurcation problems in ordinary differential equations. California Institue of Technologie, Pasadena (1986).
[8] Doedel, E.J., Jepson, A.D., and Keller, H.B.: Numerical methods for Hopf bifurcation and continuation of periodic solution paths. In: Computing Methods in Applied Sciences and Engineering, vol. VI, R. Glowinski and J.L. Lions (eds.). Amsterdam, New York: North Holland, 127-138.

[9]   Fiedler, B.: Global Bifurcation of Periodic Solutions with Symmetry,
      Lecture Notes Math. vol. 1309. Heidelberg, New York: Springer 1988.
[10] Gekeler, E.W.: ‚On the numerical solution of double-periodic elliptic
      eigenvalue problems. Computing 43 (1989), 97-114.
[11] Gekeler, E.W.: Trigonometric collocation in Hopf bifurcation. Preprint No. 91/13,
      Mathematisches Institut, Universität Stuttgart.
[12] Golubitsky, M., and Schaeffer, D.G.: Singularities and Groups in
      Bifurcation Theory I. Heidelberg, New York: Springer 1985.
[13] Golubitsky, M., Stewart, I., and Schaeffer, D.G.: Singularities and Groups
      in Bifurcation Theory II. Heidelberg, New York: Springer 1988.
[14] Hairer, E., Norsett, S.P., and Wanner, G.: Ordinary Differential Equations
      I. Heidelberg, New York: Springer 11987.
[15] Hopf, E.: Abzweigung einer periodischen Lösung von einer stationären
      Lösung eines Differentialsystems. Ber. math.-phys. Klasse Sächs. Akad.
      Wiss. 94 (1942), 1-22.
[16] Keller, H.B., and Langford, W.F.: Iterations, perturbations and
      multiplicities for nonlinear bifurcation problems. Arch. Rat. Mech. Analysis
      48 (1972), 83-108.
[17] Langford, W.F.: Numerical solution of bifurcation problems for ordinary
      differential equations. Numer. Math. 28 (1977), 171-190.
[18] Roose, D., and De Dier, B.: Numerical determination of an emanating
      branch of Hopf bifurcation points in a two-parametric problem. SIAM J.
      Sci. Stat. Comput. 10 (1989), 671-685.
[19] Sauer, R., and Szabo, I. (editors): Mathematische Hilfsmittel des
      Ingenieurs, Part III. Berlin, Heidelberg, New York: Springer 1968.
[20] Stoer, J., and Bulirsch, R.: Introduction to Numerical Analysis.
      Berlin, Heidelberg, New York: Springer 1980.
[21] Weber, H.: Numerical solution of Hopf bifurcation problems. Math.
      Methods Appl. Sciences 2 (1980), 178-190.

Exploiting Symmetry in Solving Linear Equations

Kurt Georg* and Rick Miranda*
Department of Mathematics
Colorado State University
Ft. Collins, CO 80523

Introduction

Recent efforts have shown the efficacy of applying group theoretical methods to the numerical treatment of certain partial differential equations via finite differences and finite elements, see [2, 3, 4, 5, 7, 8, 9, 12]. The articles, e.g., [4, 7], have demonstrated that the use of discretizations of partial differential equations which are suitably adapted to respect symmetry properties yield highly useful decompositions which can reduce computational effort, improve the numerical conditioning of problems, and significantly facilitate the study of bifurcation behavior at singularities. The results in the present paper extend those in [1, 4] via a systematic introduction of group representation theory.

For simplicity we deal only with the problem of solving linear equations (usually coming from discretization methods) in the presence of a group action. We find a decomposition of the given linear problem into block form which is most efficient in the sense that a general problem cannot be further decomposed. We do this in a direct way without first determining bases for the corresponding subspaces (as the method of symmetry adapted bases, see [12]). Our method can be expressed in terms of tensor products, which is described in the last section of this article; the tensor approach is more fully explored in [10].

1 Groups Acting as Permutations of Indices

Let G be a finite group with identity I.

We assume that we have a set of N indices $\mathcal{N} = \{1,\ldots,N\}$ which G permutes. We will denote this action as follows: if n is an index and g is a group element, then g sends n to gn. The action satisfies the following basic group action axioms:

$$In = n \text{ for every } n \text{ in } \mathcal{N}, \tag{1.1}$$

and

$$(gh)n = g(hn) \text{ for every } g \text{ and } h \text{ in } G \text{ and } n \text{ in } \mathcal{N}. \tag{1.2}$$

*Partially supported by NSF Grant DMS-9104058

Our basic assumption is that G acts on $\mathcal{N}$ without any fixed points: i.e.,

$$gn = n \quad \text{for some } n \text{ only if} \quad g = I. \tag{1.3}$$

The *orbit* of an index n is the set $\{gn : g \in G\}$ of indices to which n is moved by the group elements in G. Using (1.3), we see that there are exactly $\#G$ elements in each orbit. Any two orbits are either identical or disjoint, so that there are exactly $M = N/\#G$ orbits.

A *selection* of indices is a subset $\mathcal{M}$ of $\mathcal{N}$ which consists of one index in each orbit of G.

Thus, if $\mathcal{M}$ is a selection of indices, then the set $\mathcal{N}$ of all indices is equal to the set $\{gm : g \in G, m \in \mathcal{M}\}$. Moreover, with the assumption (1.3), there is no duplication in this presentation of the full set $\mathcal{N}$. There are $M = N/\#G$ elements in a selection (one element for each orbit).

2   The Induced Action on Column Vectors

Suppose now that G acts as in Section 1 on a set of indices $\mathcal{N}$. Let $V = \mathbb{C}^{\mathcal{N}}$ be the space of complex column vectors indexed by $\mathcal{N}$. If n is an index in $\mathcal{N}$ and v is a vector in V, we will denote the nth coordinate of v by $v[n]$.

Since G acts on the indices $\mathcal{N}$, we obtain an action of G on the vector space $V = \mathbb{C}^{\mathcal{N}}$ by setting $(gv)[n] = v[g^{-1}n]$. This gives a representation of G on V, which is referred to in the representation theory literature as a *permutation representation*. The axioms (1.1) and (1.2) extend to the action of G on V as:

$$Iv = v \text{ for every } v \text{ in } V, \tag{2.1}$$

and

$$(gh)v = g(hv) \text{ for every } g \text{ and } h \text{ in } G \text{ and } v \text{ in } V. \tag{2.2}$$

Note that if e_n is the standard basis vector of V with 1 in the nth position and 0 in all other positions, then the definition of the action of G on V implies that

$$g\,e_n = e_{gn}. \tag{2.3}$$

Fix an irreducible matrix representation $\rho : G \to GL(d, \mathbb{C})$; here $d = \dim\rho$ is the *dimension*, or *degree*, of the representation. We denote the ij^{th} entry of the matrix of the representation ρ on the group element g by $A_{ij}^{\rho}(g)$.

Since V is a representation of G, we have the standard projector and transfer operators P_{ij}^{ρ} defined on V, where i and j run up to the dimension of ρ. (See [11, section 2.7].) These are defined by the formula

$$P_{ij}^{\rho}(v) = \frac{\dim\rho}{\#G} \sum_{g \in G} A_{ji}^{\rho}(g^{-1})gv.$$

Note the reversal of indices and the use of the inverse element in the sum. These operators satisfy the formulas

$$gP_{ij}^{\rho} = \sum_{k} A_{ki}^{\rho}(g)P_{kj}^{\rho} \tag{2.4}$$

and, if ρ and τ are two irreducible representations of G which are either identical or non-conjugate, then

$$P_{ij}^{\rho} \circ P_{kl}^{\tau} = P_{il}^{\tau}\, \delta_{jk}\, \delta_{\rho\tau}. \tag{2.5}$$

Note that this implies that the operators P_{ii}^{ρ} are orthogonal projectors on V. The image of P_{ii}^{ρ} will be denoted by V_i^{ρ}.

Not only are the P_{ii}^{ρ} mutually orthogonal projectors, but they form a complete set of projectors in the following sense. Define *a full set of irreducible matrix representations* of G to be a set of non-conjugate irreducible matrix representations $\{\rho_\alpha\}$ such that every irreducible representation of G is conjugate to exactly one ρ_α. Since G is finite, general results from representation theory say that such a full set exists, and is finite; the number of representations in a full set equals the number of conjugacy classes of G. Moreover, one has the useful and standard formula

$$\sum_{\alpha} (\dim \rho_\alpha)^2 = \#G \tag{2.6}$$

which is proved in [6, section 31]. The completeness of the set of projectors is stated as follows.

Proposition 2.7 *Let G be a finite group acting on V. Then V splits into a direct sum of the subspaces V_i^{ρ}, as ρ ranges over a full set of irreducible matrix representations of G and i ranges from 1 to $\dim \rho$:*

$$V = \bigoplus_{\rho} \bigoplus_{i=1}^{\dim \rho} V_i^{\rho}.$$

In fact,

$$\sum_{\rho} \sum_{i=1}^{\dim \rho} P_{ii}^{\rho} \text{ is the identity on } V.$$

We will denote $P_{ij}^{\rho}(v)$ by v_{ij}^{ρ}. Explicitly,

$$v_{ij}^{\rho}[m] = \frac{\dim \rho}{\#G} \sum_{g \in G} A_{ji}^{\rho}(g^{-1}) v[g^{-1}m]. \tag{2.8}$$

By Proposition 2.7,

$$v = \sum_{\rho} \sum_{i=1}^{\dim \rho} v_{ii}^{\rho}. \tag{2.9}$$

As a consequence of (2.4) and the definitions, we have the following Lemma, whose proof is a simple computation which we leave to the reader.

Lemma 2.10 *Fix an irreducible matrix representation ρ, an index m, a group element g, and a vector v in V. Then for each j between 1 and $\dim \rho$,*

$$v_{jj}^{\rho}[gm] = \sum_{i=1}^{\dim \rho} A_{ij}^{\rho}(g^{-1}) v_{ij}^{\rho}[m].$$

We interpret this property as a condition on the coordinates of vectors in the subspaces V_j^ρ. In particular, if one knows $v_{ij}^\rho[m]$ for all $i = 1, \ldots, \dim\rho$ and all m in a selection $\mathcal{M}$, then one knows v_{jj}^ρ. Note that it then requires only $M\dim\rho = N\dim\rho/\#G$ numbers to determine the vector v_{jj}^ρ (not N); this is in fact the dimension of V_j^ρ.

Proposition 2.11 *The dimension of the subspace V_j^ρ is $M\dim\rho = N\dim\rho/\#G$.*

Proof: The above lemma shows that the dimension of V_j^ρ is at most $N\dim\rho/\#G$. However,

$$
\begin{aligned}
N &= \dim V = \dim(\bigoplus_\rho \bigoplus_{j=1}^{\dim\rho} V_j^\rho) = \sum_\rho \sum_{j=1}^{\dim\rho} \dim V_j^\rho \\
&\leq \sum_\rho \sum_{j=1}^{\dim\rho} [N\dim\rho/\#G] = \frac{N}{\#G}\sum_\rho(\dim\rho)^2 = N
\end{aligned}
$$

where the last equality uses (2.6). Therefore the inequalities must all be equalities, proving the proposition. $\Box$

3   Equivariant Linear Systems

Keeping the notation of the previous section, let $\mathcal{L} : V \to V$ be a linear transformation. We say $\mathcal{L}$ is *G-equivariant* (or simply *equivariant*) if

$$\mathcal{L}(gv) = g\mathcal{L}(v) \tag{3.1}$$

for all vectors $v \in V$ and all group elements $g \in G$.

Since V is a space of column vectors, $\mathcal{L}$ has a matrix $\mathbf{L}$, whose mn^{th} entry will be denoted by $L[m,n]$. Hence $\mathcal{L}(v) = \mathbf{L}v$, or, equivalently, $\mathcal{L}(e_n)[m] = L[m,n]$, where e_n is the nth standard basis vector for V. The equivariance of $\mathcal{L}$ is expressed in terms of its matrix $\mathbf{L}$ as follows.

Lemma 3.2 *With the above notation, $\mathcal{L}$ is equivariant if and only if $L[m,gn] = L[g^{-1}m,n]$ for every $g \in G$ and every $m,n \in \mathcal{N}$.*

Proof: Since both sides of (3.1) are linear in v, the operator $\mathcal{L}$ is equivariant if and only if $\mathcal{L}(ge_n) = g\mathcal{L}(e_n)$ for every index n. But $\mathcal{L}(ge_n)[m] = \mathcal{L}(e_{gn})[m] = L[m,gn]$ and $(g\mathcal{L}(e_n))[m] = \mathcal{L}(e_n)[g^{-1}m] = L[g^{-1}m,n]$, proving the lemma. $\Box$

Note that equivariance can also be expressed by the formula

$$L[gm,gn] = L[m,n] \tag{3.3}$$

for every $g \in G$ and every $m,n \in \mathcal{N}$.

The following lemma gives the dimension of the space of equivariant linear operators on V.

Lemma 3.4 *Let V have dimension N and G have order $\#G$. Then the dimension of the space of G-equivariant linear operators $\mathcal{L}$ on V is $N^2/\#G$.*

Proof: The formula $L[m, gn] = L[g^{-1}m, n]$ for the matrix $\mathbf{L}$ shows that $\mathbf{L}$ is determined by its entries in the columns coming from a selection. Moreover, the columns in a selection can be arbitrary. Hence the dimension of the space of equivariant operators on V is

$$(\text{\# of selected columns })(\text{\# of entries in each column})$$

which is $(N/\#G)(N) = N^2/\#G$. $\square$

Since the equivariance of $\mathcal{L}$ means that $\mathcal{L}$ commutes with the action of each group element g, it also implies that $\mathcal{L}$ commutes with all of the operators P_{ij}^ρ, since they are just linear combinations of group elements:

$$\mathcal{L} P_{ij}^\rho = P_{ij}^\rho \mathcal{L}. \tag{3.5}$$

In particular $\mathcal{L}$ commutes with the projectors P_{ii}^ρ, and so $\mathcal{L}$ respects the decomposition of Proposition 2.7 of V:

Lemma 3.6 *If $\mathcal{L}$ is equivariant, then $\mathcal{L}$ maps V_i^ρ into itself for each ρ and each i.*

We will denote the restriction of $\mathcal{L}$ to V_i^ρ by $\mathcal{L}_i^\rho$:

$$\mathcal{L}_i^\rho = \mathcal{L} \mid V_i^\rho. \tag{3.7}$$

In particular, $\mathcal{L}$ is invertible if and only if $\mathcal{L}_i^\rho$ is invertible for each ρ and each i.

4   Reducing to the Sub-Problems

In this section we fix a finite group G acting on an index set $\mathcal{N}$ without fixed points as above. We also fix a selection $\mathcal{M}$ of indices, and a full set of irreducible matrix representations $\{\rho_\alpha\}$.

Let $\mathcal{L}$ be an invertible G-equivariant linear operator $\mathcal{L} : V \to V$, with matrix $L[m, n]$ as described in the previous section. Let an arbitrary vector v in V be given. Suppose that we wish to solve the equation

$$\mathcal{L}(u) = v \tag{4.1}$$

or equivalently

$$\mathbf{L}u = v \tag{4.2}$$

where u is an unknown vector in V. Applying the projector P_{kk}^ρ to both sides of (4.1), and using the equivariance of $\mathcal{L}$, gives the sub-problem

$$\mathcal{L}_k^\rho(u_{kk}^\rho) = v_{kk}^\rho \tag{4.3}$$

which is also an invertible linear system (on the subspace V_k^ρ). If we can solve these subproblems, for each ρ and each $k = 1, \ldots, \dim \rho$, then we can reconstruct the global solution u to (4.1) as

$$u = \sum_{\rho} \sum_{k=1}^{\dim \rho} u_{kk}^{\rho}. \qquad (4.4)$$

Our job is to write down matrices for the subproblems in an efficient manner. For this we make use of the remarks following Lemma 2.10: to determine all the coordinates of the vector u_{kk}^{ρ}, it is enough to know only the transfer vectors u_{ik}^{ρ} in the coordinates from the selection $\mathcal{M}$. Moreover, the sub-problem (4.3) has as input the vector v_{kk}^{ρ}, and by Lemma 2.10, it too is determined by the selected coordinates of the transfer vectors v_{jk}^{ρ}. Therefore the heart of the sub-problem can be expressed as follows:

For a fixed ρ and k, given the selected coordinates of the vectors v_{jk}^{ρ}, determine the selected coordinates of the vectors u_{ik}^{ρ}.

For a fixed ρ and k, the scalar variables in the above set-up are therefore naturally doubly-indexed, by pairs (i, m) where $i = 1, \ldots, \dim \rho$ and $m \in \mathcal{M}$. For example, the pair (i, m) corresponds to the mth coordinate in the vector u_{ik}^{ρ}.

Therefore the matrix of the sub-problem (4.3) is a matrix

$$\mathbf{L}_k^{\rho} = L_k^{\rho}[(i, m), (j, n)]$$

such that

$$\sum_{(j,n)} L_k^{\rho}[(i, m), (j, n)] u_{jk}^{\rho}[n] = v_{ik}^{\rho}[m]. \qquad (4.5)$$

The main theorem, which gives a formula for the matrix $\mathbf{L}_k^{\rho}$, can now be stated.

Theorem 4.6 *Fix ρ and k between 1 and $\dim \rho$. Then*

$$L_k^{\rho}[(i, m), (j, n)] = \sum_{g \in G} L[g^{-1}m, n] A_{ji}^{\rho}(g^{-1}).$$

Proof: Fix the pair (i, m). Then

$$
\begin{aligned}
v_{ik}^{\rho}[m] &= (P_{ik}^{\rho}(v_{kk}^{\rho}))[m] \\
&= \frac{\dim \rho}{\#G} \sum_{h \in G} A_{ki}^{\rho}(h^{-1})(h v_{kk}^{\rho})[m] \\
&= \frac{\dim \rho}{\#G} \sum_{h \in G} A_{ki}^{\rho}(h^{-1}) v_{kk}^{\rho}[h^{-1}m] \\
&= \frac{\dim \rho}{\#G} \sum_{h \in G} A_{ki}^{\rho}(h^{-1}) \sum_{p \in \mathcal{N}} L[h^{-1}m, p] u_{kk}^{\rho}[p] \\
&\quad \text{(setting p = gn)} \\
&= \frac{\dim \rho}{\#G} \sum_{h \in G} A_{ki}^{\rho}(h^{-1}) \sum_{g \in G} \sum_{n \in \mathcal{M}} L[h^{-1}m, gn] u_{kk}^{\rho}[gn] \\
&\quad \text{(using the equivariance of L and Lemma 2.10)} \\
&= \frac{\dim \rho}{\#G} \sum_{h \in G} A_{ki}^{\rho}(h^{-1}) \sum_{g \in G} \sum_{n \in \mathcal{M}} L[g^{-1}h^{-1}m, n] \sum_{j=1}^{\dim \rho} A_{jk}^{\rho}(g^{-1}) u_{jk}^{\rho}[n]
\end{aligned}
$$

$$= \sum_{(j,n)} \left[\frac{\dim \rho}{\#G} \sum_{h \in G} \sum_{g \in G} L[g^{-1}h^{-1}m, n] A_{ki}^{\rho}(h^{-1}) A_{jk}^{\rho}(g^{-1}) \right] u_{jk}^{\rho}[n].$$

Thus, by (4.5), we have

$$\begin{aligned}
L_k^{\rho}[(i,m),(j,n)] &= \frac{\dim \rho}{\#G} \sum_{h \in G} \sum_{g \in G} L[g^{-1}h^{-1}m, n] A_{ki}^{\rho}(h^{-1}) A_{jk}^{\rho}(g^{-1}) \\
&\quad \text{(setting a = hg)} \\
&= \frac{\dim \rho}{\#G} \sum_{a \in G} \sum_{h \in G} L[a^{-1}m, n] A_{ki}^{\rho}(h^{-1}) A_{jk}^{\rho}(a^{-1}h) \\
&= \dim \rho \sum_{a \in G} L[a^{-1}m, n] [\frac{1}{\#G} \sum_{h \in G} A_{jk}^{\rho}(a^{-1}h) A_{ki}^{\rho}(h^{-1})] \\
&\quad \text{(using the orthogonality relations for the matrices } A^{\rho}) \\
&= \dim \rho \sum_{a \in G} L[a^{-1}m, n] [\frac{1}{\dim \rho} A_{ji}^{\rho}(a^{-1})] \\
&= \sum_{a \in G} L[a^{-1}m, n] A_{ji}^{\rho}(a^{-1})
\end{aligned}$$

which, after replacing a by g, is the required formula. The orthogonality relations used to obtain the second-to-last equality above can be found in [6, section 31]. $\square$

We have now the following remarkable observation:

Corollary 4.7 *The matrix L_k^{ρ} is independent of k.*

Thus there is actually only one sub-matrix for each irreducible matrix representation. We will therefore drop the subscript and define

$$L^{\rho}[(i,m),(j,n)] = \sum_{g \in G} L[g^{-1}m, n] A_{ji}^{\rho}(g^{-1}). \tag{4.8}$$

Thus the subproblems can be written in the form

$$\sum_{(j,n)} L^{\rho}[(i,m),(j,n)] u_{jk}^{\rho}[n] = v_{ik}^{\rho}[m]. \tag{4.9}$$

One final remark which should be made is that the matrices L^{ρ} for the subproblems are formed by using the entries of the full matrix $L[gm, n]$, for $g \in G$ and $m, n \in \mathcal{M}$. Therefore it is not necessary to know the entire matrix L; the reductions only require those columns of L indexed by the selection $\mathcal{M}$.

5 Putting it All Together

In this section we will give an overview of the entire process for solving the linear system (4.1), where $\mathcal{L}$ is a G-equivariant operator on the column space V, on which G acts through an action on the indices $\mathcal{N}$ for the coordinates of V without fixed points.

Given: a vector v of V.

To Solve: $\mathcal{L}(u) = v$ for the unknown vector u.

1: Choose a selection $\mathcal{M}$ of indices.

2: Find a full set of irreducible complex matrix representations for G.

3: Determine the entries $L[m, n]$ of the matrix for $\mathcal{L}$, in the selected columns.

4: For each irreducible complex matrix representation ρ in the full set,
   perform steps 4a - d:

   4a: Form the submatrices $\mathbf{L}^\rho$ using (4.8):

$$L^\rho[(i,m),(j,n)] = \sum_{g \in G} L[g^{-1}m, n] A^\rho_{ji}(g^{-1}).$$

   4b: Using the projectors P^ρ_{ik} (formula (2.8)), determine the selected coordinates
   of the vectors v^ρ_{ik}:

$$v^\rho_{ik}[m] = \frac{\dim \rho}{\#G} \sum_{g \in G} A^\rho_{ki}(g^{-1})v[g^{-1}m].$$

   4c: Solve the subproblems (4.9) for the selected coordinates of the vectors u^ρ_{jk}.

   4d: Reconstruct all the coordinates of the vector u^ρ_{kk} using Lemma 2.10:

$$u^\rho_{kk}[gm] = \sum_{j=1}^{\dim \rho} A^\rho_{jk}(g^{-1})u^\rho_{jk}[m].$$

5: Reconstruct the solution $u = \sum_\rho \sum_{k=1}^{\dim \rho} u^\rho_{kk}$.

If one has many instances of the linear system to solve with different right-hand-side vectors v, one of course only performs steps 1–4a once; then steps 4b–5 are done for each separate problem.

6   The Reduction in Complexity

Let us discuss the computational effort of the method under the assumption that the matrix is full and the linear equation solver is one of the standard direct solvers such as Gaussian elimination. The essential overhead cost of our method occurs in step 4a which amounts to N^2 flops. For large problems, the greatest part of the cost of the method is in Step 4c as outlined above: solving the sub-problems. The dimension of the subproblem for the irreducible matrix representation ρ is the dimension of V^ρ_k for any k, and this is $N \dim \rho / \#G$ by (2.11). For a direct solver as mentioned above, this gives a computational effort of

$$C \sum_\rho \left(\frac{N \dim \rho}{\#G}\right)^3 \quad \text{flops} \tag{6.1}$$

for some constant C. Without using our method, we are faced with a full size matrix of order N, which using the same direct solver gives an effort of CN^3 flops. Neglecting the lower order overhead expense and taking the quotient, we obtain the reduction factor

$$\frac{1}{(\#G)^3}\sum_\rho(\dim\rho)^3. \tag{6.2}$$

7   The Sharpness of the Method

What we have essentially done in the reduction of the full linear system (4.1) to the sub-systems (4.3) is block-diagonalize the matrix $\mathbf{L}$, without in fact finding an explicit basis for the subspaces V_k^ρ. If R is the number of irreducible matrix representations of G, (which is the number of conjugacy classes of G), we have found that $\mathbf{L}$ block-diagonalizes to R sets of blocks (one for each ρ), and the set of blocks corresponding to the irreducible matrix representation ρ is the matrix $\mathbf{L}^\rho$ repeated $\dim\rho$ times.

The reader may wonder whether any further reduction can be made. The answer is no, as a simple dimension count of available parameters shows. Each matrix $\mathbf{L}^\rho$ is of size $N\dim\rho/\#G$ (by Proposition 2.11), and so there are

$$\sum_\rho(N\dim\rho/\#G)^2 = N^2/\#G$$

total parameters (note that we have used (2.6)). Now this is precisely the dimension of the space of equivariant operators on V, by Lemma 3.4. Hence no further reduction in the problem is possible.

8   The Tensor Formulation

The reader will have noticed that the sub-matrices $\mathbf{L}^\rho$ defined in (4.8) have doubly-indexed entries. This suggests that these matrices can be conveniently expressed as tensor products. In this section we will develop this aspect of the construction. This approach is taken *ab initio* in [10].

Fix an irreducible representation ρ, and let $d = \dim\rho$ be its dimension. For any selected index m, let $\mathbf{v}^\rho[m]$ denote the d-by-d matrix whose ij^{th} entry is $v_{ij}^\rho[m]$:

$$\mathbf{v}^\rho[m] = \begin{pmatrix} v_{11}^\rho[m] & \ldots & v_{1d}^\rho[m] \\ \vdots & \ddots & \vdots \\ v_{d1}^\rho[m] & \ldots & v_{dd}^\rho[m] \end{pmatrix} \tag{8.1}$$

Using the formula (2.8), we see that the definition of $\mathbf{v}^\rho[m]$ can be expressed as a linear combination of transposes of the matrices $\mathbf{A}^\rho$ which define the representation ρ:

$$\mathbf{v}^\rho[m] = \frac{\dim\rho}{\#G}\sum_{g\in G} v[g^{-1}m]\mathbf{A}^\rho(g^{-1})^\mathsf{T}. \tag{8.2}$$

Using (8.2), we can express the matrices $\mathbf{v}^\rho[m]$ in terms of the original vector v. Inverting this procedure is also straightforward. Fix a (possibly unselected) index n, and write $n = gm$ for a unique group element g and a unique selected index m. Then by Lemma 2.10,

$$\begin{aligned} v^\rho_{jj}[gm] &= \sum_{i=1}^{d} A^\rho_{ij}(g^{-1})v^\rho_{ij}[m] \\ &= jj^{th} \text{ entry of } \mathbf{A}^\rho(g^{-1})^{\mathsf{T}}\mathbf{v}^\rho[m]. \end{aligned}$$

Hence

$$\text{trace}(\mathbf{A}^\rho(g^{-1})^{\mathsf{T}}\mathbf{v}^\rho[m]) = \sum_{j=1}^{d} v^\rho_{jj}[gm]$$

and so by (2.9)

$$v[gm] = \sum_\rho \text{trace}(\mathbf{A}^\rho(g^{-1})^{\mathsf{T}}\mathbf{v}^\rho[m]). \tag{8.3}$$

Now let $\mathbf{v}^\rho$ be the Md-by-d matrix formed by stacking the d-by-d matrices $\mathbf{v}^\rho[m]$ for the M selected indices m_α in a column:

$$\mathbf{v}^\rho = \begin{pmatrix} \mathbf{v}^\rho[m_1] \\ \mathbf{v}^\rho[m_2] \\ \vdots \\ \mathbf{v}^\rho[m_M] \end{pmatrix}. \tag{8.4}$$

We want to think of this matrix $\mathbf{v}^\rho$ as a column vector of length M, with d-by-d matrices as entries. The linear system for the subproblem (4.3) can now be written as

$$\mathbf{L}^\rho\,\mathbf{u}^\rho = \mathbf{v}^\rho \tag{8.5}$$

where $\mathbf{L}^\rho$ is an appropriately-indexed version of the matrix $L^\rho[(i,m),(j,n)]$ constructed in Section 4. This matrix $\mathbf{L}^\rho$ is an Md-by-Md matrix, which because of the double-indexing, can be viewed naturally as an M-by-M matrix whose mn^{th} entry is a d-by-d matrix. Indeed, using the definition (4.8) of $L^\rho[(i,m),(j,n)]$, we see that the mn^{th} entry is the matrix $\sum_{g\in G} L[g^{-1}m,n]\mathbf{A}^\rho(g^{-1})^{\mathsf{T}}$. Thus $\mathbf{L}^\rho$ in this form is naturally *a sum of tensor product matrices.*

Recall that the tensor product of two matrices $\mathbf{X}$ and $\mathbf{Y}$ which are d-by-d and e-by-e respectively is the de-by-de matrix $\mathbf{X} \otimes \mathbf{Y}$ which breaks naturally into a d-by-d matrix with e-by-e matrix entries: the matrix in the rs^{th} block is $X_{rs}\mathbf{Y}$.

To write our matrix $\mathbf{L}^\rho$ with this notation, one more definition is useful. Specifically, for $g \in G$ define the M-by-M matrix $\mathbf{L}_g$ by

$$\mathbf{L}_g[m,n] = L[g^{-1}m,n]. \tag{8.6}$$

Then

$$\mathbf{L}^\rho = \sum_{g\in G} \mathbf{L}_g \otimes \mathbf{A}^\rho(g^{-1})^{\mathsf{T}}. \tag{8.7}$$

The overview of the method given here can now be written with this notation as follows:

Given: a vector v of V.
To Solve: $\mathcal{L}(u) = v$ for the unknown vector u.

1: Choose a selection $\mathcal{M}$ of indices.

2: Find a full set of irreducible complex matrix representations for G.

3: Determine the entries $L[m, n]$ of the matrix for $\mathcal{L}$, in the selected columns, and find the matrices $\mathbf{L}_g$ for each group element g.

4: For each ρ in the full set of irreducible complex matrix representations, perform steps 4a - b:

    4a: Form the column vector (with matrix entries) $\mathbf{v}^\rho$ from the given vector v, using (8.2):

$$\mathbf{v}^\rho[m] = \frac{\dim \rho}{\#G} \sum_{g \in G} v[g^{-1}m]\mathbf{A}^\rho(g^{-1})^\mathsf{T}$$

    4b: Solve the subproblems (8.5) for the column vector (with matrix entries) $\mathbf{u}^\rho$:

$$\mathbf{L}^\rho\mathbf{u}^\rho = \left[\sum_{g \in G} \mathbf{L}_g \otimes \mathbf{A}^\rho(g^{-1})^\mathsf{T}\right]\mathbf{u}^\rho = \mathbf{v}^\rho.$$

5: Reconstruct the solution vector u from the $\mathbf{u}^\rho$'s using (8.3):

$$u[gm] = \sum_\rho \mathrm{trace}(\mathbf{A}^\rho(g^{-1})^\mathsf{T}\mathbf{u}^\rho[m]).$$

The reader will notice that the above outline of our method is a reorganization of the outline given in Section 5, expressed with a tensor formulation which emphasizes the block structure underlying the approach. It is our opinion that this latter organization is more suitable for a computer implementation. A small numerical example illustrating this feature has been given in [10].

References

[1] E. L. Allgower, K. Böhmer, K. Georg, and R. Miranda. Exploiting symmetry in boundary element methods. To appear in: SIAM J. Numer. Anal., 1991.

[2] E. L. Allgower, K. Böhmer, and Z. Mei. An extended equivariant branching theory. Preprint, University of Marburg, Fed. Rep. Germany, to appear in Math. Methods Appl. Sci., 1990.

[3] E. L. Allgower, K. Böhmer, and Z. Mei. On new bifurcation results for semi-linear elliptic equations with symmetries. Brunel, England, 1990. MAFELAP. To appear.

[4] E. L. Allgower, K. Böhmer, and Z. Mei. On a problem decomposition for semilinear nearly symmetric elliptic problems. In W. Hackbusch, editor, *Parallel Algorithms for PDE's*, volume 31 of *Notes on Numerical Fluid Mechanics*, pages 1–17, Braunschweig, Fed. Rep. Germany, 1991. Vieweg Verlag.

[5] A. Bossavit. Symmetry, groups, and boundary value problems. A progressive introduction to noncommutative harmonic analysis of partial differential equations in domains with geometric symmetry. *Computer Methods in Applied Mechanics and Engineering*, 56:165–215, 1986.

[6] C. W. Curtis and I. Reiner. *Representation Theory of Finite Groups and Associative Algebras*. Wiley Interscience. John Wiley and Sons, 1962.

[7] M. Dellnitz and B. Werner. Computational methods for bifurcation problems with symmetries — with special attention to steady state and Hopf bifurcation points. *J. Comput. Appl. Math.*, 26:97–123, 1989.

[8] C. C. Douglas and J. Mandel. The domain reduction method: High way reduction in three dimensions and convergence with inexact solvers. pages 149–160, Philadelphia, 1989. Fourth Copper Mountain Conference on Multigrid Methods, SIAM.

[9] C. C. Douglas and J. Mandel. A group theoretic approach to the domain reduction method: The commutative case. In preparation, 1990.

[10] K. Georg and R. Miranda. Symmetry aspects in numerical linear algebra with applications to boundary element methods. Preprint, Colorado State University, 1990.

[11] J.-P. Serre. *Linear Representations of Finite Groups*, volume 42 of *Graduate Texts in Mathematics*. Springer Verlag, Berlin, Heidelberg, New York, 1977.

[12] E. Stiefel and A. Fässler. *Gruppentheorethische Methoden und ihre Anwendung.* Teubner, Stuttgart, Fed. Rep. Germany, 1979.

International Series of Numerical Mathematics, Vol. 104, © 1992 Birkhäuser Verlag Basel

Symmetry and Preservation of Nodal Structure in Elliptic Equations Satisfying Fully Nonlinear Neumann Boundary Condtions

Timothy J. Healey & Hansjörg Kielhöfer

1.Introduction

In this paper we study a general class of quasi–linear elliptic equations of the form

$$(1.1) \qquad a_{ij}(\nabla u,u)u_{x_i x_j} + g(\lambda,\nabla u,u) = 0 \text{ in } \Omega,$$

where $\Omega \subset \mathbb{R}^2$ is a bounded domain, $u: \Omega \to \mathbb{R}$, "$\nabla(\cdot)$" denotes the gradient, and $\lambda \in \mathbb{R}$. On the boundary $\partial\Omega$ we impose fully nonlinear Neumann conditions:

$$(1.2) \qquad q(\nabla u,u) \cdot n\big|_{\partial\Omega} = 0,$$

where $q(\nabla u,u) \in \mathbb{R}^2$, n denotes the outward unit normal to $\partial\Omega$, and "$\cdot$" is the usual inner (dot) product on $\mathbb{R}^2$. Such problems arise naturally in continuum physics; (1.1) models a general class of nonlinear diffusion processes, and (1.2) is a zero–flux condition. In a recent work [5] we studied global bifurcation problems for (1.1) defined on all of $\mathbb{R}^n$ in the presence of lattice symmetry. In particular, (given certain hypotheses, which we recapitulate in Section 2) we showed that the precise nodal configuration of the eigenfunction (of the linearized problem at the trivial solution) is preserved along the associated global bifurcating solution branch.

The application of those results to homogeneous Dirichlet boundary conditions on regular polygonal domains in $\mathbb{R}^2$ is obvious. Here we establish the corresponding relationships between certain results from [5] and the fully nonlinear Neumann problem (1.1,2), which is a much more subtle task. Indeed, (1.2) renders the system fully nonlinear (even though (1.1) is quasi–linear), the direct global analysis of which is difficult. We avoid this by demonstrating that solutions obtained in [5] automatically satisfy (1.2) for particular domains. Incidently, this paper generalizes our previous work on (1.1,2) for the special case of rectangular domains [4]. We also point out that the "hidden" translational symmetry associated with *linear* Neumann conditions, $\frac{\partial u}{\partial x_i}\big|_{\partial\Omega} = 0$, is well known, e.g., [1], [2].

Throughout this paper, bold face lower case symbols denote vectors in $\mathbb{R}^2$, e.g., x, v, etc., and $|x| = |(x_1,x_2)| \equiv (x_1^2 + x_2^2)^{1/2}$ denotes the Euclidean norm of x. Lower-case Latin subscripts range from 1 to 2 (unless stated otherwise), and a repeated index (subscript) denotes summation. We frequently employ subscripts to denote partial derivatives, e.g., $g_{x_i} \equiv \frac{\partial g}{\partial x_i}$.

2. Summary of Results for Lattice Symmetries in $\mathbb{R}^2$

In this section we ignore the boundary conditions (1.2), with (1.1) defined on all of $\mathbb{R}^2$, and summarize some recent results on doubly–periodic solutions of (1.1) in the presence of specific symmetries. (We refer the reader to [5] for proofs.) Let $\mathscr{L}_4$ and $\mathscr{L}_6$ denote the square and hexagonal lattices, respectively:

$$\mathscr{L}_4 \equiv \{\alpha_1(a,0) + \alpha_2(0,a): a > 0, \; \alpha_i \in \mathbb{Z}, i = 1,2\},$$

$$(2.1) \qquad \mathscr{L}_6 \equiv \{\alpha_1(a,0) + \alpha_2(a/2,\sqrt{3}a/2): a > 0, \; \alpha_i \in \mathbb{Z}, i = 1,2\}.$$

We define doubly–periodic Hölder spaces of functions as follows:

$$X_r \equiv \{u \in C^{2,\mu}(\mathbb{R}^2): u(x + a) \equiv u(x) \; \forall \, a \in \mathscr{L}_r\},$$

$$(2.2) \qquad Y_r \equiv \{u \in C^{0,\mu}(\mathbb{R}^2): (\text{same as above})\},$$

each of which are Banach spaces when endowed with usual Hölder norms $\| \; \|_{2,\mu}$ and $\| \; \|_{0,\mu}$, respectively, where $0 < \mu < 1$, $r = 4$ & 6. (Henceforth, the subscript "r" is understood to take on the values 4 & 6.) We identify the left side of (1.1) with an operator acting on X_r:

$$G: \mathbb{R} \times X_r \to Y_r,$$

$$(2.3) \qquad G(\lambda,u) \equiv a_{ij}(\nabla u,u)u_{x_i x_j} + g(\lambda,\nabla u,u),$$

where the functions $a_{ij}: \mathbb{R}^2 \times \mathbb{R} \to \mathbb{R}$ and $g: \mathbb{R} \times \mathbb{R}^2 \times \mathbb{R} \to \mathbb{R}$ are sufficiently smooth (C^3 is enough). The principal part of the differential operator is assumed to be strongly, uniformly elliptic: $\exists$ positive constants α,β such that

$$(2.4) \qquad \alpha|\xi|^2 \leq a_{ij}(p,u) \, \xi_i \xi_j \leq \beta|\xi|^2 \; \forall \, p,\xi \in \mathbb{R}^2, u \in \mathbb{R}.$$

Our goal is to study global solution branches of the bifurcation problem

$$(2.5) \qquad G(\lambda,u) = 0.$$

We associate the lattice $\mathscr{L}_r$ with a compact Lie group of continuous linear transformations, denoted Γ_r, acting on X_r (and Y_r), as follows: Define the 2–tori

$$\mathscr{T}_4 \equiv \{\zeta_1(a,0) + \zeta_2(0,a): \zeta_i \in \mathbb{R}(\text{mod } 1), i = 1,2\},$$

$$(2.6) \qquad \mathscr{T}_6 \equiv \{\zeta_1(a,0) + \zeta_2(a/2,\sqrt{3}a/2): \zeta_i \in \mathbb{R}(\text{mod } 1), i = 1,2\},$$

and the 2 × 2 matrices

$$(2.7) \qquad R(\phi) \equiv \begin{bmatrix} \cos\phi & \sin\phi \\ -\sin\phi & \cos\phi \end{bmatrix}, \; E \equiv \begin{bmatrix} 1 & 0 \\ 0 & -1 \end{bmatrix}.$$

The dihedral group D_k is then given by

$$(2.8) \qquad D_k \cong \{R(2\pi j/k), ER(2\pi j/k): j = 1,2,...,k\},$$

and we have

(2.9) $$D_r = \{Q \in O(2): Qa \in \mathscr{L}_r \; \forall \, a \in \mathscr{L}_r\}.$$

The generators of Γ_r are:

$$\text{translations: } T_a u(x) \equiv u(x - a) \; \forall \, a \in \mathscr{L}_r.$$

$$\text{holohedry: } Qu(x) \equiv u(Q^T x) \; \forall \, Q \in D_r.$$

(2.10) $$\text{minus identity: } -Iu(x) \equiv -u(x).$$

We now give explicit conditions insuring that $u \mapsto G(\lambda,u)$ is Γ_r–equivariant, i.e.,

(2.11) $$G(\lambda,\gamma u) = \gamma G(\lambda,u) \; \forall \, \gamma \in \Gamma_r.$$

Consider first the special case when (1.1) is in divergence form:

$$\nabla \cdot q(\nabla u, u) + f(\lambda, \nabla u, u) = 0,$$

where "$\nabla \cdot$" denotes the divergence. Sufficient conditions for (2.11) are then as follows (cf. [3],[5]):

$$q(p,-u) = q(p,u), \; f(\lambda,p,-u) = -f(\lambda,p,u),$$

(2.12) $$q(Qp,u) = Qq(p,u), \; f(\lambda,Qp,u) = f(\lambda,p,u) \; \forall \, Q \in D_r.$$

In the general quasi–linear case (1.1), sufficient conditions are readily obtained by differentiations of (2.12) via

$$a_{ij}(p,u) = \partial q_i / \partial p_j (p,u),$$

(2.13) $$g(\lambda,p,u) = \partial q_i / \partial u (p,u) p_i + f(\lambda,p,u).$$

REMARK 2.1. We emphasize that the above procedure is merely a trick to obtain symmetry restrictions; (1.1) need not be of divergence form. Moreover, the field q here has nothing to do with (1.2).

One simple consequence of (2.11) is that $G(\lambda,0) \equiv 0$, i.e., we have the trivial solution branch $\{(\lambda,0): \lambda \in \mathbb{R}\}$. A necessary condition for bifurcation is that the linearized problem,

(2.14) $$A(\lambda)h \equiv G_u(\lambda,0)h = 0,$$

admit a nontrivial solution $h \in X_r$ for isolated values of λ. By virtue of (2.12,3), it is straightforward to show that (2.14) has the simple form

(2.15) $$a_0 \Delta h + c(\lambda)h = 0,$$

where $a_0 \equiv a_{11}(0,0) = a_{22}(0,0)$, "$\Delta$" denotes the two–dimensional Laplacian, and $c(\lambda) \equiv g_u(\lambda,0,0)$. If we restrict (2.15) to X_4, we find the eigenfunction

(2.16) $$h_4(x) = \sin(2\pi x_1/a)\sin(6\pi x_2/a) - \sin(6\pi x_1/a)\sin(2\pi x_2/a),$$

provided that the characteristic equation,

(2.17)
$$c(\lambda) = 40\pi^2 a_0/a^2,$$

has an isolated root, say $\lambda = \lambda_4$. On the other hand, the restriction of (2.15) to X_6 yields the nontrivial solution

(2.18)
$$h_6(x) = \sin(4\pi x_2/\sqrt{3}a) + \sin[2\pi(x_1 - x_2/\sqrt{3})/a]$$
$$- \sin[2\pi(x_1 + x_2/\sqrt{3})/a],$$

if $\lambda = \lambda_6$ is an isolated root of

(2.19)
$$c(\lambda) = 16\pi^2 a_0/3a^2.$$

REMARK 2.2. Of course, (2.15) admits many other solutions (if $c(\lambda)$ takes on appropriate values) when restricted to X_r. However, (2.16) and (2.18) are of special interest here, as we shall see in Section 3 when we return to (1.1,2).

We define the *nodal set* of $h \in X_r$ by

(2.20)
$$\mathscr{K}_h \equiv \{x \in \mathbb{R}^2 : h(x) = 0\}.$$

In particular, the nodal sets of the eigenfunctions given in (2.16) and (2.18) are depicted below. In each case, note that

(2.21)
$$\mathbb{R}^2 \setminus \mathscr{K}_{h_r} \text{ is a } \textit{monohedral tiling} \text{ of } \mathbb{R}^2,$$

i.e., there exists a bounded polygonal domain, $\Omega_0 \subset \mathbb{R}^2$, called a *tile* or *nodal domain*, and a discrete, infinite set of isometries (of $\mathbb{R}^2$ into itself), $\{g_1, g_2, ...\}$, such that $\mathbb{R}^2 \setminus \mathscr{K}_{h_r}$

$$= \bigcup_{i=0}^{\infty} \Omega_i \ (\Omega_i \equiv g_i(\Omega_0), \ i = 1,2,...) \text{ with } \Omega_i \cap \Omega_j = \emptyset \ \forall \ i \neq j.$$

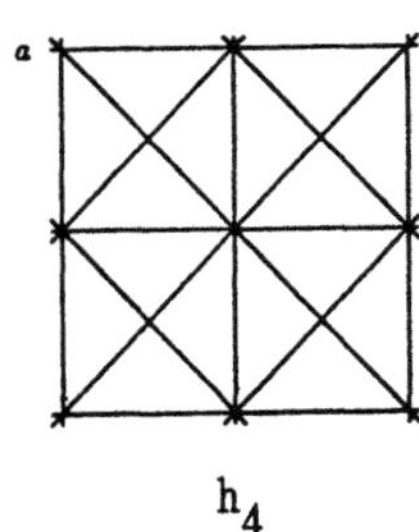

$$h_4$$

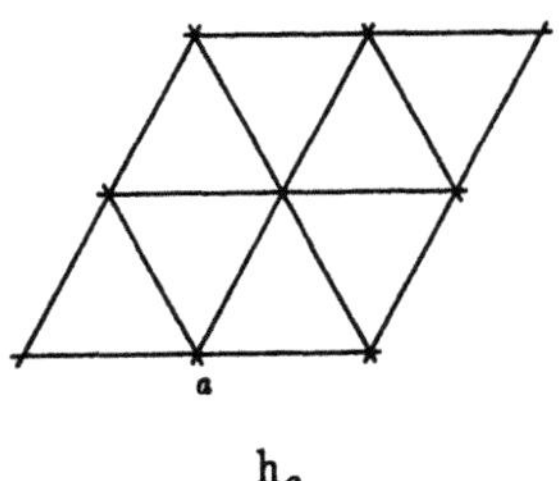

$$h_6$$

We now exploit (2.11) more fully. Recall that the *isotropy subgroup* of Γ_r at h_r is defined by

(2.22)
$$\Sigma_r \equiv \{\gamma \in \Gamma_r : \gamma h_r = h_r\}.$$

The associated fixed–point spaces are given by

$$U_r \equiv \{u \in X_r : \gamma u = u \ \forall \ \gamma \in \Sigma_r\},$$

(2.23) $$V_r \equiv \{u \in Y_r : \text{(same as above)}\},$$

each of which is a Banach space under the inherited topology of X_r and Y_r, respectively. Moreover, it is easy to show that $G: \mathbb{R} \times U_r \to V_r$, by virtue of (2.11). We henceforth refer to (2.5) under the restriction $G|_{\mathbb{R} \times U_r}$ as the Σ_r–*reduced problem*.

We now summarize pertinent information for the two cases of interest:

<u>$r = 4$</u>:

Lattice: $\mathcal{L}_4$; Point group: D_4; Eigenfunction: h_4; Nodal set:

$$\mathcal{N}_{h_4} = \{x \in \mathbb{R}^2 : x_1 = pa/2 \text{ or } x_2 = pa/2 \text{ or } x_1 - x_2 = pa/2$$
$$\text{or } x_1 + x_2 = pa/2 \ \forall \ p \in \mathbb{Z}\};$$

Generators of the isotropy subgroup Σ_4:

$$T_{(a,0)}, \ T_{(0,a)}, \ -IE, \ -IER(\pi), \ T_{(a/2,0)}ER(\pi),$$
$$T_{(0,a/2)}E, \ -IER(\pi/2), \ -IT_{(a/2,a/2)}ER(3\pi/2).$$

<u>$r = 6$</u>:

Lattice: $\mathcal{L}_6$; Point group: D_6; Eigenfunction: h_6; Nodal set:

$$\mathcal{N}_{h_6} = \{x \in \mathbb{R}^2 : x_1 = p\sqrt{3}a/2 \text{ or } x_2 = \sqrt{3}(x_1 + pa)$$
$$x_2 = \sqrt{3}(-x_1 + pa) \ \forall \ p \in \mathbb{Z}\};$$

Generators of the isotropy subgroup Σ_6:

$$T_{(a,0)}, \ T_{(a/2,\sqrt{3}a/2)}, \ -IE, \ ER(\pi/3),$$
$$-IER(2\pi/3), \ ER(\pi), \ -IER(4\pi/3), \ ER(5\pi/3).$$

In each case above, it is easy to show that

(2.24) $$\mathcal{N}_{h_r} \subset \mathcal{N}_u \ \forall \ u \in U_r,$$

i.e., all functions in the fixed–point space U_r have at least the zeros of h_r, as shown in the figure above. . As demonstrated in [5], (2.21) and (2.24) imply that for the Γ_r–reduced problem we have

$$\dim N(A(\lambda_r)|_{U_r}) = 1,$$

where λ_4 is an isolated root of the characteristic equation (2.17) and λ_6 an isolated root of (2.19). Indeed, $h_r \in N(A(\lambda_r)|_{U_r})$, which is a positive (or negative) solution of (2.15) on any tile or nodal domain. Thus, $N(A(\lambda_r)|_{U_r}) = \mathrm{span}\{h_r\}$. The central result of [5] is:

THEOREM 2.1. If $\lambda \mapsto g_u(\lambda,0,0)$ is strictly monotone in a neighborhood of λ_r, then $(\lambda_r,0)$ is a bifurcation point of a global continuum of nontrivial solutions of (2.5), denoted $C_r \subset \mathbb{R} \times U_r$, characterized by at least one of the following: (i) C_r is unbounded in $\mathbb{R} \times U_r$; (ii) $\overline{C_r}$ contains a point $(\lambda_0,0)$ with $\lambda_0 \neq \lambda_r$. Moreover,

$$(2.25) \qquad \mathcal{N}_u = \mathcal{N}_{h_r} \ \forall \ (\lambda,u) \in C_r \backslash (\lambda_r,0),$$

i.e., the *precise nodal structure of the eigenfunction* h_r *of the linearized problem at the bifurcation point is globally preserved along* C_r. In particular, if λ_r is the only root of the characteristic equation, then C_r is characterized by condition (i) above.

Outline of Proof. The existence of C_r (satisfying the above alternative) follows from a generalization [6] of a classical result of Rabinowitz [7], cf. [5]. Next we define

$$K^+ \equiv \{(\lambda,u) \in \mathbb{R} \times U_r : u > 0 \text{ in } \Omega_i\},$$
$$K^- \equiv \{(\lambda,u) : (\lambda,-u) \in K^+\},$$
$$K \equiv K^+ \cup K^-,$$

where Ω_i is any nodal domain, cf. (2.21). Of course, K is not open. Nonetheless, it can be shown [5] that

$$K \cap C_r \text{ is open relative to } C_r.$$

By the maximum principle, we then deduce

$$C_r \backslash (\lambda_r,0) \subset K. \ \square$$

The proof of Theorem 2.1 yields the following refinement:

COROLLARY 2.1. $C_r = C_r^+ \cup C_r^- \cup \{(\lambda_0,0)\}$, where C_r^+ and C_r^- are each continua characterized by $C_r^+ \subset K^+$ and $C_r^- \subset K^-$.

3. Nonlinear Neumann Boundary Conditions

In this section we demonstrate that the solutions presented in Section 2 automatically satisfy (1.1,2) on certain polygonal domains. Throughout we assume that the function q appearing in (1.2) satisfies the conditions given in (2.12). (Again we emphasize that the coeffient functions a_{ij} and g appearing in (1.1) need not be related to $\mathbf{q}$ via (2.13), cf. Remark 2.1.) We first establish a general lemma from which our results follow easily.

Let $\{\mathbf{n},\mathbf{t}\}$ be an orthonormal pair in $\mathbb{R}^2$, and let $\mathbf{Q} \in O(2)$ be the reflection that is uniquely determined by

(3.1) $\mathbf{Qt} = \mathbf{t}, \ \mathbf{Qn} = -\mathbf{n}.$

Let l denote the line

(3.2) $l \equiv \{\mathbf{x} = \alpha\mathbf{n} + s\mathbf{t}: s \in \mathbb{R}\},$

where $\alpha \in \mathbb{R}$. We say that $u: \mathbb{R}^2 \to \mathbb{R}$ has *reflection symmetry* about l if

(3.3) $u(\mathbf{Qx} + 2\alpha\mathbf{n}) = u(\mathbf{x}) \ \forall \ \mathbf{x} \in \mathbb{R}^2.$

LEMMA 3.1. If $u \in C^1(\mathbb{R}^2)$ has reflection symmetry about l and if $\mathbf{Q} \in D_r$, then

(3.4) $\mathbf{q}(\nabla u,u)|_l \cdot \mathbf{n} = 0.$

Proof. Differentiation of (3.3) yields

(3.5) $\nabla u(\mathbf{Qx} + 2\alpha\mathbf{n}) \equiv \mathbf{Q}\nabla u(\mathbf{x}).$

By virtue of (3.3), (3.5), $(2.12)_3$, and (3.1), we have

$$\begin{aligned}
\mathbf{q}(\nabla u(\mathbf{Qx} + 2\alpha\mathbf{n}),u(\mathbf{Qx} + 2\alpha\mathbf{n}))\cdot\mathbf{n} \\
= \mathbf{Q}\mathbf{q}(\nabla u(\mathbf{x}),u(\mathbf{x}))\cdot\mathbf{n} \\
= -\mathbf{q}(\nabla u(\mathbf{x}),u(\mathbf{x}))\cdot\mathbf{n}.
\end{aligned}$$

In particular, along l (cf. (3.2)) we find

$$\mathbf{q}(\nabla u,u)|_l \cdot \mathbf{n} = -\mathbf{q}(\nabla u,u)|_l \cdot \mathbf{n}. \ \square$$

Based upon Lemma 3.1, we can solve (1.1,2) for any polygonal domain whose boundary comprises line segments contained in lines about which all $u \in U_r$ possess reflection symmetry. By Theorem 2.1, $C_r \subset \mathbb{R} \times U_r$ is then a global solution branch of the boundary value problem. The lines of reflection symmetry (dashed lines) for all functions in U_4 and U_6, superimposed upon the nodal lines (solid lines) shown before, are depicted below.

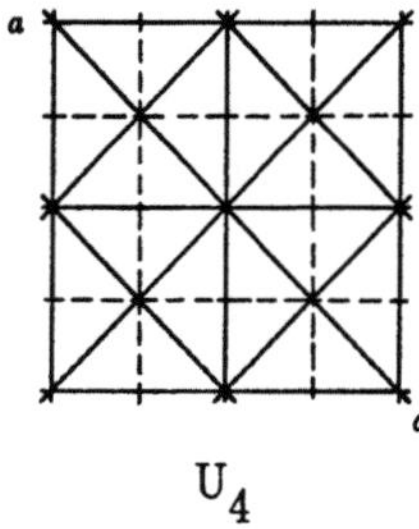

$$U_4$$

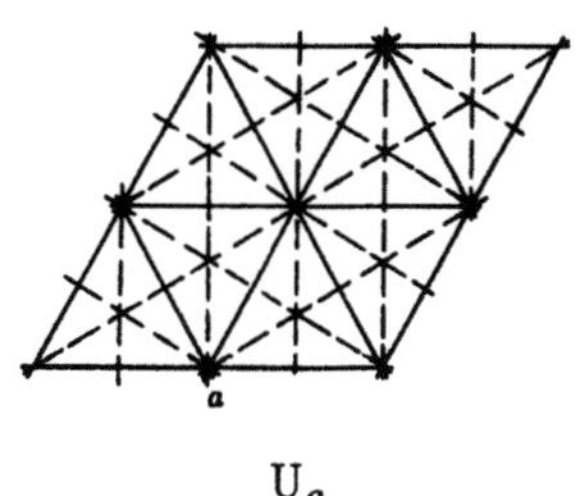

$$U_6$$

We now present some concrete examples.

Example 1 ($r = 4$): Let Ω be the square
$$\Omega \equiv (-b,b) \times (-b,b).$$
Then letting $a = 4b/m$, $m \in \mathbb{N}$, in (2.1), (2.16) and (2.17), we obtain *a countable infinity of globally separated solution branches of* (1.1,2), *along each of which the nodal pattern of* (2.16) *is preserved.*

REMARK 3.1. This particular boundary value problem also admits the same solution branches obtained in [4] for rectangular problems. These are distinct and globally separated from the branches discussed above.

Example 2 ($r = 6$): Let Ω denote the equilateral triangle
$$\Omega \equiv \{(x_1,x_2): x_1 < b \text{ and } x_1 > \sqrt{3}x_2 \text{ and } x_1 > -\sqrt{3}x_2\}.$$
Again, letting $a = 2b/m$, $m \in \mathbb{N}$, in (2.1), (2.18) and (2.19), we obtain *a countable infinity of globally separated solution branches of* (1.1,2), *along each of which the nodal pattern of* (2.19) *is preserved.*

Example 3 ($r = 6$): We can draw the same conclusions as obtained in Ex. 2 for (1.1,2) defined on a hexagonal domain Ω as follows. Let Ω_2 denote the triangular domain of

Ex. 2. Define
$$\Omega' \equiv \overline{\Omega_2} \setminus \{(x_1,x_2): x_1 = b\},$$
and then choose
$$\Omega \equiv \bigcup_{i=1}^{6} R(\pi i/3)(\Omega').$$

References

[1]   J.D. Crawford, M. Golubitsky, M.G.M. Gomes, E. Knobloch, I.N. Stewart, Boundary conditions as symmetry constraints, in Singularity Theory and its Applications Warwick 1989, Part II, M. Roberts & I. Stewart, Eds., LNM 1463, Springer–Verlag, Heidelberg (1991) 63–79.

[2]   D. Armbruster & G. Dangelmayr, Coupled stationary bifurcations in non–flux boundary value problems, Math. Proc. Camb. Phil. Soc. 101 (1987) 167–192.

[3]   T.J. Healey & H. Kielhöfer, Symmetry and nodal properties in global bifurcation analysis of quasi–linear elliptic equations, Arch. Rat. Mech. Anal. 113 (1991) 299–311.

[4]   ________________, Hidden symmetry of fully nonlinear boundary conditions in elliptic equations: global bifurcation and nodal structure, Res. Math. (1991, in press).

[5]   ________________, Preservation of nodal structure on global bifurcating solution branches of elliptic equations with symmetry, J.D.E. (1992, in press).

[6]   H. Kielhöfer: Multiple eigenvalue bifurcation for Fredholm operators, J. Reine Ang. Math. **358**, 104–124 (1985).

[7]   P.H. Rabinowitz: Some aspects of nonlinear eigenvalue problems, Rocky Mount. J. Math. **3**, 161–202 (1973).

Acknowledgements

The work of T.J.H. was supported in part by U.S. Air Force Office of Scientific Research grant AFOSR–91–0062, National Science Foundation grant DMS–9103254, and by the U.S. Army Research Office through the Mathematical Sciences Institute of Cornell University, Contract DAAG29–85–C–0018. The work of H.K. was supported by the Deutsche Forschungsgemeinschaft under Ki 227/4–1.

A New Approach for Solving Singular Nonlinear Equations

Annegret Hoy

Abstract. A new approach for constructing quadratically convergent algorithms is described. This approach covers the case of regular solutions of nonlinear equations as well as of singular solutions, so it can be viewed as a natural generalization of Newton's method.
As a consequence of the new approach, bordering and tensor methods proposed in the literature can be further developed.

1980 Mathematics Subject Classification (1985 Revision). Primary 65H10.

1. Introduction. The basic algorithm for solving nonlinear equations

$$(1.1) \quad F(x) = 0 \quad \quad F: R^n \longrightarrow R^n, \quad F \in C^2$$

is Newton's method.

Newton's method converges Q-quadratically to a solution x^* of (1.1) if the Jacobian $F'(x^*)$ is regular.

For singular $F'(x^*)$, however, the fast convergence is lost, see e.g. [1] and [2]. In the past decade several algorithms with better convergence behaviour have appeared in the literature. In [3] Griewank reviews the papers on this topic which came out in the early eighties and classified the proposed algorithms into three groups:

(i)   accelerated Newton methods

(ii)   bordering methods

(iii) tensor methods.

He compares the properties of these algorithms for the most easy type of singularity, the so-called regular singularity, which is characterized as follows:

$$(1.2) \quad \dim(\ker(F'(x^*))) = 1$$

$$(1.3) \quad v^T F''(x^*)uu \neq 0$$

where $u \in \ker(F'(x^*))$, $v \in \ker(F'(x^*)^T)$, $\|u\|_2 = \|v\|_2 = 1$.

Methods of the first group reach a regular singularity x^* only to single precision if double precision arithmetic is used. In order to achieve a high solution accuracy Griewank suggests in [3] taking bordering methods. It is interesting to note that most of the papers that have appeared after [3] fall into group (ii).

The idea of the bordering method consists of applying Newton's method not to (1.1) but to an auxiliary system

$$(1.4) \quad G(x,c) = 0, \quad G: R^{n+s} \longrightarrow R^{n+s}$$

i.e. to introduce an additional vector of variables $c \in R^s$ and additional equations so that the resulting system (1.4) has the regular solution $(x^*,0)^T$.

Several problems are connected with this approach. At first glance it is not clear

P1: whether the quadratic convergence property of the extended sequence $\{x^k,c^k\}$ carries over to the sequence $\{x^k\}$,

P2: how to organize the solution process without blowing up the algebraic costs for solving the linear Newton equations belonging to (1.4),

P3: how to approximate occurring second derivative information in an effective manner,

P4: how to form (1.4) if irregular singularities or higher dimensional null-spaces of $F'(x^*)$ are present.

To our knowledge there has been no contribution on bordering methods that answers all these questions simultaneously.

Below we will outline a new concept for solving singular equations. Using this concept the problems P1, P2, P3, P4 are answered in an easy manner. In addition more insight into bordering methods can be obtained. Furthermore it is interesting to apply the new ideas to construct an inexpensive realization of a Griewank-like tensor method.

<u>2. The new approach.</u> For the moment let us suppose that $F'(x^*)$ is regular. Additionally, p and q with $\|p\|_2 = \|q\|_2 = 1$ should be taken in such a way that $H'(x^*) = F'(x^*) + qp^T$ is regular, too.

Then the following rearrangements of the Newton-direction p_N^k give us an idea how computable directions p_D^k which lead to quadratic convergence in the singular case could look like. Applying the Sherman/Morrison rank-one updating lemma (cf. e.g. [4], p. 24) to $F'(x^k)=H_k'-qp^T$ the Newton-direction p_N^k can equivalently be expressed as

$$p_N^k = F'(x^k)^{-1}F(x^k)$$

$$= \left[[H_k']^{-1} + \frac{[H_k']^{-1}qp^T[H_k']^{-1}}{1-p^T[H_k']^{-1}q} \right] F(x^k)$$

$$= [H_k']^{-1}F_k + \frac{q^TF_k-q^TF_k+p^T[H_k']^{-1}F_k}{1-p^T[H_k']^{-1}q} [H_k']^{-1}q .$$

Using the identity

$$(2.1) \qquad F'(x^k)[H'(x^k)]^{-1} = I - qp^T[H'(x^k)]^{-1}$$

the Newton-direction p_N^k can be written as

$$(2.2) \qquad p_N^k = [H_k']^{-1}F_k + \frac{q^TF_k-q^TF_k'[H_k']^{-1}F_k}{q^TF_k'[H_k']^{-1}q} [H_k']^{-1}q .$$

From (2.2) one can observe that p_N^k is a special member of the more general class of algorithms

$$(2.3) \qquad p_D^k = [H_k']^{-1}F_k + \frac{M(x^k)- \nabla^TM(x^k)[H_k']^{-1}F_k}{\nabla^TM(x^k)[H_k']^{-1}q} [H_k']^{-1}q .$$

Algorithms of type (2.3) are of great importance for the solution of regular as well as singular equations with $\dim(\ker(F'(x^*)))=1$ as the following basic theorem shows.

THEOREM. Let x^* be a solution of (1.1) where $F'(x^*)$ is either regular or singular with $\dim(\ker(F'(x^*)))=1$. Then the sequence $\{x^k\}$ produced by $x^{k+1}=x^k-p_D^k$ with

$$(2.4) \quad p_D^k = [H_k']^{-1}F(x^k) + \frac{M(x^k) - \nabla^T M(x^k)[H_k']^{-1}F(x^k)}{\nabla^T M(x^k)[H_k']^{-1}q}[H_k']^{-1}q$$

converges Q-quadratically to x^* provided that

(i) $M(x)$ is an arbitrary function $M: R^n \to R^1$ that satisfies
$M(x^*)=0$, $\nabla^T M(x^*)H'(x^*)^{-1}q \neq 0$
where $\nabla^T M(x)$ denotes the gradient of the function M,

(ii) p and q are chosen to ensure $H'(x^*)=F'(x^*)+qp^T$ to
be regular,

(iii) x^0 is chosen to satisfy $\|x^0-x^*\| \leq \delta$ with $\delta>0$ and δ
sufficiently small.

Proof. Our proof is based on Taylor's Theorem and on the
identity (2.1).
Taylor's Theorem and identity (2.1) together yield

$$(2.5) \quad [H_k']^{-1}F(x^k) = (x^k-x^*)-[H_k']^{-1}qp^T(x^k-x^*)+\beta_2(x^k-x^*).$$

Furthermore, we know from Taylor's Theorem and (i) that

$$(2.6) \quad M(x^k) = \nabla^T M(x^k)(x^k-x^*) + \beta_2(x^k-x^*).$$

Inserting (2.5) and (2.6) into (2.4) we obtain

$$p_D^k = [H_k']^{-1}F(x^k) + \frac{\nabla^T M(x^k)[H_k']^{-1}qp^T(x^k-x^*)}{\nabla^T M(x^k)[H_k']^{-1}q}[H_k']^{-1}q + \beta_2(x^k-x^*)$$

$$= [H_k']^{-1}F(x^k) + p^T(x^k-x^*)[H_k']^{-1}q + \beta_2(x^k-x^*)$$

$$= [H_k']^{-1}H(x^k) + \beta_2(x^k-x^*)$$

where the function $H(x)$ is defined as

$$(2.7) \quad H(x) = F(x) + qp^T(x-x^*) .$$

As a consequence the search direction p_D^k can be interpreted
as a β_2-perturbed Newton-direction belonging to the system
$H(x)=0$. From (ii) we conclude that $H(x)$ has x^* as a regular
solution. This proves the quadratic convergence of the
sequence $\{x^k\}$ which is produced by p_D^k .

In the next paragraph we will show how the search direction of type (2.4) can be adapted to several types of singularities by appropriate choices of the function $M(x)$.

3. Construction of special algorithms. Our aim is to construct special algorithms for different types of singularities. To this end we define a family of functions by

$$(3.1) \quad \begin{aligned} M_0(x) &= q^T F(x) \\ M_m(x) &= \nabla^T M_{m-1}(x) H'(x)^{-1} q \quad \text{für } m > 0. \end{aligned}$$

The algorithm belonging to the starting function $M=M_0$ is just Newton's method itself.

The Theorem reproduces the well-known quadratic convergence property of Newton's method if

$$\nabla^T M_0(x^*) H'(x^*)^{-1} q = 1 - p^T H'(x^*)^{-1} q \neq 0,$$

i.e. for regular $F'(x^*)$.

Next we will study the case $m=1$.

We have

$$M_1(x) = \nabla^T M_0(x) H'(x)^{-1} q = 1 - p^T H'(x)^{-1} q .$$

Above we have realized that $M_1(x^*)=1-p^T H'(x^*)^{-1} q=0$ if and only if $F'(x^*)$ is singular. Therefore we conclude that M_1 is an appropriate candidate in the singular case. Note that the natural hypothesis on F which belongs to the choice M_1 is $F \in C^3$. But what is the meaning of the second important condition

$$(3.2) \quad \nabla^T M_1(x^*) H'(x^*)^{-1} q \neq 0 ?$$

We get

$$\nabla^T M_1(x^*) H'(x^*)^{-1} q = p^T H'(x^*)^{-1} F''(x^*) H'(x^*)^{-1} q H'(x^*)^{-1} q .$$

From

$$F'(x^*) H'(x^*)^{-1} q = [I - q p^T H'(x^*)^{-1}] q = q(1 - p^T H'(x^*)^{-1} q) = 0$$

we can see that

$$\nabla^T M_1(x^*) H'(x^*)^{-1} q = \lambda v^T F''(x^*) uu$$

where $\lambda \neq 0$, i.e. condition (3.2) is fulfilled if and only
if the regularity condition (1.3) is satisfied. Therefore the
choice $M(x)=M_1(x)$ is appropriate for regular singularities.
Now, we will proceed with the case $m=2$. We have

$$M_2(x) = p^T H'(x)^{-1} F''(x) H'(x)^{-1} q H'(x)^{-1} q .$$

The natural hypothesis on F is now $F \in C^4$.
Above we have realized that $M_2(x^*)=0$ if and only if x^* is
an irregular solution of (1.1).
The second condition $v^T M_2(x^*) H'(x^*)^{-1} q \neq 0$ can equivalently
be expressed as

$$(3.3) \quad v^T F'''(x^*)uuu - 3v^T F''(x^*)uy \neq 0$$

where y is the unique solution to
$$F'(x^*)y = F''(x^*)uu, \quad u^T y = 0.$$
Condition (3.3) occurs in the literature at several places.
So we find it as the characterization of cusp points in para-
meter-dependent problems, cf. [5], and as condition for geometric
isolation of an irregular singularity, cf. [6].
The Theorem and the construction rule (3.1) together answer
the four problems which were outlined at the beginning in
connection with bordering methods.
For solving singular problems it is not necessary to introduce
new variables. As a consequence the problems P1 and P2 do not
arise.
The special structure (2.4) of the algorithms has the useful
advantage that the higher derivative information needed in
$v^T M(x) H'(x)^{-1} F$ and $v^T M(x) H'(x)^{-1} q$ occurs in form of
directional derivatives. Directional derivatives can effectively
be approximated by a few function values only, compare [7],
[8].
Irregular singularities can be handled with the choice
$M(x)=M_2(x)$. A generalization of the Theorem to higher-dimensional null-spaces of the Jacobian $F'(x^*)$ is possible,
compare [9]. More detailed discussions on higher-dimensional
null-spaces with consequences to bifurcation theory will be
published in a jpint work with Allgower, Böhmer and Janowski.

In the following example it should be illustrated how symmetries in the system $F(x)=0$ can be exploited by special choices of $M(x)$.

Consider the polynomial

$$(3.4) \quad y = x^n + a_1 x^{n-1} + \ldots + a_{n-1} x + a_n .$$

One approach to determine all polynomial roots simultaneously is to solve Vieta's equations

$$F(x) = 0 \quad F: R^n \longrightarrow R^n \quad \text{with}$$
$$F_1 = x_1 + x_2 + \ldots + x_n + a_1$$
$$F_2 = x_1 x_2 + \ldots + x_{n-1} x_n - a_2$$
$$\vdots$$
$$F_n = x_1 x_2 \ldots x_n - (-1)^n a_n$$

by Newton's method, compare e.g. [10], [11]. It is interesting to observe that neither in [10] nor in the recent paper [11] the case of multiple zeros is incorporated into the discussion. For the sake of simplicity we restrict ourselves to polynomials (3.4) which possess exactly one double zero $x_1^* = x_2^*$ and $(n-2)$ simple zeros $x_3^*, \ldots, x_n^*$.

Then the Jacobian $F'(x^*)$ has a regular singularity at the solution $x^* = (x_1^*, \ldots , x_n^*)^T$,
i.e. the algorithm

$$x^{k+1} = x^k - p_D^k$$

with p_D^k according to (2.4) and $M(x)=M_1(x)$ leads to quadratic convergence.

However, because of the special symmetric structure of Vieta's equations there exists a more appropriate choice of $M(x)$ giving us an algorithm which works without second derivative information. Take

$$M(x) = \widetilde{M}(x) = x_2 - x_1 .$$

The two assumptions of the Theorem concerning $\widetilde{M}(x)$ are fulfilled, cf. [12] .

4. Consequences for bordering and tensor methods. The history
of the bordering approach is outlined in [3].
Consider the most easy type of a bordered system

$$(4.1) \qquad \begin{aligned} F(x) + \mu q &= 0 \\ \mathcal{Y}(x) &= 0 \ . \end{aligned}$$

Suppose that $q^T v \neq 0$ and $\nabla^T \mathcal{Y}(x^*)u \neq 0$. Then the Jacobian of
(4.1) is regular for a regular singularity x^* .
In the following we will analyze the Newton step $p_B = (t^T, \tau)^T$
which belongs to (4.1). The Newton step is defined as the
solution of the linear system

$$(4.2) \qquad \begin{aligned} F'(x)t + \tau q &= F(x) + \mu q \\ \nabla^T \mathcal{Y}(x)t &= \mathcal{Y}(x) \ . \end{aligned}$$

Premultiplication of the first equation by $H'(x)^{-1}$ yields

$$H'(x)^{-1}F'(x)t + \tau H'(x)^{-1}q = H'(x)^{-1}F(x) + \mu H'(x)^{-1}q \ .$$

Utilizing (2.1) we further obtain

$$(4.3) \qquad t = H'(x)^{-1}F(x) + [\mu - \tau + p^T t]H'(x)^{-1}q \ .$$

Substitution of (4.3) into the second equation of (4.2) leads
to

$$\mu - \tau + p^T t = \frac{\mathcal{Y}(x) - \nabla^T \mathcal{Y}(x)H'(x)^{-1}F(x)}{\nabla^T \mathcal{Y}(x)H'(x)^{-1}q} \ ,$$

i.e. we have

$$(4.4) \qquad t = H'(x)^{-1}F(x) + \frac{\mathcal{Y}(x) - \nabla^T \mathcal{Y}(x)H'(x)^{-1}F(x)}{\nabla^T \mathcal{Y}(x)H'(x)^{-1}q} H'(x)^{-1}q \ .$$

Indeed, (4.4) is equivalent to (2.4) with $\mathcal{Y}(x) = M(x)$.
The previous rearrangements show that the x-correction t of
p_B can be expressed independently of τ . This fact has
apparently not been pointed out in contributions to bordering
methods, compare [3], [13], [14], [15], [16], [17]. Our special
algorithmic proposals with $\mathcal{Y}(x) = M_m(x)$ seem to be new and
they can apparently be handled more easily than other algorithms
described in the literature.

A tensor step s_T^k is defined as solution of the quadratic model

$$(4.5) \quad F(x^k) + F'(x^k)s_T^k + 0.5 T^k s_T^k s_T^k = 0 .$$

The new iterate is then calculated according to $x^{k+1} = x^k + s_T^k$. In [18] the tensors T^k are built up using function values at previous iterates.

As is reported in [19] the resulting algorithm is 3-step convergent with Q-order 7/6 towards regular singularities. In [3] Griewank proposes a tensor method which converges with Q-order 1.5 in the singular case. However, he does not give an inexpensive computational realization of his method. In [9] we study the special choice of tensors

$$(4.6) \quad T^k = qp^T [H_k']^{-1} F''(x^k) .$$

Then the solution s_T^k of (4.5) is given by

$$(4.7) \quad s_T^k = -[H_k']^{-1}F(x^k) - \alpha_k[H_k']^{-1}q$$

$$= -p_D^k \pm \sqrt{A_k^2 - B_k} .$$

One can show, see [9], that $A_k^2 - B_k = \beta_3(x^k - x^*)$ if $F'(x^*)$ is a regular singularity. Consequently, the tensor method based on (4.6) converges with Q-order 1.5 in the singular case independently of the chosen sign in front of the square root. The terms A_k and B_k are formed using second order directional derivatives only. This means that the algorithm which is based on (4.6) is inexpensively realizable.

<u>References.</u>

[1] A. Griewank & M.R. Osborne, "Newton's method for singular problems when the dimension of the null space is > 1," SIAM J. Numer. Anal., v.18, 1981, pp. 145-149.
[2] A. Griewank & M.R. Osborne, "Analysis of Newton's method at irregular singularities," SIAM J. Numer. Anal. v.20, 1983, pp. 747-773.

[3] A. Griewank, "On solving nonlinear equations with simple
singularities or nearly singular solutions," SIAM Rev., v.27,
1985, pp. 537-563.

[4] H. Schwetlick, Numerische Lösung nichtlinearer Gleichungen.
VEB Deutscher Verlag der Wissenschaften, Berlin, 1979.

[5] A. Spence & A.D. Jepson, "The numerical calculation of
cusps, bifurcation points and isola formation points in two-
parameter problems," in ISNM 70 (T. Küpper, H.D. Mittelmann
and H. Weber, eds.), Birkhäuser, Basel, 1984, pp. 459-479.

[6] H.B. Keller, "Geometrically isolated nonisolated solutions
and their approximation," SIAM J. Numer. Anal., v.18, 1981,
pp. 822-838.

[7] G. Pönisch & H. Schwetlick, "Computing turning points of
curves implicitly defined by nonlinear equations depending on
a parameter," Computing v.26, 1981, pp. 107-121.

[8] A. Hoy & H. Schwetlick, "Some superlinearly convergent
methods for solving singular nonlinear equations," in Lectures
in Applied Mathematics v.26 (E.L. Allgower and K. Georg, eds.),
Providence, 1990, pp. 285-300.

[9] A. Hoy, "Numerische Lösung singulärer nichtlinearer
Gleichungssysteme," Habilschrift, Martin-Luther-Universität
Halle, 1991.

[10] I. Kerner, "Ein Gesamtschrittverfahren zur Berechnung von
Nullstellen von Polynomen," Numer. Math., v.8, 1966, pp. 290-294.

[11] A. Frommer, "A unified approach to methods for the simultaneous computation of all zeros of generalized polynomials,"
Numer. Math., v.54, 1988, pp. 105-116.

[12] A. Hoy, "A note on the computation of double zeros of
polynomials via Vieta's Equations," to appear in Wiss. Z.
Univ. Halle, v.40, 1991, pp. 143-146.

[13] H. Weber & W. Werner, "On the accurate determination of
nonisolated solutions of nonlinear equations," Computing, v.26,
1981, pp. 315-326.

[14] R. Menzel & G. Pönisch, "A quadratically convergent method
for computing simple singular roots and its application to
determining simple bifurcation points," Computing, v.32, 1984,
pp. 127-138.

[15] N. Yamamoto, "Regularization of solutions of nonlinear equations with singular Jacobian matrices," J. Inf. Process., v.7, 1984, pp. 16-21.

[16] T. Tsuchiya, "Enlargement procedure for resolution of singularities at simple singular solutions of nonlinear equations," Numer. Math., v.52, 1988, pp. 401-411.

[17] Mei Zhen, "Splitting iteration method for simple singular points and simple bifurcation points," Computing, v.41, 1989, pp. 87-96.

[18] R.B. Schnabel & P.D. Frank, "Tensor methods for nonlinear equations," SIAM J. Numer. Anal., v.21, 1984, pp. 815-843.

[19] R.B. Schnabel & P.D. Frank, "Solving systems of nonlinear equations by tensor methods," Technical report CU-CS-334-86, Department of Computer Science, University of Colorado at Boulder, 1986.

Fachbereich Mathematik und Informatik
Martin-Luther-Universität Halle-Wittenberg
Postfach
O-4010 Halle/Saale
Germany

Quasiperiodic drift flow in the Couette-Taylor problem

by

P. Laure[*],   J. Menck[†],   and J. Scheurle [†]

1. Introduction

The Couette-Taylor problem deals with the flow of an incompressible, viscous fluid between two coaxial rotating cylinders. Depending on the angular velocities of the cylinders, different flow patterns are observed in experiments. Mathematically, transitions between different flow patterns can be described by instabilities and bifurcations of certain solutions of the Navier-Stokes equations for this problem. In this paper we consider the counterrotating case, i.e. the cylinders rotate in opposite directions. We describe a sequence of three successive instabilities and corresponding bifurcations which occur in a certain parameter regime when the Reynolds number is increased. The primary bifurcation is the classical bifurcation from Couette flow to Görtler-Taylor vortex flow (cf. Taylor [1923]), the secondary one leads to wavy vortex flow, and the tertiary one leads to what we call quasiperiodic drift flow. We also indicate how this result is linked to related work on the Couette-Taylor problem and give an outline of the method by which we have obtained it. A key step of our method is the reduction of the Navier-Stokes equations to a system of ordinary differential equations. This is achieved by invariant manifold theory and ideas from the theory of dynamical systems with symmetry.

The plan of the paper is as follows. In section 2 we introduce the basic equations. In section 3 we discuss stability of the Couette flow. Section 4 is devoted to the reduction procedure. Finally, in section 5 we analyse the reduced equation and state the main result.

Acknowledgements. The authors thank the European Community and Gerard Ioos for supporting their collaboration for this paper through the European Stimulation Program ST2J-0316 (EDB). The third author is grateful to Jerry Marsden, who invited him to spend some time in Berkeley in fall of 1991. Most of this paper has been written up during that pleasant stay. Also, we thank Marty Golubitsky for some helpful comments.

---

[*] Institut Non-Lineaire de Nice, UMR 129,Université de Nice, Parc Valrose, 06034 Nice Cedex, France
[†] Institut für Angewandte Mathematik, Universität Hamburg, Bundesstrasse 55, W-2000 Hamburg 13, Germany

2. The basic equations and solutions

Let R_i and R_0 be the radii of the inner and outer cylinders, respectively, and denote their angular velocities by Ω_i and Ω_0. According to the geometry of the experimental apparatus, we choose cylindrical coordinates r, Θ, z and denote the velocity field in the fluid by $V = (V_r, V_\Theta, V_z)^T$ and the pressure field by p. The superscript T denotes the corresponding column vector. Both are functions of the spatial coordinates and time t. Also, we introduce the dimensionless parameters

$$(2.1) \qquad \Omega = \Omega_0/\Omega_i, \quad \eta = R_i/R_0, \quad r_1 = \frac{\eta}{1-\eta}, \quad r_2 = \frac{1}{1-\eta}$$

and
$$(2.2) \qquad R = \frac{R_i \Omega_i d}{\nu},$$

where R is the Reynolds number, $d = R_0 - R_i$ is the width of the gap between the cylinders, and ν is the kinematic viscosity of the fluid. As usual, we impose non-slip boundary conditions at the cylinder walls and assume $2\pi/\alpha$-periodicity of V and p in the axial direction (infinitely long cylinders). Here the wave number α will be fixed eventually. Then the Navier-Stokes equations for V and p in dimensionless form read as follows:

$$(2.3) \qquad
\begin{aligned}
&\left.\begin{aligned}
\frac{\partial V}{\partial t} &= \Delta V - R(V \cdot \nabla)V - \nabla p \\
\nabla \cdot V &= 0
\end{aligned}\right\} (r_1 \leq r \leq r_2, \Theta \in \mathbb{R}, z \in \mathbb{R}) \\
&V_r = V_z = 0 \quad (r = r_1, r = r_2) \\
&V_\Theta = 1 \quad (r = r_1) \\
&V_\Theta = \Omega/\eta \quad (r = r_2) \\
&V \quad \text{and} \quad p \quad \text{are} \quad 2\pi/\alpha - \text{periodic in} \quad z \quad \text{and} \quad 2\pi - \text{periodic in} \quad \Theta.
\end{aligned}$$

Here Δ is the Laplace operator, and ∇ the nabla operator.

These equations are covariant with respect to the symmetry group $\Gamma = SO(2) \times O(2)$, where $SO(2)$ acts by rotations $R_\varphi : \Theta \mapsto \Theta + \varphi$ with angle $\varphi \in [0, 2\pi)$ around the z-axis, and $O(2)$ acts by translations $T_a : z \mapsto z + a$ along the z-axis and through the flip $S : z \mapsto -z$ (see Golubitsky and Stewart [1986]).

One advantage of the symmetry is that one can describe the transitions between different flow patterns by symmetry breaking bifurcations and characterize different solutions (V, p) by their symmetry, or more precisely by the isotropy subgroup of V

$$\Sigma_V = \{\gamma \in \Gamma \mid \gamma V = V\}.$$

There is a stationary solution, namely the <u>Couette flow</u> (COU), for which one has the explicit analytic expression

(2.4)
$$V^0 = (0, V_\Theta^0, 0)^T, \quad p^0 = R \int (V_\Theta^0(r)^2/r)\,dr$$

$$V_\Theta^0(r) = Ar + B/r, \quad A = \frac{\Omega - \eta^2}{\eta(1+\eta)}, \quad B = \frac{\eta(1-\Omega)}{(1-\eta)(1-\eta^2)}.$$

It has the symmetry of the full group Γ, i.e. $\Sigma_{COU} = \Gamma$, and represents an azimuthal flow. Another type of stationary solution which is going to play a role in the discussion below is the Görtler-Taylor vortex flow (GTV). Its isotropy subgroup is $\Sigma_{GTV} = SO(2) \times \mathbb{Z}_2(S)$ where $\mathbb{Z}_2(S)$ is the subgroup of $O(2)$ generated by the flip S. Consequently, this flow is not invariant under the translations T_a. Here flat flow cells form in the fluid. In contrast to those flows, for the wavy vortex flow (WV) also the $SO(2)$-symmetry is broken. This flow is time periodic and for fixed time invariant under $\Sigma_{WV} = \mathbb{Z}_2(\mathbb{R}_\pi, S)$ only, i.e. under a rotation R_π followed by the flip S. Here wavy flow cells form in the fluid. Actually, the wavy vortex solution is a rotating wave or relative equilibrium with respect to the group $SO(2)$, i.e. its trajectory in (V, p)-space is also a group orbit. The corresponding flow cells rotate periodically around the axis of the cylinders. Since the translational symmetry is broken, these solutions occur in families of conjugate trajectories which just differ by a translation T_a.

3. Stability of the Couette flow

Next we discuss stability of the Couette flow depending on the parameters R, Ω and η. To this end we introduce relative variables U and q via

(3.1)
$$V = V^0 + U, \quad p = p^0 + q$$

and write the basic equations (2.3) as an evolution equation

(3.2)
$$\frac{dU}{dt} = L(R, \Omega, \eta)U + N(R, \Omega, \eta)(U) \qquad (U \in H)$$

for U in a suitable Hilbert space $H \subset [L_2((r_1, r_2) \times \mathbb{R} \times \mathbb{R})]^3$ of solenoidal vector fields which are periodic in Θ and z. It is well known that this can be achieved using the Weyl projection operator to eliminate the pressure field q (see Ladyshenskaya [1963], Iudovich [1965], Iooss [1971], and Témam [1977]). Here L is a closed linear operator with dense domain of definition D in H. The elements of D have zero trace at $r = r_1$ and $r = r_2$. By $\|\cdot\|_D$ we denote the graph norm in D with respect to L. The resolvent of L is compact, i.e. L has pure point spectrum, and L generates a holomorphic compact semigroup $\exp(Lt)_{t\geq 0}$ in H. The operator $N : D \to H$ is quadratic and continuous.

There is a local existence and uniqueness theorem for the initial value problem corresponding to (3.2):

Theorem. For all $T > 0$, there exists a δ such that (3.2) has an unique (classical) solution $U \in C^0([0,T], D) \cap C^1((0,T], H)$ with $U(0) = U_0$ for all $U_0 \in D$ with $\|U\|_D < \delta$.

Hence, the evolution equation (3.2) generates a local semiflow S_t in D. Moreover, it has been shown, e.g. by Sattinger [1969/70] and Kirchgässner and Kielhöfer [1973], that the principle of linearized stability holds true for the trivial solution $U = 0$ which represents the Couette flow. It says the following: if all eigenvalues of L have negative real parts, then $U = 0$ is asymptotically stable with respect to the semiflow S_t (in the sense of Liapunov). If there is at least one eigenvalue of L with positive real part, then $U = 0$ is unstable. Note, that here in Liapunov's notion of stability the norm $\| \cdot \|_D$ is used to measure initial values and the H-norm is used to measure $U(t)$ for $t > 0$.

Using this principle, it is not difficult to prove that for sufficiently small values of R the Couette flow is asymptotically stable. Indeed, in this case L is just a small perturbation of the Laplace operator which is negative definite. Hence, all eigenvalues of L have negative real parts. However, for all Ω and η, there exists a critical Reynolds number $R_c = R_c(\Omega, \eta)$ such that for $R > R_c(\Omega, \eta)$ the Couette flow is unstable. How many eigenvalues cross the imaginary axis right at $R = R_c$ and how the corresponding eigenfunctions which we call critical modes, look like depends on Ω and η. By definition of H, the general complex form of the critical modes is

$$U = U(r)e^{i(k\alpha z + m\Theta)} \qquad (k, m \in \mathbb{Z}).$$

Numerically one finds a curve in the rectangle $-1.2 \leq \Omega \leq -0.4$, $0.4 \leq \eta \leq 1.0$, along which critical modes with $k = 1$ and two different azimuthal wave numbers $m = 0$ and $m = 1$ occur simultaneously at $R = R_c(\Omega, \eta)$ (cf. Langford et al. [1988]). This is called a bicritical instability of the Couette flow. We now fix a point (Ω_c, η_c) on this curve, choose α appropriately, and consider (3.2) for $\eta = \eta_c$ and (R, Ω) near the corresponding critical point $P_c = (R_c(\Omega_c, \eta_c), \Omega_c)$ in the (R, Ω)-plane. There L has a real eigenvalue $\mu(R, \Omega)$ and a pair of complex conjugate eigenvalues $\gamma(R, \Omega) \pm i\omega(R, \Omega)$. All these eigenvalues have multiplicity two. For $R = R_c$ and $\Omega = \Omega_c$ they simultaneously sit on the imaginary axis. Hence, the corresponding critical eigenspace E is six-dimensional. The other eigenvalues of L are strictly bounded away from the imaginary axis for (R, Ω) near P_c.

4. Reduction

Because of the above properties of L and N, one can use center manifold theory (see e.g. Henry [1981]) to reduce (3.2) to a six-dimensional system of first order ordinary differential equations for (R, Ω) near P_c and $U \in D$ near 0

$$(4.1) \qquad \frac{dx}{dt} = X(R, \Omega)(x) \qquad (x \quad \text{near} \quad 0 \quad \text{in} \quad \mathbb{R}^6),$$

where x is a vector of coordinates in the critical eigenspace E. The vector field $X = X(R, \Omega, \cdot)$ is equivariant with respect to the symmetry group Γ, i.e. it commutes with a certain representation of Γ on E. This system fully describes all solutions of (3.2) which exist and stay close to $U = 0$ (in the H-norm) for all t, including stability properties. It is obtained by restricting (3.2) to a six-dimensional invariant manifold $M \subset D$, the <u>center manifold</u>, which is represented as the graph of a smooth map $\Psi = \Psi(R, \Omega, \cdot)$. The latter is defined for x near 0 in E and has values in the complementary eigenspace of $L(R_c, \Omega_c, \eta_c)$. The vector field X and the map Ψ satisfy a so-called <u>homological equation</u>

$$(4.2) \qquad [id + D\Psi]X = L(\cdot + \Psi) + N(\cdot + \Psi)$$

which can be used to compute Taylor expansions for both X and Ψ at $x = 0$. This leads to linear elliptic boundary value problems for the Taylor coefficients of Ψ. The results in this paper only depend on terms of order up through two of Ψ and up through three of X.

A further reduction can be achieved using the symmetry of X. For example, following Menck [1991], one can devide out the center manifold M by the action of the subgroup $\tilde{\Gamma} = SO(2) \times S^1$ of Γ, which leads to the <u>orbit space</u> $M_{|\tilde{\Gamma}}$. Here different points on the group orbits of $\tilde{\Gamma}$ are identified. Globally, this is not a manifold, rather it is an algebraic variety with cone-like singularities. But corresponding to a certain region on M, where the group orbits of $\tilde{\Gamma}$ are two-tori, $M_{|\tilde{\Gamma}}$ has a stratum which is a four-dimensional manifold locally. There the motion of (4.1) transverse to the group orbits of $\tilde{\Gamma}$ is described by a smooth four-dimensional system

$$(4.3) \qquad \frac{d\xi}{dt} = Y(R, \Omega)(\xi) \qquad (\xi \quad \text{near} \quad 0 \quad \text{in} \quad \mathbb{R}^4),$$

where ξ is a vector of suitably choosen invariant coordinates ξ_1, ξ_2, ξ_3 and ξ_4. To describe the relation between ξ and the coordinates on the center manifold M we introduce complex coordinates $\zeta_j \ (j = 0, 1, 2)$ in E defined by

$$(4.4) \qquad x = \sum_{j=0}^{2} (\zeta_j V_j + \bar{\zeta}_j W_j),$$

where $V_0 = \hat{U}_0(r)e^{i\alpha z}, W_0 = SV_0$ and $V_1 = \hat{U}_1(r)e^{i(\alpha z + \theta)}, V_2 = SV_1, W_1 = \bar{V}_1, W_2 = \bar{V}_2$ denote critical modes corresponding to the zero eigenvalue and the pair of complex conjugate eigenvalues, respectively. Then we have

$$(4.5) \qquad \begin{aligned} \xi_1 &= |\zeta_0|^2, \quad \xi_2 = \frac{1}{2}(|\zeta_1|^2 + |\zeta_2|^2) \\ \xi_3 &= \frac{1}{2}(|\zeta_2|^2 - |\zeta_1|^2), \quad \xi_4 = Im(\bar{\zeta}_0^2 \zeta_1 \bar{\zeta}_2). \end{aligned}$$

It turns out, that the $\mathbb{Z}_2$-symmetry $\xi_1 \mapsto \xi_1$, $\xi_2 \mapsto \xi_2$, $\xi_3 \mapsto -\xi_3$ and $\xi_4 \mapsto -\xi_4$ still acts nontrivially on this system. Hence, the restriction of (4.3) to the fixed point subspace $Fix(\mathbb{Z}_2) = \{\xi \in \mathbb{R}^4 | \xi_3 = \xi_4 = 0\}$ finally leads to a two-dimensional system for ξ_1 and ξ_2. Setting

$$(4.6) \qquad \begin{aligned} \mu(R, \Omega) &= \lambda \\ \gamma(R, \Omega) &= \lambda - \sigma, \end{aligned}$$

this system has the form

$$(4.7) \qquad \begin{aligned} \dot{\xi}_1 &= \lambda \xi_1 + b_1 \xi_1 \xi_2 + b_2 \xi_1^2 + \text{h.o.t.} \\ \dot{\xi}_2 &= (\lambda - \sigma)\xi_2 + a_1 \xi_1 \xi_2 + a_2 \xi_2^2 + \text{h.o.t..} \end{aligned}$$

The coefficients depend on the parameters λ and σ. Note, that the terms which are quadratic in (ξ_1, ξ_2) involve third order terms of the vector field X. This is a consequence of the fact that the orbit space $M_{|\hat{\Gamma}}$ is globally nonlinear.

5. Analysis of the reduced system

Numerical computations show that roughly speaking, we can think of λ and σ given by (4.6) as being

$$(5.1) \qquad \lambda \approx R - R_c(\Omega_c, \eta_c) \quad \text{and} \quad \sigma \approx \Omega - \Omega_c \quad (\text{as} \quad (R, \Omega) \to P_c).$$

To compute the relevant Taylor coefficients of X and Ψ from (4.2), we used a combination of symbolic computations and a numerical boundary value problem solver (cf. Laure and Demay [1988]).

Here is a table of results for X which we obtained for differnt values of $\eta_c \in [0.4, 0.85]$, $\alpha \sim 3.6, R = R_c$ and $\Omega = \Omega_c$:

Table

η_c	c_1^1	p_2^1	p_0^3	c_0^3	$p_1^1 - p_0^3$	$c_2^1 - c_0^3$	p_0^2	c_0^2	q_0^3	q_0^2
0.85	-7.918	-152.5	-109.5	-191.2	2.8	-88.2	-44.17	-182.5	-62.16	-117
0.825	-18.43	-188.7	-118.8	-208.1	-6.5	-122.5	-46.96	-204.8	-68.44	-130.4
0.8	-30.04	-228.2	-129	-227.1	-16.5	-160.1	-50.02	-227.9	-74.47	-144.2
0.775	-42.71	-270.9	-139.8	-248.2	-27.5	-201.4	-53.35	-252.2	-80.44	-158.5
0.75	-57.01	-318.4	-152	-272.4	-39.6	-247.4	-57.2	-277.7	-86.15	-173.4
0.725	-72.83	-370.4	-165.3	-299.4	-52.6	-298.8	-61.4	-305.3	-91.92	-189.3
0.7	-90.89	-428.7	-180.4	-330.7	-67.1	-357.1	-66.35	-334.9	-97.5	-206.4
0.675	-111.4	-494	-197.3	-366.5	-83	-423.3	-72	-367.6	-103.1	-224.9
0.65	-134.8	-567.2	-216.3	-407.4	-100.5	-499.2	-78.5	-404	-109	-245.4
0.625	-162.1	-651	-238.4	-455.6	-120.2	-587	-86.2	-444.7	-114.8	-268.1
0.6	-194.1	-747.2	-263.9	-512.1	-142.5	-689.1	-95.2	-491.2	-120.9	-294
0.575	-232.2	-858.7	-293.8	-579.2	-168	-809.2	-105.9	-545	-127.4	-323.7
0.55	-278.5	-989.9	-329.6	-660.5	-197.9	-951.9	-118.7	-608.2	-134.5	-358.6
0.525	-335.6	-1146	-373.2	-760.5	-233.3	-1124	-134.1	-684.4	-142.5	-400.7
0.5	-408.2	-1337	-427.8	-885.7	-276.9	-1334	-152.9	-777.8	-151.9	-452.5
0.475	-503.2	-1573	-498.6	-1047	-332.4	-1595	-176.4	-896.1	-163.6	-518.8
0.45	-633	-1877	-594.8	-1260	-405.2	-1932	-206.4	-1053	-180.2	-608.1
0.425	-822	-2288	-735.6	-1557	-509.4	-2383	-246.9	-1272	-206.6	-736.7
0.4	-1125	-2890	-965.9	-2001	-674.1	-3041	-306	-1610	-257.4	-942.5

This table refers to the following representation of X as a Γ-equivariant vector-field (cf. Golubitsky and Langford [1988]):

$$X = (c^1 + i2\xi_4 c^2)\begin{pmatrix}\zeta_1 \\ 0 \\ 0\end{pmatrix} + (c^3 + i2\xi_4 c^4)\begin{pmatrix}\bar{\zeta}_0\bar{\zeta}_1\bar{\zeta}_2 \\ 0 \\ 0\end{pmatrix}$$

$$(5.2) \qquad + (p^1 + iq^1)\begin{pmatrix}0 \\ \zeta_1 \\ \zeta_2\end{pmatrix} + 2\xi_4(p^2 + iq^2)\begin{pmatrix}0 \\ \zeta_1 \\ -\zeta_2\end{pmatrix}$$

$$+ (p^3 + iq^3)\begin{pmatrix}0 \\ \zeta_0^2\zeta_2 \\ \bar{\zeta}_0^2\zeta_1\end{pmatrix} + 2\xi_4(p^4 + iq^4)\begin{pmatrix}0 \\ \zeta_0^2\zeta_2 \\ -\bar{\zeta}_0^2\zeta_1\end{pmatrix}$$

Here c^j, p^j, q^j $(j = 1, 2, 3, 4)$ denote functions of R, Ω and the Γ-invariants $\xi_1, \xi_2, \xi_3^2, \xi_3\xi_4$ and $\xi_5 = Re(\bar{\zeta}_0^2\zeta_1\bar{\zeta}_2)$. A subscript 0 denotes the value of these functions for $R = R_c, \Omega = \Omega_c$ and all the other arguments equal to zero. A subscript 1 or 2 denotes the corresponding value of the partial derivative with respect to ξ_1 and ξ_2, respectively.

The correspondence between the values of the coefficients in (4.7) at $\lambda = \sigma = 0$ and the entries of the table is as follows:

$$(5.3) \qquad \begin{aligned} b_1 &= c_2^1 - c_0^3, & b_2 &= c_1^1 \\ a_1 &= p_1^1 - p_0^3, & a_2 &= p_2^1 \end{aligned}$$

Therefore, the following inequalities are satisfied:

$$(5.4) \qquad \begin{aligned} &b_1, b_2, b_2 - a_1 < 0 \\ &b_1 - a_2 \lessgtr 0 \quad \text{for} \quad \eta_c \lessgtr 0.49 \\ &d = a_2 b_2 - a_1 b_1 > 0 \end{aligned}$$

We now analyse (4.7) under these conditions. Neglecting the higher order terms and choosing the values of the coefficients at $\lambda = \sigma = 0$, by simple algebraic computations one obtains the following approximations. The curve given by $\lambda = 0$ in the (λ, σ)-plane is a curve of primary bifurcations of the trivial solution $\xi_1 = \xi_2 = 0$. This corresponds to the classical primary instability of the Couette flow which has already been studied by Taylor [1923]. Early rigorous treatments of this instability are in Iudovich [1965], Velte [1966], and Kirchgässner and Sorger [1969]. The bifurcating Taylor vortex flow is represented by the equilibria

$$(5.5) \qquad \xi_1 = \pm\sqrt{-\lambda/b_2}, \quad \xi_2 = 0 \qquad (\lambda > 0, \quad \beta \text{ arbitrary})$$

of the truncated system (4.7). Furthermore, one finds the curve of secondary bifurcations $\lambda = \sigma b_2/(b_2 - a_1)$, along which another family of equilibria given by

$$(5.6) \qquad \xi_1 = \frac{(b_1 - a_2)\lambda - b_1\sigma}{d}, \quad \xi_2 = \frac{(a_1 - b_2)\lambda + b_2\sigma}{d}$$

$$\left(\lambda > 0, \rho_2\lambda > \sigma > \rho_1\lambda, \quad \text{where} \quad \rho_1 = \frac{b_1 - a_2}{b_1}, \quad \rho_2 = \frac{b_2 - a_1}{b_2}\right)$$

branches off from the previous one. This corresponds to a curve of secondary instabilities in the Couette-Taylor problem where the Taylor vortex flow looses stability to wavy vortex flow. This happens actually through a Hopf bifurcation (see Davey, DiPrima and Stewart [1968], Chossat and Iooss [1985], Golubitsky and Stewart [1986] and Golubitsky and Langford [1988]). We point out, that the numerical value of ρ_1 in (5.6) is positive for $0.4 \leq \eta_c \leq 0.475$ and negative for $0.5 \leq \eta_c \leq 0.85$ according to our computations. Also note, that the stability of the approximate Couette and Görtler-Taylor vortex flows inside $Fix(\mathbb{Z}_2)$ is consistent with their actual stability. However, to really prove stability by the present method, one has to take into account the full orbit space $M_{|\tilde{\Gamma}}$. Equation (4.3) is not adequate to do this in case of the Couette and Görtler-Taylor vortex flows. Since those are $SO(2)$-symmetric, their $\tilde{\Gamma}$-orbits in M are not two-tori. Menck [1992] uses an extended system to overcome this difficulty.

To analyse the stability of the wavy vortex flow, we can use equation (4.3). Inside $Fix(\mathbb{Z}_2)$ the corresponding equilibria (5.6) are asymptotically stable for all values of λ and σ for which they exist. According to Menck [1991], their stability in $M_{|\tilde{\Gamma}}$ is therefore determined by the eigenvalues of the 2×2 matrix

$$(5.7) \qquad B = \begin{pmatrix} -2p_0^2\xi_2 + p_0^3\xi_1 & -q_0^3\xi_1 \\ q_0^3\xi_1 + 2(q_0^2 - c_0^2)\xi_2 & c_0^3\xi_2 + p_0^3\xi_1 \end{pmatrix}.$$

In points of $Fix(\mathbb{Z}_2)$ the matrix of the linearization of (4.3) block-diagonalizes in a (ξ_1,ξ_2)- and a (ξ_3,ξ_4)-block. The matrix B is equivalent to the latter. If we evaluate B along the branch of equilibria (5.6), then trace $B(\lambda,\sigma)$ becomes a linear function of σ and det $B(\lambda,\sigma)$ becomes a quadratic function of σ. According to our computations,

$$(5.8)$$
$$\text{trace} \quad B(\lambda,\sigma) = 2p_0^3\xi_1 < 0, \quad \det B(\lambda,\sigma) = \xi_1[(p_0^3)^2 + (q_0^3)^2] > 0 \quad \text{for} \quad \sigma = \lambda\rho_2$$
$$\text{trace} \quad B(\lambda,\sigma) = \xi_2(c_0^3 - 2p_0^2) < 0, \quad \det B(\lambda,\sigma) = \xi_2^2 p_0^2 c_0^3 < 0 \quad \text{for} \quad \sigma = \lambda\rho_1.$$

Therefore, trace $B(\lambda,\sigma) < 0$ for all λ and σ as in (5.6) and the function det $B(\lambda,\cdot)$ has a unique simple zero $\sigma = \sigma_0(\lambda) \in (\lambda\rho_1, \lambda\rho_2)$ for all $\lambda > 0$,

$$(5.9) \qquad \det B(\lambda,\rho_0(\lambda)) = 0.$$

This implies that $B(\lambda,\sigma)$ has a simple zero eigenvalue and a negative real eigenvalue along the curve $\sigma = \sigma_0(\lambda)$ $(\lambda > 0)$ in the (λ,σ)-plane. Hence, along such a curve the equilibria of (4.3) corresponding to (5.6) loose their stability through a bifurcation of still another family of equilibria when either σ is decreased or λ is increased. Because of the $\mathbb{Z}_2$-symmetry, this is actually a pitchfork bifurcation. Note, that the eigenvector belonging to the zero eigenvalue is antisymmetric with respect to this symmetry. This follows from the block structure of the corresponding matrix. Consequently, the bifurcating equilibria are not $\mathbb{Z}_2$-symmetric.

We also mention, that $\sigma_0(\lambda)$ turns out to be positive for $0.4 \leq \eta_c \leq 0.55$ and negative for $0.6 \leq \eta_c \leq 0.85$.

A more careful analysis shows that this tertiary bifurcation is subcritical, and the bifurcating equilibria are unstable near the bifurcation point. It is an interesting open question whether the bifurcating solution branch turns to the right and attains stability somewhere away from the bifurcation point. Correspondingly, in the Couette-Taylor problem there is a curve of tertiary instabilities starting at the origin in the (λ, σ)-plane. There the wavy vortex flow looses stability through a bifurcation of <u>quasiperiodic drift solutions</u> (QD). These do not have any obvious spatial symmetry and, therefore, from a generic point of view they fill the $\tilde{\Gamma}$-orbits, i.e. 2-tori, corresponding to the bifurcating equilibria of (4.3) densely. Actually, they are quasiperiodic rotating waves with two frequencies. In a still photograph the corresponding fluid flow almost looks like the wavy vortex flow. But the flow cells do not have any spatial symmetry. As time increases, they do not only rotate in the azimuthal direction, but also slowly drift in the axial direction of the cylinders.

We summarize our results in the following theorem.

Theorem. There is a parameter regime in the Couette-Taylor problem (2.3), in particular $0.4 \leq \eta \leq 0.55, \alpha \sim 3.6$, where the following sequence of successive bifurcations occurs when the Reynolds number R is increased quasistatically:

$$COU \longrightarrow GTV \longrightarrow WV \longrightarrow QD.$$

At the primary and secondary bifurcations asymptotic stability (in the sense of Liapunov) is exchanged to the bifurcating solutions. The tertiary bifurcation is a subcritical pitchfork bifurcation, through which the wavy vortex flow looses stability. Here the bifurcating quasiperiodic drift flow is unstable close to the bifurcation point.

Remark. The quasiperiodic drift flow is not to be confused with the modulated wavy vortex flow which usually is observed after the tertiary instability in the standard Couette-Taylor experiment, where the outer cylinder is held fixed. That flow is also quasiperiodic, but still has a spatial $\mathbb{Z}_2$-symmetry. In the parameter regime, which we have studied, there appear to be no such solutions. As far as the identification of the quasiperiodic solutions as drift states and a theoretical and numerical computation of the curve of bifurcation, the direction of bifurcation and stability of that drift state is concerned, our theorem provides a supplement to the general bifurcation picture for this parameter regime developed by Golubitsky and Stewart [1986] and Golubitsky and Langford [1988] (see also Chossat and Iooss [1992]). For an analysis of bifurcations to drift states in related contexts, cf. Chossat and Golubistky [1988] and Golubitsky, Krupa and Lin [1991].

References

P. Chossat and G. Iooss [1985], Primary and secondary bifurcations in the Couette-Taylor problem, Jap. J. Appl. Math.2, 37–68.

P. Chossat and G. Iooss [1992], The Couette-Taylor problem, Monograph to appear.

P. Chossat and M. Golubitsky [1988], Iterates of maps with symmetry, SIAM J. Math. Anal. 19 (6), 1259–1270.

A. Devay, R. C. DiPrima and Y. T. Stewart [1968], On the instability of Taylor vortices, J. Fluid Mech. 31, 17–52.

M. Golubitsky, M. Krupa and Ch. C. Lin [1991], Time-reversibility and particle sedimentation, SIAM J. Appl. Math. 51 (1), 49–72.

M. Golubitsky and W. F. Langford [1988], Pattern formation and bistability in flow between counterrotating cylinders, Physica D32, 362–392.

M. Golubitsky and I. Stewart [1986], Symmetry and stability in Taylor-Couette flows, SIAM J. Math. Anal. 17 (2), 249–288.

D. Henry [1981], Geometrical theory of semilinear parabolic equations, Springer Lecture notes in Math. 840, New York.

G. Iooss [1971], Théorie non linéaire de la stabilité des écoulements laminaires dans le cas de l'echange des stabilités, Arch. Rat. Mech. Anal. 40(3), 166–208.

V. I. Iudovich [1965], On the stability of steady flow of a viscous incompressible fluid, Dokl. Akad. Nauk. SSSR 161(5), 1037–1040.

K. Kirchgässner and H.-J. Kielhöfer [1973], Stability and bifurcation in fluid dynamics, Rocky Mountain J. Math. 3(2), 275–318.

K. Kirchgässner and P. Sorger [1969], Branching analysis for the Taylor problem, Quart. J. Mech. Appl. Math. 22, 183–209.

O. A. Ladyshenskaya [1963], The mathematical theory of viscous incompressible fluid flow, Gordon and Breach, New York.

W. F. Langford, R. Tagg, E. Kostelich, H. L. Swinney and M. Golubitsky [1988], Primary instabilities and bicriticality in flow between counterrotating cylinders, Phys. Fluids 31, 776–785.

P. Laure and Y. Demay [1988], Symbolic computations and equation on the center manifold: application to the Couette-Taylor problem, Computers and Fluids 16(3), 229–238.

J. Menck [1991], A tertiary Hopf bifurcation with applications to problems with symmetry, to appear in Dynamics and Stability of Systems 7 (1992).

J. Menck [1992], Analyse nicht-hyperbolischer Gleichgewichtspunkte in dynamischen Systemen unter Ausnutzung von Symmetrien, mit Anwendung von Computeralgebra, Doctorial thesis, Universität Hamburg.

D. H. Sattinger [1969/70], The mathematical problem of hydrodynamic stability, J. Math. Mech. 19, 154–166.

G. I. Taylor [1923], Stability of a viscous liquid contained between two rotating cylinders, Phil. Trans. Roy. Soc. (London) A 223, 289–343.

R. Témam [1977], Navier-Stokes equations, North-Holland, Amsterdam.

W. Velte [1966], Stabilität und Verzweigung stationärer Lösungen der Navier-Stokesschen Gleichungen beim Taylorproblem, Arch. Rat. Mech. Anal. 22, 1–14.

Numerical applications of equivariant reduction techniques

VLADIMÍR JANOVSKÝ AND PETR PLECHÁČ

Abstract. A new numerical method for detection and continuation of symmetry breaking bifurcation points is proposed. It avoids the construction of a symmetry adapted basis of the state space and the relevant block diagonalisation of the Jacobian. The method is related to the well–known techniques of augmented (bordered) Jacobians. The idea is to make a symmetry adapted choice of the bordering matrices which shoud induce a given representation of the symmetry group.

1. INTRODUCTION

Let $F : \mathbf{R}^N \times \mathbf{R}^p \to \mathbf{R}^N$ be a smooth mapping. We consider the equation $F(u, \beta) = 0$ for the state variable $u \in \mathbf{R}^N$ which depends on parameters $\beta \in \mathbf{R}^p$.

Let $(u^*, \beta^*) \in \mathbf{R}^N \times \mathbf{R}^p$ be a singular point of F with $corank = m$, i.e., $F(u^*, \beta^*) = 0$ and $dim\, Ker\, F_u(u^*, \beta^*) = m \geq 1$. The classical Liapunov–Schmidt reduction (see e.g. [4]) is an algorithm which reduces the problem $F(u, \beta) = 0$ in a neighbourhood of the given (u^*, β^*) to a set of m nonlinear equations $g(x, y) = 0$, where $g : \mathbf{R}^m \times \mathbf{R}^p \to \mathbf{R}^m$. Singular points can be classified by a finite set of conditions (equations and inequalities) imposed upon partial derivatives of the relevant g at the origin $0 \in \mathbf{R}^m \times \mathbf{R}^p$.

If the singular point is not known *a priori* and is the object of a numerical computation then one needs to derive an *extended system* for singular points of the particular class. The aim is to define each class of singular points as regular roots of a new mapping, which consists of F augmented by further nonlinear conditions. In fact, one has to reformulate the defining conditions for g as direct conditions upon partial derivatives of F at the singular point. There are technical difficulties due to the fact that the classical reduction projects the solution set into the kernel of F_u at (u^*, β^*) which is not known: one has to introduce auxiliary variables to span this kernel, etc.

Generalised Liapunov–Schmidt reduction techniques, see e.g. [10], [1], [11] and [6] can be performed at *any* point $(u, \beta) \in \mathbf{R}^N \times \mathbf{R}^p$. They make it possible to formulate extended systems for specified singular points from a classification list in a systematic and consistent way without auxiliary variables (*minimally extended* defining equations). The generalised reductions can also compete with the classical reduction as far as the local aposteriori analysis of $F(u, \beta) = 0$ is concerned: Let us mention applications to the local computer aided analysis of F, see [8], [7], [9].

We recall the reduction following essentially [6]. The value of m ($= corank$ of the singular point under question) is a part of the input data. We make a choice of two full–rank matrices $M \in \mathcal{L}(\mathbf{R}^m, \mathbf{R}^N)$ and $L \in \mathcal{L}(\mathbf{R}^N, \mathbf{R}^m)$. Given $(u, \beta) \in \mathbf{R}^N \times \mathbf{R}^p$, we define $g : \mathbf{R}^m \times \mathbf{R}^p \to \mathbf{R}^m$ and $v : \mathbf{R}^m \times \mathbf{R}^p \to \mathbf{R}^N$ requiring $g = g(x, y) \in \mathbf{R}^m$ and $v = v(x, y) \in \mathbf{R}^N$ to satisfy

$$
\begin{aligned}
F(u + v, \beta + y) + M\, g &= F(u, \beta) \\
L\, v &= x
\end{aligned}
$$
(1.1)

for $(x, y) \in \mathbf{R}^m \times \mathbf{R}^p$. Obviously, $g(0,0) = 0$ and $v(0,0) = 0$. Assuming that the augmented (bordered) Jacobian

$$(1.2) \qquad \mathcal{J}(u,\beta) \equiv \begin{pmatrix} F_u(u,\beta) & M \\ L & 0 \end{pmatrix} \in \mathcal{L}(\mathbf{R}^{N+m}, \mathbf{R}^{N+m})$$

of F at (u,β) is *regular*, both g and v are well–defined as germs of smooth mappings centred at the origin. Note that if the necessary condition $dim\, Ker\, F_u(u,\beta) \leq m$ is satisfied then the above augmented Jacobian $\mathcal{J}(u,\beta)$ is regular *generically* (i.e., assuming a generic choice of L and M). In what follows, we consider L and M to be independent of (u,β); we can assume this at least locally. Then, both g and v depend smoothly on (u,β) as a parameter i.e., $g = g(x,y;u,\beta)$ and $v = v(x,y;u,\beta)$.

The reduction techniques proposed in [11], [10] or in [1] can be shown equivalent to (1.1).

Just to give an example of an extended system, let us consider *pitchfork bifurcation point* as a bifurcation singularity with *codim* $= 2$, see e.g. [4]. To this end, we distinguish one component of β (say, the first one) as a controle parameter λ. Thus, $\beta = (\lambda, \alpha) \in \mathbf{R}^p = \mathbf{R}^1 \times \mathbf{R}^k$. Consequently, see (1.1), $y = (t, z) \in \mathbf{R}^1 \times \mathbf{R}^k$. Setting $k = 2$ $(= codim)$ and $m = 1$ $(= corank)$, and choosing $L \in \mathcal{L}(\mathbf{R}^N, \mathbf{R}^1)$ and $M \in \mathcal{L}(\mathbf{R}^1, \mathbf{R}^N)$ generically, we may define $(u, \lambda, \alpha) \in \mathbf{R}^N \times \mathbf{R}^{1+k}$ to be pitchfork bifurcation point iff (u, λ, α) is a *regular root* of the system

$$(1.3) \qquad \begin{aligned} F(u,\lambda,\alpha) = 0 \quad&, \quad g_x(x,t,z;u,\lambda,\alpha) = g_{xx}(x,t,z;u,\lambda,\alpha) = \\ g_t(x,t,z;u,\lambda,\alpha) = 0 \quad&, \quad (x,t,z) = 0 \in \mathbf{R}^4 \end{aligned}$$

of $N + 3$ equations for $(u, \lambda, \alpha) \in \mathbf{R}^{N+3}$. Differentials of g w.r.t. x, t, z and u, λ, α (the latter are needed for a numerical treatment of the extended system, e.g. for a Newton–like method) can be calculated by the chain rule from (1.1).

The aim of this paper is to adapt the mentined techniques to *equivariant mappings*. Namely, we shall assume that there exists a compact Lie group Γ acting on $\mathbf{R}^N$ such that

$$(1.4) \qquad F(\gamma u, \beta) = \gamma\, F(u,\beta)$$

for each $(u,\beta) \in \mathbf{R}^N \times \mathbf{R}^p$ and each $\gamma \in \Gamma$.

Using a symmetry adapted choice of L and M, we show that $g : \mathbf{R}^m \times \mathbf{R}^p \to \mathbf{R}^m$ constructed via (1.1) will be also equivariant w.r.t. a representation of Γ on $\mathbf{R}^m$. It makes it possible to define symmetry breaking bifurcation points in a similar way as in the example (1.3). We can take any steady–state bifurcation case classified in terms of the (classical) Liapunov–Schmidt reduction of F, see e.g. [5], and write down immediatelly the relevant defining equations on $\mathbf{R}^N \times \mathbf{R}^p$.

We shall give examples how to apply this technique to

- continuation of a branch with a given symmetry,
- detection of a symmetry breaking bifurcation point on the branch,
- branch-switching at a bifurcation point,
- computation of bifurcation points with nonlinear degeneracies and
- detection of mode-interaction points.

We also try to compare the presented approach with the methods using the symmetry adapted basis of the whole state space $\mathbf{R}^N$ as it was suggested in [2]; see also [13], [3].

2. The construction of an equivariant reduction

We assume F to be Γ–equivariant in the sense of (1.4). We also assume w.l.o.g. that $\Gamma \subset \mathbf{O}(N)$. Let $(u,\beta) \in \mathbf{R}^N \times \mathbf{R}^p$ satisfy $dim\,Ker\,F_u(u,\beta) \leq m$. We consider the reduction (1.1) at the chosen point (u,β).

DEFINITION 2.1. *We say that $M \in \mathcal{L}(\mathbf{R}^m,\mathbf{R}^N)$ and $L \in \mathcal{L}(\mathbf{R}^N,\mathbf{R}^m)$ are symmetry adapted bordering matrices provided that*
(i) the images $Im\,M$ and $Im\,L^t$ spanned by the columns of M and the columns of the transposed L respectively, are m-dimensional invariant subspaces w.r.t. the action of Γ on $\mathbf{R}^N$
(ii) Γ acts on $Im\,M$ in the same way as it acts on $Im\,L^t$ precisely, given $\gamma \in \Gamma$ there exists the unique $\sigma \in \mathbf{O}(m)$ such that

$$(2.1) \qquad \gamma\,M = M\,\sigma \quad , \quad \gamma\,L^t = L^t\,\sigma \quad , \quad (i.e.,\ L\gamma = \sigma\,L).$$

We say that the representation $\vartheta \in Hom\,(\Gamma,\mathbf{O}(m))$ defined by (2.1) is induced by the matrices M, L. The image of ϑ is denoted by Γ_0; the group Γ_0 acts on $\mathbf{R}^m$.

Let us note that algoritms constructing (irreducible) Γ–invariant subspaces in a prescribed isotypic component are available, see e.g. [12, p 104].

The following statement is just a simple observation:

THEOREM 2.2. *If M and L are symmetry adapted bordering matrices and if the relevant augmented Jacobian $\mathcal{J}(u,\beta)$, see(1.2), is regular then the operators g and v constructed via (1.1) satisfy*

$$(2.2) \qquad \sigma g(x,y;u,\beta) = g(\sigma x,y;\gamma u,\beta) \quad , \quad \gamma v(x,y;u,\beta) = v(\sigma x,y;\gamma u,\beta)$$

for $(x,y) \in \mathbf{R}^m \times \mathbf{R}^p$ and $\gamma \in \Gamma$, $\sigma = \vartheta(\gamma)$.

Let $Fix\,\Gamma = \{u \in \mathbf{R}^N : \gamma u = u$ for each $\gamma \in \Gamma\}$ denote the fixed–point space of Γ.

COROLLARY 2.3. *If $u \in Fix\,\Gamma$ then $g : \mathbf{R}^m \times \mathbf{R}^p \to \mathbf{R}^m$ is Γ_0–equivariant.*

COROLLARY 2.4. *Let $\{e_m^i\}_{i=1}^m$ denote the canonical basis of $\mathbf{R}^m$. If $u \in Fix\,\Gamma$ then span $\{v_x(0,0;u,\beta)e_m^i\}_{i=1}^m$ is Γ–invariant and Γ acts on this space in the same way as it acts on $Im\,M$ i.e., $\gamma\,v_x(0,0;u,\beta) = v_x(0,0;u,\beta)\sigma$ for $\sigma = \vartheta(\gamma)$, $\gamma \in \Gamma$.*

COROLLARY 2.5. *If $u \in Fix\,\Gamma$ and the action of Γ on $Im\,M$ is absolutely irreducible then the differential $g_x(0,0;u,\beta) \in \mathcal{L}(\mathbf{R}^m,\mathbf{R}^m)$ is a real multiple of identity.*

Given a pair of symmetry adapted bordering matrices M and L, we define the set

$$\mathcal{D} = \{(u,\beta) \in \mathbf{R}^N \times \mathbf{R}^p : det\,\mathcal{J}(u,\beta) \neq 0\}$$

Assuming $\mathcal{D}$ nonempty, we may claim $\mathcal{D}$ to be open. Suppose that $(u,\beta) \in \mathcal{D}$ solves

$$(2.3) \qquad F(u,\beta) = 0 \quad , \quad g_x(0,0;u,\beta) = 0$$

under the linear constraint $u \in Fix\,\Gamma$.

We conclude (without using a symmetry) that (u, β) is a singular point with $corank = m$ namely, $Ker\, F_u(u, \beta) = span\, \{v_x(0, 0; u, \beta)\, e_m^i\}_{i=1}^m$. Let us make two important remarks:

a) The choice of M and L *imposes* (up to an equivalence) the representation ϑ of Γ on $Ker\, F_u(u, \beta)$.

b) $g_x(0, 0; u, \beta) = 0$ yields substantially less then m^2 different conditions upon (u, β) (e.g., if ϑ is absolutely irreducible then it gives just one condition, if ϑ is a sum of two nonconjugate absolutely irreducible representations then $g_x = 0$ yields precisely two scalar conditions upon (u, β), etc.).

It motivates the way, the conditions (2.3) may serve as defining equations for symmetry breaking bifurcation points. We also show in the next Section that the awkward constraint $u \in Fix\, \Gamma$ may be replaced by m conditions $Lu = 0$ when solving (2.3).

3. Pathfollowing symmetric solutions

We start with a local description of the manifold of symmetric solutions to $F = 0$. Let M and L be symmetry adapted bordering matrices. Let a point $(u^0, \beta^0) \in \mathcal{D}$ satisfy $u_0 \in Fix\, \Gamma$. Note that the case $F(u^0, \beta^0) = 0$ appeals to our aim. By virtue of (1.1) and Theorem 2.2 ,

THEOREM 3.1. *It holds in the obvious local sense:*
$F(u, \beta) = F(u^0, \beta^0)$, $u \in Fix\, \Gamma$ iff $u = u^0 + v(x, y; u^0, \beta^0)$, $\beta = \beta^0 + y$, where (x, y) satisfies $g(x, y; u^0, \beta^0) = 0$, $x \in Fix\, \Gamma_0$.

In other words, the condition $u \in Fix\, \Gamma$ can be replaced by $Lu \in Fix\, \Gamma_0$ due to the assumption $F(u, \beta) = F(u^0, \beta^0)$. Linearising this assumption, we expect the same statement:

COROLLARY 3.2. *A vector* $(\delta u, \delta \beta) \in \mathbf{R}^N \times \mathbf{R}^p$ *satisfies* $F_u(u^0, \beta^0)\, \delta u + F_\beta(u^0, \beta^0)\, \delta \beta = 0$, $\delta u \in Fix\, \Gamma$ *iff* $\delta u = v_x(0, 0; u^0, \beta^0)\, \delta x + v_y(0, 0; u^0, \beta^0)\, \delta y$, $\delta \beta = \delta y$, *where* $(\delta x, \delta y)$ *solves* $g_x(0, 0; u^0, \beta^0)\, \delta x + g_y(0, 0; u^0, \beta^0)\, \delta y = 0 \in \mathbf{R}^m$, $\delta x \in Fix\, \Gamma_0$.

Note that Corollary 3.2 can be applied directly to a numerical continuation of a symmetric solution branch.

Remark 3.3. Solving the equation $F(u, \beta) = 0$ on $Fix\, \Gamma \times \mathbf{R}^p$ via *damped Newton iterations*, we may also relax the constraint upon u: If $F_u(u^0, \beta^0)\, \delta u + F_\beta(u^0, \beta^0)\, \delta \beta = -\omega F(u^0, \beta^0)$, ω is a parameter, then $\delta u \in Fix\, \Gamma$ iff $L\, \delta u = \delta x \in \Gamma_0$. The formulae for $(\delta u, \delta \beta)$ are omited here. The above Newton iteration step is formally equivalent to the linearisation of an auxiliary equation $(1 - \omega)\, F(u, \beta) - F(u^0, \beta^0) = 0$ for $(u, \beta, \omega) \in \mathbf{R}^N \times \mathbf{R}^p \times \mathbf{R}^1$ at $(u^0, \beta^0, 0)$ and Corollary 3.2 can be applied.

Manifolds of bifurcation points can be treated similarly. For example, let (u^0, β^0) satisfy (2.3) in addition. Obviously, the tangent vectors $(\delta u, \delta \beta) \in Fix\, \Gamma \times \mathbf{R}^p$ to the manifold (2.3) at (u^0, β^0) are given by Corollary 3.2 where the conditions upon $(\delta x, \delta y)$ are to be appended by $g_{xx}(0, 0; u^0, \beta^0)\, \delta x + g_{xy}(0, 0; u^0, \beta^0)\, \delta y = 0 \in \mathbf{R}^{m^2}$. Most of the additional $\mathbf{R}^{m^2}$ conditions are redundant due to the symmetry. For example, if ϑ is absolutely irreducible then they yield just one independent condition, etc.

4. Numerical example

In order to illustrate numerical applications of the equivivariant reduction from Section 2, we consider the 6–box Brusselator model for the steady state $u \in \mathbf{R}^{12}$ of two reactants in six coupled cells: Setting $u = (w^{(1)}, ..., w^{(6)})$, $w^{(i)} \in \mathbf{R}^2$, we require $D\left(w^{(j+1)} - 2w^{(j)} + w^{(j-1)}\right) + \lambda f(w^{(j)}) = 0$ for $j = 1, ..., 6$ and $w^{(0)} = w^{(7)}$. D is a 2×2 diagonal diffusion matrix taken as $D = diag(1, 10)$; $\lambda \in \mathbf{R}^1$ is a control parameter. The function $f : \mathbf{R}^2 \to \mathbf{R}^1$, $f(r, s) = \begin{pmatrix} A - (B+1)r + r^2 s \\ Br - r^2 s \end{pmatrix}$ models a nonlinear reaction of substances r and s. It depends on two constants A and B which are considered as an imperfection $\alpha \in \mathbf{R}^2$. Thus, $\beta = (\lambda, \alpha) \in \mathbf{R}^{1+2}$ in this model.

The problem has $\mathbf{D}_6$–symmetry, see [2]. The flip κ and the rotation ρ which generate $\mathbf{D}_6$ act on $\mathbf{R}^{12}$ as permutations namely, $\kappa u = (w^{(1)}, w^{(6)}, w^{(5)}, w^{(4)}, w^{(3)}, w^{(2)})$, $\rho u = (w^{(6)}, w^{(1)}, w^{(2)}, w^{(3)}, w^{(4)}, w^{(5)})$. This particular representation ϑ of $\mathbf{D}_6$ can be decomposed by means of irreducible representations (irreps) of $\mathbf{D}_6$ as follows: $\vartheta = 2\vartheta_1^{(1)} + 2\vartheta_3^{(1)} + 2\vartheta_1^{(2)} + 2\vartheta_2^{(2)}$ where $\vartheta_1^{(1)} : \kappa \to 1$, $\rho \to 1$, $\vartheta_3^{(1)} : \kappa \to 1$, $\rho \to -1$ are two (out of four) irreps of $\mathbf{D}_6$ on $\mathbf{R}^1$ and

$$\vartheta_p^{(2)} : \kappa \to \begin{pmatrix} 1 & 0 \\ 0 & -1 \end{pmatrix}, \rho \to \begin{pmatrix} cos(2p\pi/6) & -sin(2p\pi/6) \\ sin(2p\pi/6) & cos(2p\pi/6) \end{pmatrix} \quad, \quad p = 1, 2.$$

are two irreps of $\mathbf{D}_6$ on $\mathbf{R}^2$.

Let us comment on the construction of symmetry adapted bordering matrices M and L. Obviously, it suffices to construct bordering matrices $M_p^{(l)}$ and $L_p^{(l)}$ which induce each particular irrep $\vartheta_p^{(l)}$ from the above list. Since $\mathbf{D}_6$ already acts as permutations, the construction is easy (see e.g. [12, p 56]): Setting $L_3^{(1)} = (-c_1, c_2, c_1, -c_2, -c_1, c_2, c_1, -c_2, -c_1, c_2, c_1, -c_2)$ for an arbitrary choice of constants c_1, c_2 and setting $M_3^{(1)} = (L_3^{(1)})^t$ (for another arbitrary choice of c_1, c_2), we obtain bordering matrices inducing $\vartheta_3^{(1)}$. The formulae for $M_1^{(1)}$ and $L_1^{(1)}$ are similar. Given constants c_1 and c_2, we define

$$a_i^{(p)} = \begin{pmatrix} c_1 cos(2p(i-1)\pi/6) & c_2 cos(2p(i-1)\pi/6) \\ c_1 sin(2p(i-1)\pi/6) & c_2 sin(2p(i-1)\pi/6) \end{pmatrix} \in \mathcal{L}(\mathbf{R}^2, \mathbf{R}^2)$$

for $i = 1, ..., 6$. Let us set $L_p^{(2)} = (a_1^{(p)}, ..., a_6^{(p)}) \in \mathcal{L}(\mathbf{R}^N, \mathbf{R}^2)$ and, taking a different pair of constants c_1 and c_2 , $M_p^{(2)} = (a_1^{(p)}, ..., a_6^{(p)})^t \in \mathcal{L}(\mathbf{R}^2, \mathbf{R}^N)$. It can be readily verified that $L_p^{(2)}$ and $M_p^{(2)}$ induce the representation $\vartheta_p^{(2)}$.

5. Detection of singular points

Refering to the above example, let α be fixed. In the particular numerical experiment reported on Figs 1–2, there is taken $A = 1$ and $B = 2$ (the same data as in [3]).

We consider a numerical continuation of the symmetric solution branch $\mathcal{S}_0 = \{(u, \lambda) \in Fix\mathbf{D}_6 \times \mathbf{R}^1 : F((u, \lambda, \alpha)) = 0\}$ in a neighbourhood of a symmetry breaking bifurcation point (u^*, λ^*). Let us discuss the numerical procedure based on Corollary 3.2 . The problem is a propper choice of the bordering matrices which would guarantee the regularity of $\mathcal{J}(u^*, \lambda^*)$. Due to the generic scenario of a symmetry breaking bifurcation (see [5]), the representation of $\mathbf{D}_6$ on $Ker F_u(u^*, \lambda^*)$ is absolutely irreducible. In our case, this representation is equivalent either to $\vartheta_1^{(3)}$ or $\vartheta_2^{(1)}$ or $\vartheta_2^{(2)}$.

If we know the classification (i.e. the particular irrep $\vartheta_p^{(l)}$) then we simply choose $M_p^{(l)}$ and $L_p^{(l)}$ to be the bordering matrices inducing the irrep. The resulting $\mathcal{J}(u^*, \lambda^*)$ will be regular generically (i.e., for a generic choice ot the constants c_1 and c_2) and the continuation will work. Moreover, the diagonal entry (see Corollary 2.5) of $g_x(0, 0; u, \lambda)$ gives a scalar *test function* indicating (by the change of its sign) the particular bifurcation point along the solution branch. Direct techniques can be also applied namely, Newton–like methods solving the extended system (2.3) will converge locally to (u^*, λ^*).

The described strategy breaks down when the information concerning the type of irrep is not a part of data. In this case it is possible to take $M = (M_1^{(1)}, M_3^{(1)}, M_1^{(2)}, M_2^{(2)})$ $\in \mathcal{L}(\mathbf{R}^6, \mathbf{R}^N)$ and $L \in \mathcal{L}(\mathbf{R}^N, \mathbf{R}^6)$ composed row–wise by $L_1^{(1)}$, $L_3^{(1)}$, $L_1^{(2)}$, $L_2^{(2)}$. The bordering matrices are symmetry adapted and induce the representation $\vartheta = \vartheta_1^{(1)} + \vartheta_3^{(1)} + \vartheta_1^{(2)} + \vartheta_2^{(2)}$ on $\mathbf{R}^6$. The corresponding bordered Jacobian $\mathcal{J}(u^*, \lambda^*)$ will be regular generically at any symmetry breaking bifurcation point. Given a point $(u, \lambda) \in \mathcal{S}_0$, the differential $g_x(0, 0; u, \lambda) \in \mathcal{L}(\mathbf{R}^6, \mathbf{R}^6)$ will be a diagonal matrix $diag(f_1, f_2, f_3, f_3, f_4, f_4)$ with four different entries. These entries f_i (considered as functions of (u, λ)) serve as test functions of singular points. Refering to the *bifurcation graph* of $\mathbf{D}_6$ for 6–box Brusselator in [2] where the *maximal isotropy subgroups* of $\mathbf{D}_6$ are listed, we claim that f_2 and f_4 detect respectively $\mathbf{D}_6 \to \mathbf{D}_3$ and $\mathbf{D}_6 \to \mathbf{Z}_2 \oplus \mathbf{Z}_2$ symmetry breaking bifurcation points. The test function f_3 indicates the simultaneous appearence of $\mathbf{D}_6 \to \mathbf{Z}_2(\kappa)$ and $\mathbf{D}_6 \to \mathbf{Z}_2(\rho\kappa)$ bifurcation. Finally, f_1 should detect a fold on the symmetric branch.

Remark 5.2. Let us relax one of the components of the unfolding parameter $\alpha = (A, B)$. The above described choice of $M \in \mathcal{L}(\mathbf{R}^6, \mathbf{R}^N)$ and $L \in \mathcal{L}(\mathbf{R}^N, \mathbf{R}^6)$ makes it possible to continue globally the solution curve to (2.3) using the technique sketched at the end of Section 3. One of the test functions f_i will be zero along the path. The sign changes of the remaining three test functions will indicate mode interaction points.

Let us come back to the problem of pathfolowing $\mathcal{S}_0$. We can preserve the robustness of the bordering matrices $M \in \mathcal{L}(\mathbf{R}^6, \mathbf{R}^N)$ and $L \in \mathcal{L}(\mathbf{R}^N, \mathbf{R}^6)$ while shrinking their size by a sort of lumping. Since dimension of irreps is bounded by two, it is necessary to require $M \in \mathcal{L}(\mathbf{R}^2, \mathbf{R}^N)$ and $L \in \mathcal{L}(\mathbf{R}^N, \mathbf{R}^2)$. For example, we define $M = M_1^{(2)} + M_2^{(2)} + (M_1^{(1)}, M_3^{(1)})$. In the same spirit, $L = L_1^{(2)} + L_2^{(2)} +$ the matrix composed row–wise from $L_1^{(1)}$ and $L_3^{(1)}$. Again, $\mathcal{J}(u^*, \lambda^*)$ will be regular generically. On the other hand, M and L are no longer symmetry adapted as far as $\mathbf{D}_6$ is concerned. Nevertheless, they are symmetry adapted w.r.t. the subgroup $\mathbf{Z}_2(\kappa)$ of $\mathbf{D}_6$. One can easily verify that $L\kappa = \begin{pmatrix} 1 & 0 \\ 0 & -1 \end{pmatrix} L$. As a consequence, the matrix $g_x(0, 0; u, \lambda)$ is a diagonal matrix $diag(t_1, t_2)$ at each point (u, λ) of the path $\mathcal{S}_0$. The values of t_1 and t_2 (when considered as functions of (u, λ)) may also serve as test functions. The information they yield is not that strict: $t_1 = 0$ and $t_2 \neq 0$ indicate a fold on $\mathcal{S}_0$, $t_1 \neq 0$ and $t_2 = 0$ correspond to $\mathbf{D}_6 \to \mathbf{D}_3$ symmetry breaking bifurcation and the case $t_1 = t_2 = 0$ detects either $\mathbf{D}_6 \to \mathbf{Z}_2 \oplus \mathbf{Z}_2$ bifurcation or the simultaneous $\mathbf{D}_6 \to \mathbf{Z}_2(\kappa)$ and $\mathbf{D}_6 \to \mathbf{Z}_2(\rho\kappa)$

symmetry breaking bifurcation.

The branch $\mathcal{S}_0$ is depicted on Fig 1 by the solid horizontal line as the projection into the (u_1, λ)-plain. Using the techniques explained above, six primary symmetry breaking bifurcation points P_i with $\lambda > 0$ were detected. The type of symmetry breaking at P_i is listed in the legend.

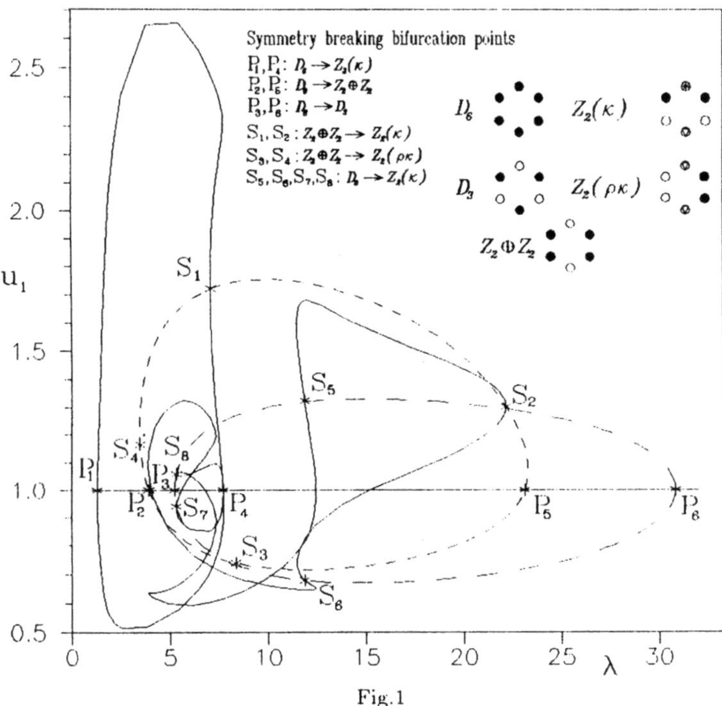

Fig.1

Suppose a bifurcation point (u^*, λ^*) is detected and a numerical pathfollowig of the bifurcating branches with the symmetry of the relevant maximal isotropy subgroup is required. We describe the branch–switching procedure on an example: Let us consider, say, the point P_5. The particular type of symmetry breaking at this point motivates to choose $M_2^{(2)}$ and $L_2^{(2)}$ as bordering matrices. Computing $v_x(0, 0; u^*, \lambda^*) \in \mathcal{L}(\mathbf{R}^2, \mathbf{R}^N)$ (which might be considered as a by–product of the pathfollowig $\mathcal{S}_0$ anyway), we claim that the vector τ

$$(5.1) \qquad \tau = (v_x(0, 0; u^*, \lambda^*)d, 1) \in \mathbf{R}^N \times \mathbf{R}^1 \quad , \quad d = \begin{pmatrix} 1 \\ 0 \end{pmatrix}$$

is tangent to the bifurcated branch with $\mathbf{Z}_2(\kappa) \oplus \mathbf{Z}_2(\rho^3 \kappa)$ symmetry. The reasoning is based on the fact that the direction $d \in \mathbf{R}^2$ in (5.1) spans the fixed point space of the maximal isotropy subgroup $Z_2\left(\begin{pmatrix} 1 & 0 \\ 0 & -1 \end{pmatrix}\right)$ of the 2-dim representation $\vartheta_2^{(2)}$. The tangents to the remaining conjugate branches can be found replacing d by its $2\pi/3$–rotations in $\mathbf{R}^2$. Higher–order predictors of the bifurcated branch (specified by the choice of the isotropy direction d) can be obtained at the cost of computation of higher differentials of v and g along d at $(0,0;u^*,\lambda^*)$.

The continuation of a bifurcated branch and the detection of (secondary) bifurcation points follows the same routine as the continuation of the primary branch S_0 with the highest symmetry $\mathbf{D}_6$. We stick to the above example again and we consider the continuation of the $\mathbf{Z}_2(\kappa) \oplus \mathbf{Z}_2(\rho^3\kappa)$–branch.

The group $\mathbf{Z}_2(\kappa) \oplus \mathbf{Z}_2(\rho^3\kappa) = \{id,\,\kappa,\,\rho^3,\,\rho^3\kappa\}$ has four 1–*dim* irreducible representations $\widetilde{\vartheta}_1^{(1)}$, $\widetilde{\vartheta}_2^{(1)}$, $\widetilde{\vartheta}_3^{(1)}$ and $\widetilde{\vartheta}_4^{(1)}$ (with multiplicities 4, 2, 4 and 2 respectively). In particular, $\widetilde{\vartheta}_1^{(1)} : \kappa \to 1,\, \rho^3 \to 1$, $\widetilde{\vartheta}_2^{(1)} : \kappa \to -1,\, \rho^3 \to 1$, $\widetilde{\vartheta}_3^{(1)} : \kappa \to 1,\, \rho^3 \to -1$, $\widetilde{\vartheta}_4^{(1)} : \kappa \to -1,\, \rho^3 \to -1$. The bordering matrices $M_p^{(l)}$, $L_p^{(l)}$ generating the irreps of the higher symmetry $\mathbf{D}_6$ can be naturally exploited to a construction of bordering matrices $\widetilde{M}_p^{(l)}$, $\widetilde{L}_p^{(l)}$ which induce the above irreps $\widetilde{\vartheta}_p^{(l)}$ of $\mathbf{Z}_2 \oplus \mathbf{Z}_2$. Obviously, the matrices $M_1^{(l)}$, $L_1^{(l)}$ and $M_3^{(1)}$, $L_3^{(1)}$ also induce $\widetilde{\vartheta}_1^{(1)}$ and $\widetilde{\vartheta}_3^{(1)}$. Each column of $M_p^{(2)}$ spans now an invariant subspace of $\mathbf{Z}_2 \oplus \mathbf{Z}_2$. As a consequence, we may set (as an alternative) $\widetilde{M}_1^{(1)} = M_2^{(2)}(1,0)^t$, $\widetilde{L}_1^{(1)} = (1,0)L_2^{(2)}$ and $\widetilde{M}_3^{(1)} = M_1^{(2)}(1,0)^t$, $\widetilde{L}_3^{(1)} = (1,0)L_1^{(2)}$. The remaining irreps are induced as follows: $\widetilde{M}_2^{(1)} = M_2^{(2)}(0,1)^t$, $\widetilde{L}_2^{(1)} = (0,1)L_2^{(2)}$ and $\widetilde{M}_4^{(1)} = M_1^{(2)}(0,1)^t$, $\widetilde{L}_4^{(1)} = (0,1)L_1^{(2)}$.

We refer to Fig 1 for the secondary branches emanating from the bifurcation points P_i which break the symmetry of $\mathbf{D}_6$. All these branches have (at least) $\mathbf{Z}_2(\kappa)$–symmetry. This was the choice eliminating all conjugate branches (with $\mathbf{Z}_2(\rho^2\kappa)$, etc. symmetries) and the branches with $\mathbf{Z}_2(\rho^3\kappa)$ (and cojugate) symmetries. The type of the symmetry breaking at all detected secondary bifurcation points S_i is listed in the legend. Let us note that the points P_1 and S_4 (and, similarly P_4 and S_3) are connected via a $\mathbf{Z}_2(\rho^3\kappa)$–branch. The $\mathbf{Z}_2(\kappa)$–branch connecting P_1 with S_1 does not intersect P_4 as Fig 1 might suggest (the confusion is due to the particular projection). The (rather messy) $\mathbf{Z}_2(\kappa)$–branch emanating from P_4 is also depicted in another projection on Fig 2. Here $\xi_1 = L_3^{(1)}u$, choosing $c_1 = -1, c_2 = 1$ respectively. The fact that this curve

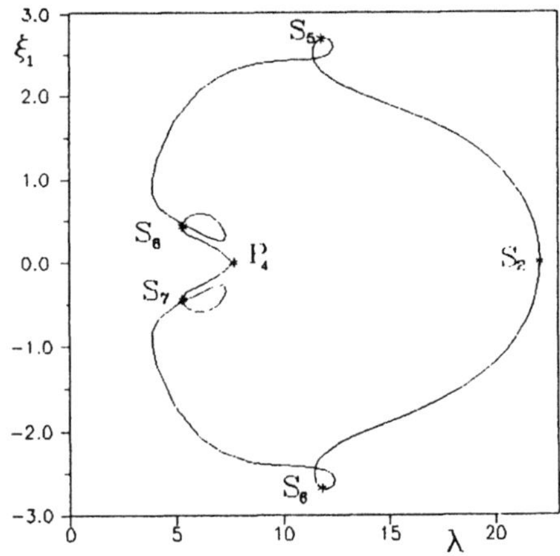

Fig.2

connects P_4 with the secondary branches emanating from P_3, P_6 and P_2, P_5 can be explained by a sequence of mode interactions. For example, relaxing the parameter B one can find (using the technique described in Remark 5.2) three mode interaction points close to $B = 2$, see the following

Table. Mode interaction on the $\mathbf{D}_6$–solution manifold, $A = 1.$ is fixed.

B	λ	f_2	f_3	f_4	$modes$
1.8905695	6.32455488	$1.e - 16$	$8.6e - 8$	$.18353$	$\mathbf{D}_3$ vs $\mathbf{Z}_2(\kappa), \mathbf{Z}_2(\rho\,\kappa)$
1.8302967	5.47722571	$-.10106$	$-2.1e - 8$	$1.e - 16$	$\mathbf{Z}_2 \oplus \mathbf{Z}_2$ vs $\mathbf{Z}_2(\kappa), \mathbf{Z}_2(\rho\,\kappa)$
1.7390097	10.9544512	$-1.e - 16$	$-.95445$	$1.e - 16$	$\mathbf{D}_3$ vs $\mathbf{Z}_2 \oplus \mathbf{Z}_2$

(For the meaning of f_i we refer to Remark 5.2 .) Trying to interprete the Table we may say that, decreasing $B \leq 2$, we reshuffle the primary bifurcation points three times: At first P_3 and P_4 coalesce and exchange positions, and the same happens later with the couples P_2, P_4 and P_3, P_5.

The already quoted *bifurcation graph* for the 6–box Brusselator [2] admites a $\mathbf{Z}_2^c$–branch bifurcating from the $\mathbf{Z}_2 \oplus \mathbf{Z}_2$–branch. Such a case was not detected for the set of parameters $A = 1$, $B = 2$ (Fig 1). Following [5, pp 218–223], the $\mathbf{Z}_2^c$–branch can be organized by a $\mathbf{D}_6 \to \mathbf{Z}_2 \oplus \mathbf{Z}_2$ bifurcation point with a nonlinear degeneracy. The defining equations for this point $(u^*, \lambda^*, \alpha^*)$ are as follows: Choosing $M_2^{(2)}$ and $L_2^{(2)}$ as bordering matrices (i.e., $\mathbf{D}_6$ will act as $\mathbf{D}_3$ on $Ker F_u(u^*, \lambda^*, \alpha^*)$), the point $(u^*, \lambda^*, \alpha^*)$ is a root of

$$F(u, \lambda, \alpha) = 0, \quad d^t g_x(0,0,0; u, \lambda, \alpha) d = 0, \quad (d^t g_x(0,0,0; u, \lambda, \alpha) d)_x d = 0$$

where $d = (1,0)^t$, see (5.1).

We fixed $A = 1$ and found $u^* = (1, 2.2, ...)^t \in \mathbf{R}^{12}$, $\lambda^* = 30$, $B^* = 2.2$ to be the organizing centre (degenerated P_5). Its asymptotic analysis yields that a secondary bifurcated branch which connects all three conjugate $\mathbf{Z}_2 \oplus \mathbf{Z}_2$–branches bifurcating from the organizing centre can be expected for $B > B^*$. Fig 3 is an example: It depicts a phase projection of a $\mathbf{Z}_2^c$–branch connecting three conjugate $\mathbf{Z}_2 \oplus \mathbf{Z}_2$–branches; the unfolding parameters are $A = 1$ and $B = 2.5$.

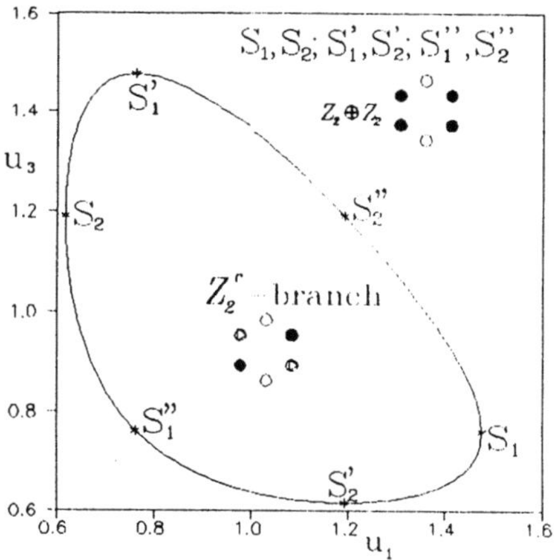

Fig.3

6. Conclusions and comparisons

The method by *Dellnitz* and *Werner*, see [2] and [3], requires the construction of a symmetry adapted basis (see e.g. [12, p 43]) and the relevant block–diagonalisation of the Jacobian. This complex but very expensive information concerning the action of the symmetry group on the whole state space $\mathbf{R}^N$ is not needed for the purpose of a classification and analysis of the symmetry breaking bifurcation. What really matters (see [5]) is the representation of the symmetry group on the kernel of the Jacobian at the bifurcation point. This representation is a part of classification data (together with nonlinear degeneracies of the mapping). Our numerical approach is based on this philosophy. The particular representation required on the kernel can be encoded in defining equations by a choice of bordering matrices M and L which induce (in the sense of Definition 2.1) the required representation. The existing "nonsymmetric" bifurcation packages can be simply converted to a symmetry breaking bifurcation analysis.

The unbeatable advantage of [2] is the reduction of numerical manipulations (e.g. the continuation of symmetric solutions) from the state space $\mathbf{R}^N$ to the substantionally less dimensional $Fix\,\Gamma$ that has to be constructed explicitly (compare with Section 3). The computation of particular blocks of the diagonalised Jacobian in [2] needs the evaluation of the full scaled F_u anyway (unless using symbolic differentiation, see [3]). Going down the isotropy lattice (=the bifurcation graph), one needs to change the basis for each isotropy subgroup. In the proposed method one changes only the bordering matrices and the working state–space is always the same $\mathbf{R}^N$.

Acknowledgement. We wish to express our thanks to Bodo Werner for his friendly encouragement and for sharing many ideas which stimulated this paper.

REFERENCES

[1]. Beyn W.J., *Defining equations for singular solutions and numerical applications*, In: Numerical Methods for Bifurcation Problems (eds. Küpper T., Mittelmann H., D., Weber H.), Birkhäuser Verlag, Basel 1984.

[2]. Dellnitz M., Werner B., *Computational methods for bifurcation problems with symmetries - with a special attention to steady state and Hopf bifurcation points*, J. Comp. Appl. Math. 26 (1989), 97-123.

[3]. Gaterman K., Hohmann A., *Symbolic exploatation of symmetry in numerical pathfollowing*, Preprint SC 90-11, Konrad–Zuse–Zentrum für Informationstechnik Berlin 1990.

[4]. Golubitsky M., Schaeffer D., "Singularities and Groups in Bifurcation Theory Vol. 1," Springer Verlag, New York, 1985.

[5]. Golubitsky M., Stewart I., Schaeffer D., "Singularities and Groups in Bifurcation Theory Vol. 2," Springer Verlag, New York, 1988.

[6]. Griewank A., Reddien G.,W., *Computation of cusp singularities for operator equations and their discretisation*, J. Comp. Appl. Math. 26 (1989), 133-153.

[7]. Janovsky V., Plechac P., *Asymptotic analysis of Takens–Bogdanov points*, J. Comp. Appl. Math. 36 (1991), 349-359.

[8]. Janovsky V., Plechac P., *Computer aided analysis of imperfect bifurcation diagrams I. Simple bifurcation point and isola formation centre.*, to appear in SIAM J.Numer.Math. Vol 29, 1992.

[9]. Janovsky V., Werner B., *Constructive analysis of Takens–Bogdanov points with Z_2-symmetry*, submitted to IMA J. Numer. Anal..

[10]. Jepson A.D., Spence A., *Singular points and their computation*, In: Numerical Methods for Bifurcation Problems (eds. Küpper T., Mittelmann H., D., Weber H.), Birkhäuser Verlag, Basel 1984.

[11]. ______________, *A reduction process for nonlinear equations*, SIAM J. Math. Anal. 20 (1989), 39-56.

[12]. Stiefel E., Fässler A., "Gruppentheoretische Methoden und ihre Anwendung," Teubner, Stuttgart, 1979.

[13]. Werner B., *Eigenvalue problems with the symmetry of a group and bifurcations*, In: Continuation and Bifurcations: Numerical Techniques and Applications (eds. Roose D., de Dier B., Spence A.), Kluver Academic Publishers, Dordrecht 1990.

Department of Numerical Analysis, Faculty of Mathematics and Physics, Charles University, Malostranské nám. 25, Prague, Czechoslovakia

International Series of Numerical Mathematics, Vol. 104, © 1992 Birkhäuser Verlag Basel

NUMERICAL BIFURCATION ANALYSIS OF A MODEL
OF COUPLED NEURAL OSCILLATORS

Alexander I. Khibnik †‡,   Roman M. Borisyuk †,   Dirk Roose ‡

†Research Computing Centre, USSR Academy of Sciences,
Pushchino Moscow Region, 142292 USSR

‡Dept. of Computer Science, Katholieke Universiteit Leuven,
B-3001 Leuven, Belgium

1. Introduction

Biological and artificial neural networks that consist of coupled neural oscillators receive much attention in neural network modelling nowadays. The construction of a network of this type requires the selection of a mathematical model for a single neural oscillator and the choice of a mechanism to couple the oscillators. Furthermore, one can consider neural oscillators that are all identical to each other or oscillators with different parameters (e.g., different intrinsic frequencies).

The present paper is devoted to the study of a rather simple and basic neural network model, which involves two identical Wilson-Cowan neural oscillators. Each oscillator is represented by a model of two differential equations, which describes the average activity of a population of neurons, assuming that it consists of an excitatory and an inhibitory subpopulations.

Due to the simple and fixed structure of this neural network, we can investigate how the architecture and the strength of the coupling between the oscillators influences the dynamical behavior. In this paper we present a systematic numerical bifurcation study of four typical schemes for the connections between the excitatory and inhibitory subpopulations of the oscillators. The bifurcation diagrams that we present for each case show the existence and the stability of different types of periodic solutions. These diagrams also allow to predict quasi-periodic and chaotic behavior, clearly demonstrated by numerical simulations. Such a behavior is rather typical for this neural network model and points to some interesting mathematical problems concerning symmetry-breaking and symmetry-increasing of complex oscillatory regimes (tori and strange attractors).

Note that many papers on neural networks are devoted to the study of phase-locking mode in the case of a weak coupling between the oscillators (see, e.g., [6]). Our results are related mainly to an intermediate strength of connections between neural oscillators. Besides typical phase-locking (i.e. when both oscillators are oscillating in-phase), we observe also a situation when difference between the phases is approximately zero most of the time. Such a more general type of phase-locking arises under certain conditions for quasi-periodical regime and becomes visible on a large time scale only (for other applications of quasi-periodical regime see also [2]).

The results that we present here are entirely numerical, produced with the help of *interactive* software for bifurcation analysis and for simulation of dynamical systems. All the bifurcation diagrams below are computed by using LOCBIF [11,12], a program for continua-

tion and bifurcation analysis, with special attention to multi-parameter problems and higher codimension singularities. It allows to study:
- equilibrium points and periodic solutions (limit cycles) of autonomous ODEs,
- periodic solutions of nonautonomous time-periodic ODEs,
- fixed points and periodic orbits of iterated maps.

For studying quasi-periodic and chaotic behavior we used TraX [10,14], a program for simulation of continuous and discrete-time dynamical systems. We refer the reader to [8,9,10,11,12,14] for more information concerning the software.

2. A neural network model based on neural oscillators

We consider a system of coupled neural oscillators. For a single oscillator we take the Wilson-Cowan system of two autonomous differential equations which describes the dynamics of activity of a neural net consisting of excitatory and inhibitory subpopulations [17]:

$$
\begin{aligned}
dE/dt &= -E + (k_e - E) \cdot S_e(c_1 E - c_2 I + P), \\
dI/dt &= -I + (k_i - I) \cdot S_i(c_3 E - c_4 I + Q).
\end{aligned}
\tag{1}
$$

Here $E(t)$ and $I(t)$ are the average activities of the excitatory and the inhibitory subpopulations, $S_e(x)$ and $S_i(x)$ are monotonically increasing sigmoid-type functions, k_e and k_i are constants, $k_e = S_e(+\infty)$, $k_i = S_i(+\infty)$, P and Q are the external inputs for the excitatory and inhibitory populations, and c_1, c_2, c_3, c_4 are the strengths of connections between the populations.

The nonlinearity of sigmoid type is given by

$$
S(x; b, q) = 1/(1 + \exp(-b(x - \theta))) - 1/(1 + \exp(b\theta)),
\tag{2}
$$

with $S_e(x) = S(x; b_e, \theta_e)$ and $S_i(x) = S(x; b_i, \theta_i)$, where $b_e, \theta_e, b_i, \theta_i$ are parameters. The values E and I may be negative which means that the activity of the network is lower than that of the background.

A two-parameter bifurcation analysis of model (1) with P and c_3 as control parameters, due to [3], shows that the model exhibits several bifurcations of codimension one: fold, Hopf, multiple limit cycle and homoclinic bifurcations (and also several bifurcations of codimension two). It is important to observe that there exists a parameter region, for which the model has an unstable equilibrium point surrounded by a stable limit cycle. We take the parameter values in this region, fixing them for the rest of the paper as: $q_e = 4, b_e = 1.3, q_i = 3.7, b_i = 2, c_1 = 16, c_2 = 12, c_3 = 15, c_4 = 3, P = 1.5$ and $Q = 0$.

Consider now two identical neural oscillators of the form (1), which are connected by means of the external input terms:

$$
\begin{aligned}
dE_1/dt &= -E_1 + (k_e - E_1) \cdot S_e(c_1 E_1 - c_2 I_1 + a_1 E_2 - a_2 I_2 + P), \\
dI_1/dt &= -I_1 + (k_i - I_1) \cdot S_i(c_3 E_1 - c_4 I_1 + a_3 E_2 - a_4 I_2 + Q), \\
dE_2/dt &= -E_2 + (k_e - E_2) \cdot S_e(c_1 E_2 - c_2 I_2 + a_1 E_1 - a_2 I_1 + P), \\
dI_2/dt &= -I_2 + (k_i - I_2) \cdot S_i(c_3 E_2 - c_4 I_2 + a_3 E_1 - a_4 I_1 + Q).
\end{aligned}
\tag{3}
$$

Here E_1, I_1 and E_2, I_2 describe the first and second oscillators respectively, and a_1, a_2, a_3, a_4 represent the strength of the connections between subpopulations related to different oscillators. Since the oscillators are identical and the coupling is symmetric, the model (3) has reflection symmetry where reflection means exchange of the first and second oscillator.

We use model (3) to investigate *separately* four types of connections assuming that only one of parameters $a_j, j = 1, 2, 3, 4$, has nonzero value. These cases are:

1. *Connections between the excitatory populations* ($\alpha = a_1$ *is the control parameter*).
2. *Connections from the inhibitory population of one oscillator to the excitatory population of the other oscillator* ($\alpha = a_2$ *is the control parameter*).
3. *Connections from the excitatory population of one oscillator to the inhibitory population of the other oscillator* ($\alpha = a_3$ *is the control parameter*).
4. *Connections between the inhibitory populations* ($\alpha = a_4$ *is the control parameter*).

We are interested in the analysis of possible oscillatory regimes in the model, their stability and bifurcations under variation of the strength of the connections a_j. In this context, the relationships between different types of periodic oscillations are the most interesting. According to the symmetry of the system (3), there exist three types of limit cycles in the phase space: *symmetric* (C_s), *antisymmetric* (C_a), *and nonsymmetric* (C_n). The symmetric limit cycle is such that $E_1(t) \equiv E_2(t)$, $I_1(t) \equiv I_2(t)$; it corresponds to *in-phase* oscillations. The antisymmetric limit cycle is characterized by a phase shift of half a period between the oscillators: $E_1(t + T/2) \equiv E_2(t)$, $I_1(t + T/2) \equiv I_2(t)$, where T is a period. We call this *anti-phase* oscillations based on the observations of the corresponding time series. A limit cycle which is neither symmetric nor antisymmetric is called nonsymmetric.

If all $a_j = 0$, we can consider the system (3) to be a product of two oscillatory systems, each of them having an unstable equilibrium point and a stable limit cycle, O_1, C_1 and O_2, C_2 respectively. Therefore, the following invariant sets exist, being a product of these regimes:
- an absolutely unstable equilibrium point $O_s = O_1 \times O_2$;
- two saddle nonsymmetric limit cycles $C_{n,1} = C_1 \times O_2$ and $C_{n,2} = C_2 \times O_1$;
- a stable invariant two-torus $T_0 = C_1 \times C_2$; it consists of a continuum of periodic orbits, and therefore it is structurally unstable; note that two of these orbits are C_s and C_a, and all the others are nonsymmetric (i.e. of C_n-type).

If one or more of the connection parameters are perturbed, all these invariant sets persist and remain of the same stability type, but typically only a finite (and even) number of periodic orbits "survive" on the torus T_0. The symmetric and antisymmetric periodic orbits C_s and C_a always persist under perturbations due to their internal symmetry. One of these orbits is stable, while the remaining is unstable (of saddle type).

Remark 1. The model (3) is not similar to a discrete variant of a reaction-diffusion system. Indeed, a coupling of neural oscillators is not of "diffusion-type", but involves essentially the nonlinear function (2) characterizing partial oscillators. This implies that the variation of the strength of coupling may cause the appearance or disappearance of symmetric equilibrium points and limit cycles.

Remark 2. Let L denote an invariant symmetry plane of system (3), $L : E_1 = E_2, I_1 = I_2$.

It can be proved that in the particular case $a_1 = a_4$ (e.g., $a_1 = a_4 = 0$) the symmetric equilibrium point $O_s \subset L$ and the symmetric limit cycle $C_s \subset L$ satisfy the following "degeneracy" conditions:

$$\lambda_1 + \lambda_2 = \lambda_3 + \lambda_4 \tag{4}$$

$$\mu_1 \cdot \mu_2 = \mu_3 \cdot \mu_4 \tag{5}$$

Here $\lambda_1, \lambda_2, \lambda_3, \lambda_4$ are eigenvalues of O_s, $\mu_1 = 1, \mu_2, \mu_3, \mu_4$ are multipliers of C_s, with λ_1, λ_2 and μ_1, μ_2 corresponding to the symmetry plane L. These degeneracy conditions imply that the codimension two singularities, characterized e.g. by $(\lambda_1, \lambda_2, \lambda_3, \lambda_4) = (\pm i\omega_1, \pm i\omega_2)$ and $(\mu_2, \mu_3, \mu_4) = (1, \exp(\pm i\omega))$, may arise generically in a one-parameter study of system (3). Furthermore, if the equilibrium point O_s loses stability via Hopf bifurcation, then necessarily one has two pairs of purely imaginary eigenvalues at the critical parameter value, instead of one pair as in the case of "ordinary" Hopf bifurcation. This leads, under certain generic conditions, to the simultaneous appearance of two stable limit cycles, symmetric C_s and antisymmetric C_a. Indeed, such a situation occurs in our model (see below). Roughly speaking, if the connections between excitatory populations of both oscillators have the same strength as those between the inhibitory populations, then typically the small-amplitude phase and antiphase oscillatory regimes appear (or disappear) together.

Remark 3. In the sequel we will use the terminology "symmetric", "antisymmetric" and "nonsymmetric" also with respect to arbitrary invariant sets. This can be defined as follows. Let M be an invariant set of system (3). Assume M is invariant under reflection, then it is called symmetric if $M \subset L$ (L is the symmetry plane), and antisymmetric in the other case. If M is not invariant under reflection, we call it nonsymmetric.

3. Numerical bifurcation analysis

In schematic solution diagrams (Fig.1,4,6,7), we present the results of the one-parameter continuation of the periodic solutions of system (3) when $\alpha = a_j$ varies for the *cases 1-4* described above. For the most complex case (Fig.4), we show in Fig.5 a diagram of periodic solutions by plotting an integral norm of the solution versus the strength of connections α on actual scale. We now discuss each of the cases separately. Note that the presentation of this section refers to some hypotheses concerning bifurcations of invariant tori and strange attractors, which are discussed in more details in Section 4.

Case 1. Connections between the excitatory populations (Fig.1). For small values of $\alpha = a_1$, the antisymmetric limit cycle C_a is the only stable regime. It loses stability via a torus bifurcation at $\alpha = 0.25$ and disappears by merging with the equilibrium point O_s at the Hopf bifurcation point for $\alpha = 0.50$. The symmetric cycle C_s is of saddle type when α is small, and becomes stable via a symmetry-breaking bifurcation, occurring at $\alpha = 1.72$, when two saddle nonsymmetric cycles branch off from it. The cycle C_s disappears at $\alpha = 5.34$ on a homoclinic trajectory of a saddle symmetric equilibrium point.

When α increases, the saddle nonsymmetric limit cycles $C_{n,1}$ and $C_{n,2}$, which have one multiplier inside and two outside the unit circle at $\alpha = 0$, undergo at $\alpha = 1.02$ a (backward) torus (Neimark-Sacker) bifurcation and become stable. They disappear at a fold point for $\alpha = 1.76$, by coalescing with the saddle nonsymmetric limit cycles mentioned above.

For parameter values between the two torus bifurcations the system exhibits quasi-periodic behavior, suggesting that a two-dimensional torus exists in the phase space. Notice that the torus bifurcation of C_a leads to the appearance of an antisymmetric torus T_a, while the torus bifurcations of $C_{n,1}$ and $C_{n,2}$ give rise to two nonsymmetric tori $T_{n,1}$ and $T_{n,2}$, which are symmetric to each other. Thus, two different types of quasi-periodic behavior become possible in the system (see, e.g., Fig.3 in [2]).

When the parameter α increases, one can expect a phenomenon of *torus symmetry-breaking*, i.e. a transition from an antisymmetric torus to two nonsymmetric tori. Indeed, this phenomenon is revealed between $\alpha = 0.691$ and $\alpha = 0.692$. It appears to be associated with a homoclinic bifurcation of system (3), at which a homoclinic trajectory of the saddle symmetric limit cycle C_s arises (because of symmetry we have two such trajectories simultaneously). The phenomenon of torus symmetry-breaking is illustrated in Fig.2, which also shows the rearrangement of the invariant closed curve of the Poincare map. The origin is the saddle fixed point of the map that corresponds to the limit cycle C_s. Evidently, this rearrangement suggests that two (due to symmetry) homoclinic trajectories of the fixed point arise (see also the discussion in Section 4).

It should be emphasized that near the torus symmetry-breaking point both oscillators are oscillating most of the time in-phase (Fig.3), since a trajectory traces out very close to the symmetric periodic orbit. But because this orbit is unstable, eventually a phase shift appears. The final picture demonstrates the alternation of (approximately) in-phase and out-of-phase oscillations, with apparent domination in time of the in-phase oscillations.

We conclude our discussion of the "excitatory to excitatory" connections case by presenting how a stable asymptotical regime evolves, as α increases from zero: 1) antisymmetric limit cycle (anti-phase periodic oscillations) $\rightarrow$ 2) antisymmetric torus $\rightarrow$ 3) nonsymmetric tori $\rightarrow$ 4) nonsymmetric limit cycles $\rightarrow$ 5) symmetric limit cycle (in-phase periodic oscillations) $\rightarrow$ 6) symmetric equilibrium point. Note that the symmetric limit cycle becomes stable even earlier than the nonsymmetric stable limit cycles disappear. Therefore for some parameter values three stable periodic regimes exist simultaneously, and a transition from nonsymmetric to in-phase periodic regime appears to be sharp.

Case 2: Connections from the inhibitory population of one oscillator to the excitatory population of the other oscillator (Fig.4,5). This case exhibits a much more complicated bifurcation diagram of periodic solutions as in the previous case. This complexity is linked with the fact that a chaotic dynamical behavior becomes possible and is observed for a large range of parameter values. The diagram (Fig.4) presents four "principal" branches of limit cycles. Their existence follows from a perturbation analysis of the trivial case $\alpha = 0$. These branches C_s, C_a, $C_{n,1}$ and $C_{n,2}$ have received their names according to the original limit cycles they contain. Additionally, one can see on the diagram at least five more branches, which do not exist for small α; some of them exhibit "wiggles", similar to those occurring near homoclinic or heteroclinic bifurcations involving a saddle-focus equilibrium point [16]. Note that, for example, the Lorenz system presents the same kind of wiggles [15].

As α increases, the symmetric limit cycle C_s, being originally stable, changes stability three times, respectively at $\alpha = 1.16$, $\alpha = 5.46$, and $\alpha = 5.98$. The first and second cases

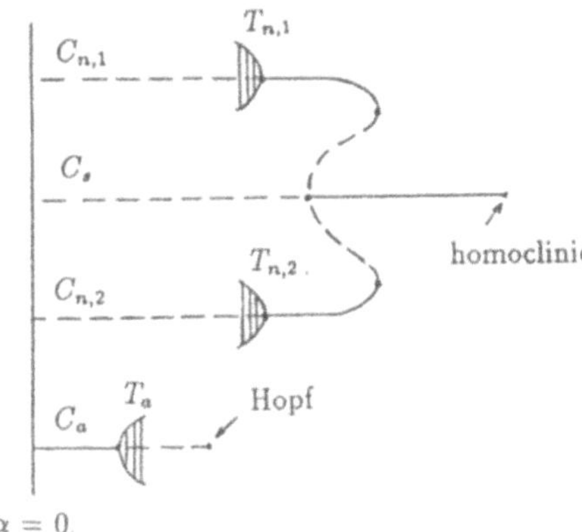

Figure 1: Schematic diagram of periodic solutions for *case 1* - mutual connections between the excitatory populations of two neural oscillators. Solid lines shows stable solutions, dashed lines - unstable ones. The shadowed parabolic region represents the appearance of a stable two-torus. C_s - symmetric, C_a - antisymmetric, $C_{n,1}, C_{n,2}$ - nonsymmetric limit cycle, T_a - antisymmetric, $T_{n,1}, T_{n,2}$ - nonsymmetric two-torus.

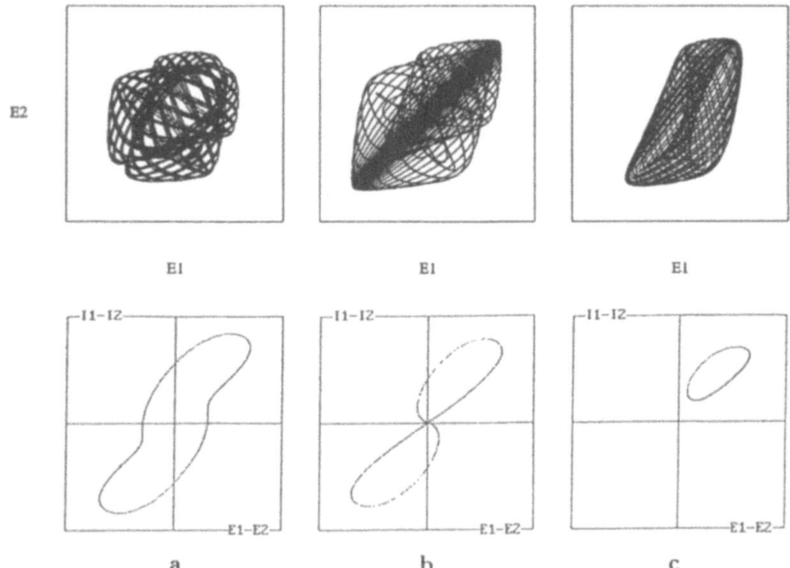

Figure 2: Antisymmetric (*a,b*) and nonsymmetric (*c*) two-tori and their Poincaré sections. *a:* $\alpha = 0.5$, *b:* $\alpha = 0.691$, *c:* $\alpha = 0.85$. *Secant plane:* $E_1 + E_2 = 0.44$.

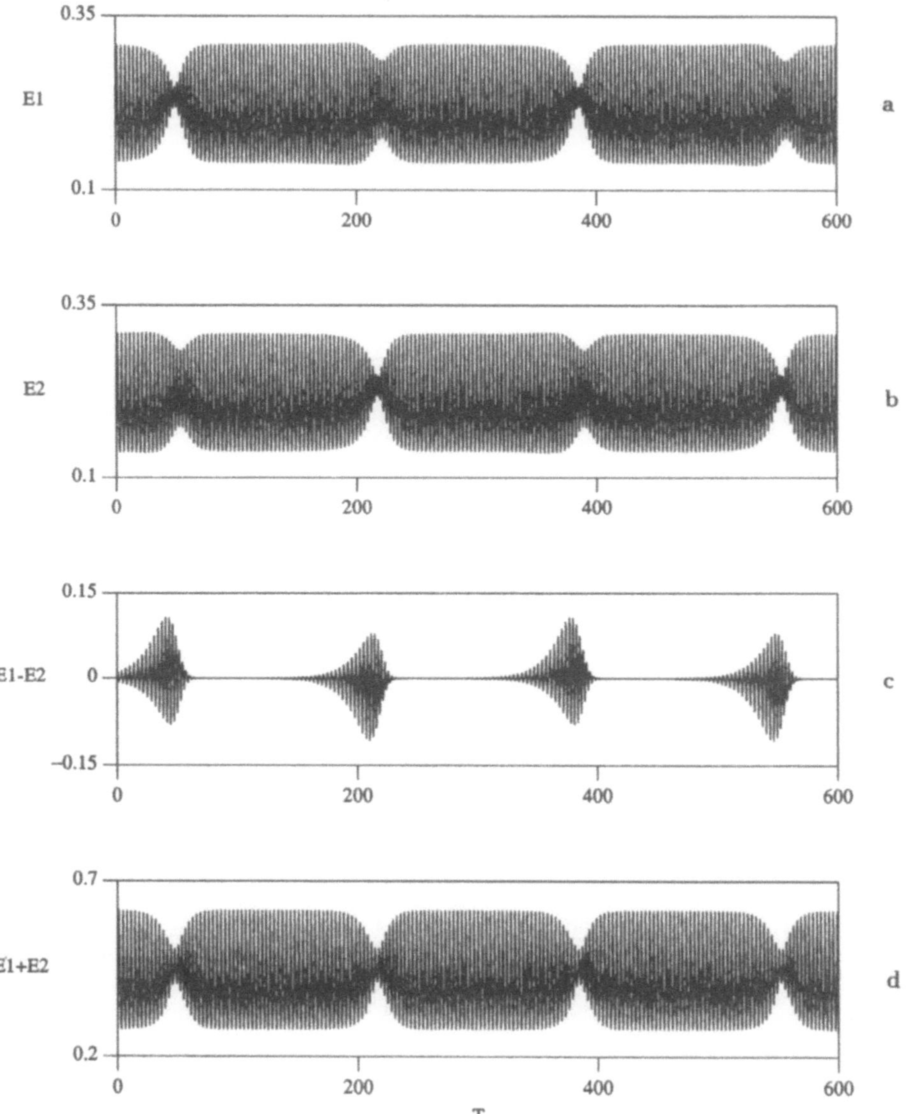

Figure 3: Quasi-periodic regime near the torus symmetry-breaking ($\alpha = 0.691$). The difference between the activities (c) shows clearly that both oscillators are oscillating most of the time in-phase. The sum of the activities (d) demonstrates the halving of the period of modulation, which is typical for an antisymmetric torus.

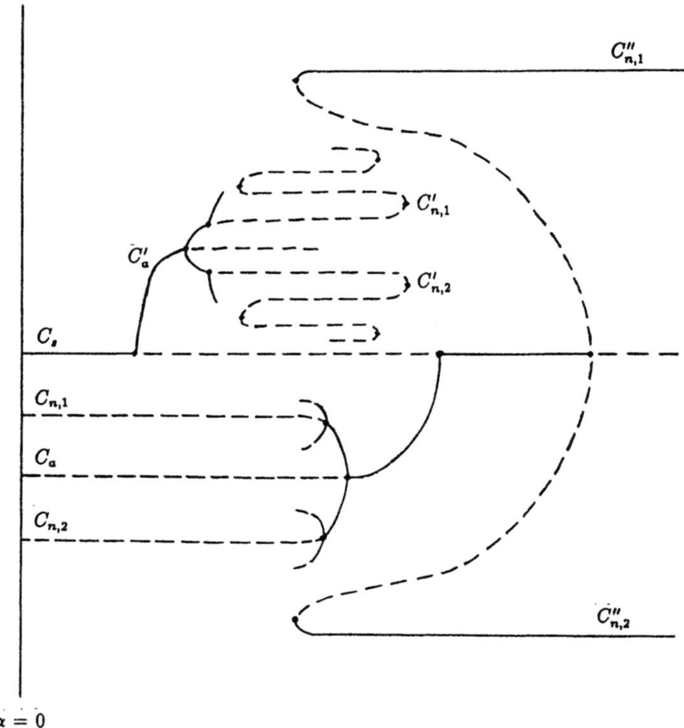

Figure 4: Schematic diagram of periodic solutions for *case 2* - connections from the inhibitory to the excitatory population. Solid line shows stable solutions, dashed line - unstable ones.

correspond to a period doubling bifurcation point of symmetry-breaking type, giving rise to a stable antisymmetric limit cycle. The point $\alpha = 5.46$ is the end point of the branch C_a, and at $\alpha = 1.16$ a "new" branch C'_a emanates from C_s. Although it is not shown on Fig.4, the branch C'_a when being continued further, exhibits the same wiggly behavior as that of the branches of nonsymmetric limit cycles (see Fig.4,5 and the discussion below).

Both branches C_a and C'_a contain a pitchfork bifurcation point. This corresponds to a loss of stability and symmetry-breaking of the antisymmetric limit cycle C_a (C'_a), as α decreases (increases). A pair of nonsymmetric stable limit cycles then arises. The bifurcation point $\alpha = 5.03$ turns out to be the end point of the branches $C_{n,1}$ and $C_{n,2}$, while at the bifurcation point $\alpha = 1.90$ "new" branches $C'_{n,1}$ and $C'_{n,2}$ emanate from C'_a.

The nonsymmetric limit cycles $C'_{n,1}$ and $C'_{n,2}$ become unstable at $\alpha = 1.94$ via a period doubling bifurcation. A cascade of period doubling bifurcations for the nonsymmetric limit cycles then follows, and, eventually, chaotic oscillations appear, which probably can be associated with two different strange attractors. Both attractors are nonsymmetric by itself,

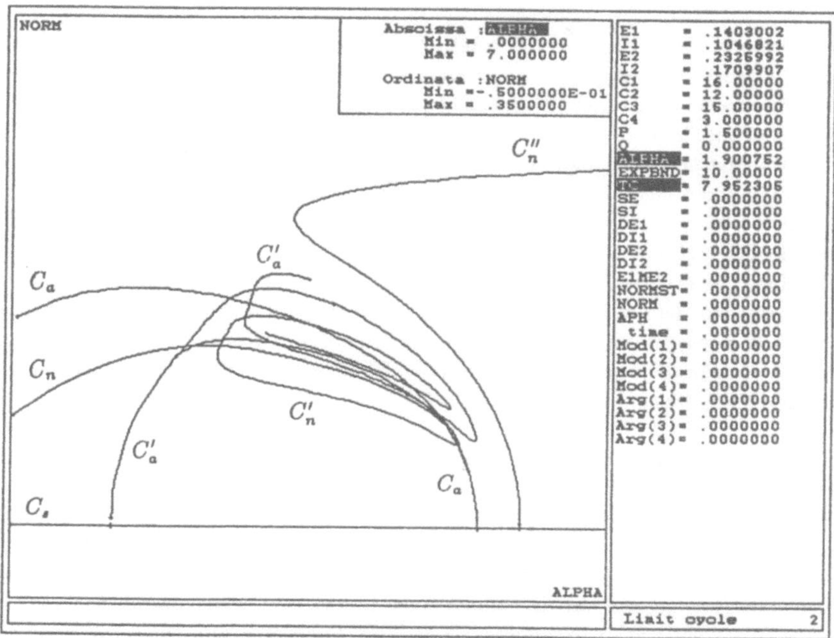

Figure 5: LOCBIF screen with the computed diagram of periodic solutions corresponding to the schematic diagram of Fig.4. The integral norm of a solution vs α is plotted, where for the norm we use an average difference between the oscillators: $norm^2 = \frac{1}{T}\int_0^T ((E_1 - E_2)^2 + (I_1 - I_2)^2)\,dt$.

being symmetric to each other under reflection. (Note that as α decreases, the nonsymmetric limit cycle $C_{n,1}$ $(C_{n,2})$ also becomes unstable via a period doubling bifurcation at $\alpha = 4.99$.)

Let us keep track of the nonsymmetric strange attractors, when α increases further. One can observe that just after their arising the two nonsymmetric attractors "glue", giving rise to an antisymmetric strange attractor. This phenomenon may be called a "symmetry-increasing of the strange attractor" (for motivation of such a terminology see the papers [4,5]). The antisymmetric strange attractor dies at $\alpha = 3.27$. This critical parameter value corresponds to the fold bifurcation point on the branches $C_{n,1}''$ and $C_{n,2}''$, i.e. two branches of nonsymmetric limit cycles which emanate from the branch of symmetric limit cycles C_s at a pitchfork symmetry-breaking point for $\alpha = 5.98$. The nonsymmetric cycles, being initially unstable (near the symmetry-breaking point), become stable when the fold bifurcation point is passed. In some sense, the antisymmetric strange attractor, when dying, transfers its stability to the nonsymmetric limit cycles, which appear at the fold point. For smaller values of α but near such a critical value, an intermittency phenomenon is present in the system.

The scenario of disappearance of the antisymmetric strange attractor that we observed numerically suggests that at a critical parameter value, two heteroclinic trajectories exist that are symmetric to each other. Each one of them connects two nonhyperbolic limit cycles $C''_{n,1}$ and $C''_{n,2}$ of saddle-node type.

We remark that the continuation of the branches of the nonsymmetric limit cycles $C'_{n,1}$ and $C'_{n,2}$ reveals their wiggly behavior (Fig.4,5). The subsequent fold bifurcation points are located at $\alpha = 5.25$, $\alpha = 2.38$, $\alpha = 5.15$. The period T of the limit cycle is increasing along the branches, although not monotonically.

We conclude the discussion of this case by presenting the sequence of stable regimes and transitions between them when α increases: 1) symmetric limit cycle $\to$ 2) antisymmetric limit cycle $\to$ 3) nonsymmetric limit cycles which undergo a series of period doubling bifurcations $\to$ 4) nonsymmetric strange attractors $\to$ 5) antisymmetric strange attractor $\to$ 6) nonsymmetric limit cycles.

Case 3: Connections from the excitatory population of one oscillator to the inhibitory population of the other oscillator (Fig.6). In this case we have only the basic periodic solution branches which correspond to the limit cycles C_s, C_a, $C_{n,1}$ and $C_{n,2}$. The symmetric limit cycle C_s is stable for small α and remains stable until it shrinks and merges with the equilibrium O_s at $\alpha = 2.49$. This critical parameter value corresponds to a double Hopf bifurcation point (two pairs of pure imaginary eigenvalues), from which both branches C_s and C_a emanate. Near the critical parameter value both symmetric and antisymmetric cycles are stable, but C_a becomes unstable, as α decreases, via a pitchfork symmetry-breaking bifurcation at $\alpha = 1.67$. This gives rise to two stable nonsymmetric limit cycles $C_{n,1}$ and $C_{n,2}$, which, in turn, become unstable at $\alpha = 1.59$ via a torus bifurcation, giving rise to two stable nonsymmetric tori $T_{n,1}$ and $T_{n,2}$.

These tori die just after their appearance, close to $\alpha = 1.58$, giving rise to one stable antisymmetric torus T_a . This symmetry-increasing torus bifurcation probably is associated with homoclinicity of the antisymmetric limit cycle. The antisymmetric torus disappears quite suddenly, as α decreases, at $\alpha = 1.38$.

In this case of "excitatory to inhibitory" coupling the evolution of stable asymptotical regimes, as α increases, is rather simple: 1) symmetric limit cycle $\to$ 2) symmetric equilibrium point $\to$ 3) nonsymmetric equilibrium point. The last transition is due to symmetry-breaking of the equilibrium point O_s at $\alpha = 7.43$.

Note that the local bifurcation theory [7] predicts that if two stable limit cycles arise near a double Hopf bifurcation point, simultaneously a small unstable torus of saddle type exists. Indeed, in our numerical experiments we observed long-transient behavior having the form of small-amplitude quasiperiodic oscillations. This made such a saddle torus "visible". According to the numerical results, the torus seems to be of antisymmetric type.

Case 4: Connections between the inhibitory populations (Fig.7). This case is the simplest one. Its specific feature is that the antisymmetric limit cycle C_a is always stable. Therefore, no transitions between asymptotic regimes may be expected.

All other regimes are unstable, but nevertheless they undergo bifurcations when the parameter α varies. There are four bifurcation points: pitchfork symmetry-breaking bifurcation

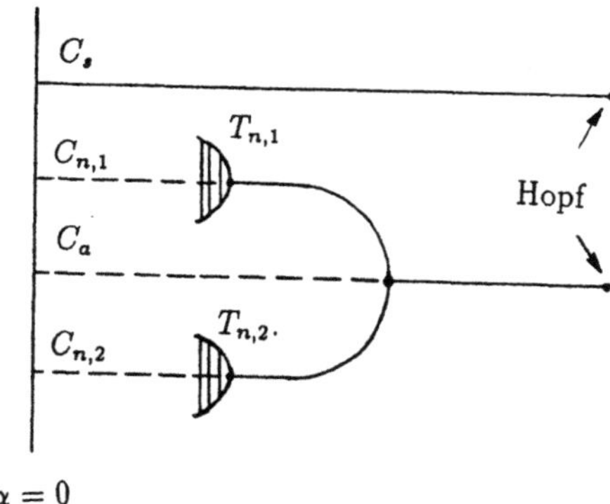

Figure 6: Schematic diagram of periodic solutions for *case 3* - connections from the excitatory to the inhibitory population. Solid line shows stable solutions, dashed line - unstable ones. Shadowed parabolic region shows appearance of a stable two-torus.

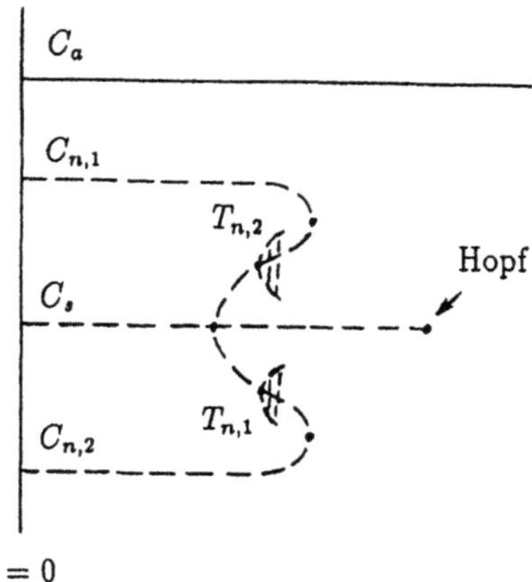

Figure 7: Schematic diagram of periodic solutions for *case 4* - mutual connections between the inhibitory populations. Solid line shows stable solutions, dashed line - unstable ones. Parabolic region shows appearance of an unstable two-torus.

of C_s at $\alpha = 0.49$, torus bifurcation of (unstable) nonsymmetric cycle C_n at $\alpha = 0.54$, fold bifurcation of C_a at $\alpha = 0.55$, and shrinking of C_s and collision with the equilibrium O_s at the Hopf bifurcation point at $\alpha = 0.61$. Notice that in this case neither torus nor fold bifurcations lead to the appearance of a stable torus or a stable limit cycle respectively (compare with the previous cases).

4. Discussion of global bifurcation phenomena

The numerical results described above give a rather complete picture with respect to the bifurcation analysis of (basic) periodic solutions for the mathematical model under study. However, these results point to some open mathematical and numerical questions concerning symmetry-breaking and symmetry-increasing of invariant tori and strange attractors. Note that our study of these phenomena in model (3) has a preliminary character, being solely based on a numerical simulation approach.

In *case 1* we observe a transition from one antisymmetric torus to two nonsymmetric tori, when the parameter α varies. We call this phenomenon a torus symmetry-breaking since the antisymmetric torus can be "traced" up to a certain parameter value, after which it disappears, but instead two nonsymmetric tori appear "at the same place". We also know that the torus symmetry-breaking involves in this case a homoclinic bifurcation of symmetric limit cycle.

This leads to the question whether we should consider this phenomenon as one particular bifurcation or as a bifurcation scenario (bifurcation sequence). The first possibility implies that a corresponding critical parameter value must exist, while the second possibility requires a parameter interval in which the whole scenario happens. In practice, such an interval may be so small that it is not recognized in the numerical simulations.

Based on the theory [1,16], the appearance of the homoclinic trajectory of the saddle limit cycle implies in generic situations a bifurcation scenario. This is completely due to the fact that stable and unstable manifolds of the limit cycle intersect transversally, which gives rise to a homoclinic bundle and makes it possible to obtain chaotic behavior. On the other hand, when the torus symmetry-breaking occurs at one critical parameter value, this would suggest the simultaneous appearance of a continuum of homoclinic trajectories. These trajectories form the invariant manifold with an edge at C_s, homeomorphic to a product of a circle and a "figure eight". The last situation appears to be highly degenerate, and expecting it for the system under consideration requires strong theoretical arguments, related apparently to the symmetry of the system. (For a similar example see, e.g., [13], where heteroclinic bifurcation in a presence of $O(2)$ symmetry is treated. We should notice however that continuous and discrete symmetry groups may lead to completely different answers concerning the possibility of the degeneration described above.)

Observations of the torus symmetry-breaking (or symmetry-increasing) phenomenon in *case 3* also refer to a key role of a homoclinic bifurcation, related now to an antisymmetric limit cycle of saddle type.

Our study of *case 2* reveals a transition from two nonsymmetric strange attractors to one strange attractor of antisymmetric type (a so called symmetry-increasing of a strange attractor). Apparently homoclinic or heteroclinic bifurcations should be involved in such a phenomenon too. Whether there exists any relation between bifurcations of the strange attractor and the wiggly form of the branches of the periodic solutions is an interesting question.

5. Conclusion

The results clearly demonstrate that different types of connections between neural populations lead to substantial differences in the dynamical behavior. Each case is characterized by a specific type of bifurcation diagram and by a number of stable asymptotic regimes which one can observe when the strength of connections changes.

Quasi-periodic oscillations appear to be quite typical for the model (3). These oscillations possess different types of symmetry that are related to the symmetry of the corresponding torus. Transitions between quasi-periodic oscillations with different types of symmetry show that the period of modulation grows in an unlimited way. This phenomenon is closely related to the fact that both oscillators are oscillating in-phase almost all the time.

For a discussion of the possible role of modulated oscillations in the context of information processing by brain structures and the binding problem we refer to [2].

Some mathematical and numerical problems which come from our numerical study of the system (3), still remain open. They are mainly related to the phenomenon of symmetry-breaking and symmetry-increasing of quasiperiodic and chaotic regimes, involving homoclinic and heteroclinic bifurcations of the saddle limit cycles.

With respect to modelling of neural networks, it is important to study the behavior of a large ensemble of neural oscillators described by (1) with different types of connections, with different (geometric) locations of the oscillators and with taking into account possible delays in connections. We refer the reader to [2] for some results and discussions of such more general neural network schemes.

Acknowledgments

We are deeply grateful to Prof. Wolf Singer, Prof. Bodo Werner and Dr. Michael Dellnitz for helpful discussions.

References

[1] V.S. Afrajmovich, V.I. Arnold, Yu.S. Il'yashenko and L.P. Shil'nikov (1991). Theory of Bifurcations. In *Dynamical Systems 5*, V.I. Arnold (ed.). Encyclopaedia of Mathematical Sciences, Springer-Verlag, New York.

[2] G.N. Borisyuk, R.M. Borisyuk, A.I. Khibnik. Analysis of oscillatory regimes in a coupled neural oscillator system with application to visual cortex modeling. In *Proceedings of Workshop on Complex Dynamics in Neural Networks*, (IIASS, Vietri sul Mare, June 1991), Springer series Approaches to Neurocomputing, Springer-Verlag.

[3] R.M. Borisyuk, A.B. Kirillov (1991). Bifurcation analysis of neural network model. *Biological Cybernetic* (in press).

[4] M. Dellnitz, M. Golubitsky, I. Melbourne (1991). Mechanisms of symmetry creation. *Preprint*, University of Houston (see also this volume).

[5] M. Dellnitz, M. Golubitsky, I. Melbourne (1991). The structure of symmetric attractors. *Preprint*, University of Houston.

[6] G.B. Ermentrout, N.Kopell (1991). Multiple pulse interactions and averaging in systems of coupled neural oscillators. *J. of Mathematical Biology*, **29**, pp. 195-217.

[7] J. Guckenheimer and P. Holmes (1983). *Nonlinear Oscillations, Dynamical Systems, and Bifurcations of Vector Fields*. Springer-Verlag, New York.

[8] A.I. Khibnik (1990). Numerical methods in bifurcation analysis of dynamical systems: a continuation approach. In *Mathematics and Modelling*, A.D. Bazykin and Yu.G. Zarhin (eds.), pp.162-197. USSR Academy of Sciences, Pushchino (in Russian).

[9] A.I. Khibnik (1990). LINLBF: A program for continuation and bifurcation analysis of equilibria up to codimension three. In *Continuation and Bifurcations: Numerical Techniques and Applications*, D. Roose, B. de Dier, A. Spence (eds.), pp. 283-296. Kluwer.

[10] A.I. Khibnik (1990). *Using TraX: A Tutorial to Accompany TraX, A Program for Simulation and Analysis of Dynamical Systems*. Exeter Software, New York.

[11] A. Khibnik, Yu. Kuznetsov, V. Levitin, E. Nikolaev (1990). LOCBIF: Interactive Local Bifurcation Analyser, version 1.1. *Report*, Research Computing Centre, USSR Academy of Sciences, Pushchino.

[12] A.I. Khibnik, Yu.A. Kuznetsov, V.V. Levitin, E.N. Nikolaev. Continuation techniques and interactive software for bifurcation analysis of ODEs and iterated maps. In *Proceedings of Advanced NATO Workshop on Homoclinic Chaos* (Brussels, May 1991), *Physica D* (submitted for publication).

[13] E. Knobloch, D. Moore (1991). Chaotic travelling wave convection. In *Proc. IUTAM Symposium on Nonlinear Hydrodynamic Stability and Transition*.

[14] V.V. Levitin (1989). *TraX: Simulation and Analysis of Dynamical Systems*. Exeter Software, New York.

[15] C. Sparrow (1982). *The Lorenz Equations*. Springer-Verlag, New York.

[16] S. Wiggins (1988). *Global Bifurcations ans Chaos*. Springer-Verlag, New York.

[17] H.R. Wilson, J.D. Cowan (1972). Excitatory and inhibitory interactions in localized populations of model neurons. *Biophys. J.*, **12**, pp. 1-24.

Numerical Exploration of Bifurcations and Chaos in Coupled Oscillators

Seunghwan Kim and Won Gyu Choe
Depts. of Physics and Mathematics,
POSTECH, Pohang, Korea 790-600

January 6, 1992

Abstract

A system of coupled oscillators has recently been of great interests since it serves as a paradigm for a variety of problems of competing periods and associated phenomena occurring in nature. A common approach is to study a system of coupled oscillators through models of circle and torus mappings and attempt to emulate rich phenomena of quasiperiodicity, mode-locking, and the transition to chaos often observed in these systems. We attempt to classify types of chaos, which are often observed in three coupled oscillations, and explain some mechanisms for their appearance. Through an interactive use of computational and visual tools especially developed for studying torus maps, we have explored interesting cross-sections of the bifurcation set in the parameter space of the rotation parameter and the nonlinearity. We also present some preliminary numerical results on the structure of the parameter space of the torus maps with an exchange symmetry.

1 Introduction

Many physical systems possess competing temporal periods and can be modeled by a system of coupled oscillators. Examples include a swing, the periodically forced damped pendulum, and Josephson junctions in microwave fields.[4]

For problems of coupled oscillations, the ratios of the periods of oscillations in the system, whether they are intrinsically generated or externally imposed by periodic forcing, are often used to characterize dynamics in the phase space of the torus. For example, suppose that the system has two competing periods. Then we can define the *rotation number*, ρ, which is the ratio of periods.

If ρ is rational, the oscillators are said to be *commensurate* (or *resonant*) and the corresponding phase space motion is periodic with the period given by the denominator of ρ. For nonlinear systems with dissipation, this resonance often persists under small perturbations so that the rotation number is locked to a rational value over a range of the system parameters. This phenomena is called the *mode-locking*.

If ρ is irrational, the system is said to be *incommensurate*. The corresponding motion in the phase space is not closed, covering the two dimensional phase space densely, which is called *quasiperiodic*.

When the nonlinearity in the system is sufficiently large, ρ may not be well-defined and fluctuate. In this case, phase space dynamics is called *chaotic* with a characteristic sensitive dependence on initial conditions.

It is now well established that the study of a system of weakly coupled oscillators with dissipation can be reduced to the study of discrete dynamical models of circle and torus mappings which contain a few control parameters encoding the essential ingredients of the incommensurate problem: competing periods and nonlinearity. Though these models are relatively simple and low-dimensional, it has been shown that they can accurately emulate the common phenomena of quasiperiodicity, mode-locking, the transition to chaos occurring frequently in incommensurate systems[11]. It has also been shown that the success of these models largely hinges on the fact that after long time strong dissipation reduces the dimension of the relevant phase space to a low-dimensional subspace and that some of their predictions are "model independent."[6, 19]

However, most research activities have so far focused on the circle maps and an extension of the circle map theory to torus maps has only recently been investigated systematically.[3] The study of torus maps has been quite difficult because toroidal dynamics was not well known, requiring explorations and experimentations, and because analytic methods have been scarce for nonlinear problems in the two-dimension. Moreover, toroidal dynamics and the bifurcation set are, though intriguing, quite complicated due to the torus topology and the large phase and parameter spaces. In spite of these difficulties, a judicious combination of some analytic results, extensive computation and an interactive use of graphics and visual representations has helped to make steady progresses in understanding the torus map. In this paper, we present a brief summary of the theory of circle and torus maps and some recent results on torus maps focusing on exploring the parameter space by visual-numerical methods.

2 Two Coupled Oscillations

Suppose that for weak coupling and dissipation the asymptotic dynamics of a system with two coupled oscillations occur on the invariant torus, T^2, points on which are completely described by two phase angles, $x_0, x_1 \in S^1$. The phase variable x_0 typically denotes the time variable, t. Rather than studying the continuous dynamics it is more efficient to study the discrete problem by watching the system with "strobe lights" at discrete times $x_{0,n} = nT_0$, with $T_0 > 0$. This strobing gives a mapping $f : S^1 \to S^1$, called the *return map*, which maps $x_1(x_0 = nT_0)$ to $x_1(x_0 = (n+1)T_0)$.

2.1 Definitions

We assume that the state of the system with two competing periods is given by a sequence of an angle coordinate, $\{x_n\}$ with $x_n \in [0,1]$, generated by the one-dimensional recursion relation

$$x_{n+1} = f(x_n), \tag{1}$$

where f is assumed to be a smooth mapping of a circle onto itself. Let F be a lift of f onto R. One of the most widely studied is the "sine map"[6, 19] given by

$$F(x) = x + \Omega - \frac{a}{2\pi} sin(2\pi x) \tag{2}$$

where a is the nonlinear parameter and Ω the rotation parameter.

An important number characterizing dynamics is called the *rotation number*, and is defined by

$$\rho = \lim_{n\to\infty}(x_n - x_0)/n, \tag{3}$$

which represents the ratio of periods of two coupled oscillators. This number is well-defined and independent of x_0 for $a \le 1$.

2.2 Mode-locked Tongues and Devil's Staircase

Typically one want to understand the structure of the phase diagram in the parameter space of (Ω, a) and, in particular, how the transition to chaos occurs as the nonlinearity increases. A schematic phase diagram for the sine map family is shown in Figure 1. The shaded regions denote mode-locked regions, where only a few of them are drawn and the curve denotes the set of the parameter values for which motion is quasiperiodic with a given irrational winding number, ρ.

When the map is linear ($a = 0$), we get $\rho = \Omega$. Since a set of all irrational numbers occupies full measure, quasiperiodicity occurs with probability one, whereas the mode-locking occurs with probability zero. As the nonlinearity is turned on ($a \ne 0$), the mode-locked states acquire nonlinear stability and become stable under small but finite perturbations. The rotation number ρ is still related to Ω but in a rather nontrivial way, as shown in the plot of Ω versus ρ with $a = .7$ in Figure 2. This mode-locking structure with an infinite number of steps representing mode-locking to rational values of ρ is called the *devil's staircase*[11]. However, it has been numerically observed that for $a < 1$, the set of all mode-locked steps still does not occupy the entire interval in Ω since the size of steps decreases sufficiently fast as a function of the period. Furthermore, Herman has shown that most quasiperiodic orbits persist under nonlinear perturbations and for a given a the set of parameter values of Ω for which quasiperiodic orbits exist has a positive measure.[1, 10] Therefore, both the quasiperiodicity and mode-locking occurs with non-zero but finite probability for $a < 1$.

As the nonlinearity increases, the mode-locked intervals increases in size and at a critical nonlinearity, $a = 1$ for the sine map, the mode-locked intervals occupy full measure so that quasiperiodicity occurs with probability zero. It can be shown that this singular change in the prevalence of quasiperiodicity has been induced by a noninvertibility of the sine map for $a = 1$. Beyond the critical nonlinearity, quasiperiodicity finds no room for its existence and has to give way to chaos.

The self-similarity of the Devil's staircase and the phase diagram below and at the criticality can be and has been analyzed by numerical methods and renormalization group theory to show that certain aspects of the staircase and the transition to chaos are universal and independent of details of the model [6, 19].

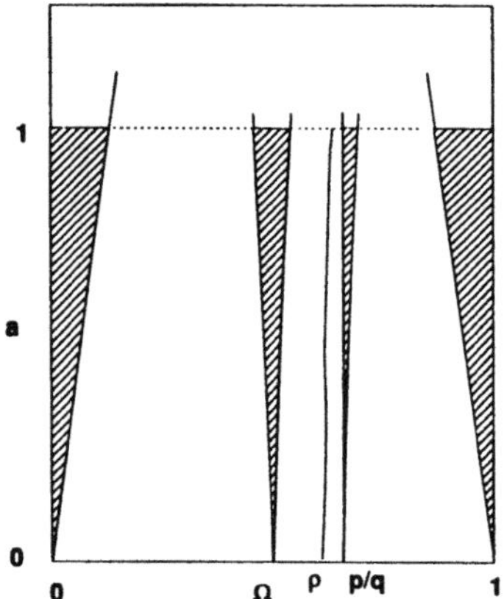

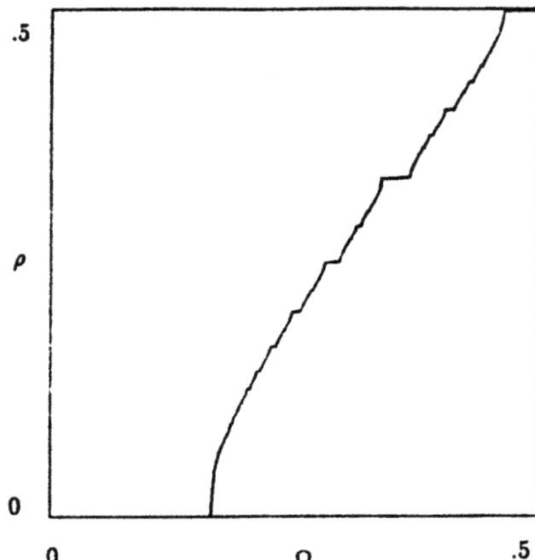

Figure 1. A schematic phase diagram for the sine map family. The shaded regions denote a few dominant mode-locked regions.

Figure 2. The Devil's staircase: A plot of the rotation parameter Ω versus the rotation number ρ for a sine map ($a = .7$). Only the left half of the staircase with a few dominant steps is shown.

3   Three Coupled Oscillations

The study of circle maps has yielded extensive information about the resonances and the mode-locking structure for systems with two interacting periods. However, the problem of three coupled oscillations is still not well understood, though dynamics with more than two competing periods have been observed in a variety of experiments.[7, 16] Moreover, in spite of extensive efforts to understand the mechanism through which the nonlinear systems exhibit chaos, the full quasiperiodic route to chaos remains to be the least understood. Ruelle-Takens and Newhouse-Ruelle-Takens[20, 18] have shown rigorously that *triply quasiperiodic motion* (quasiperiodic motion with three incommensurate periods) is not stable in the sense that it can be perturbed into a chaotic one by an arbitrarily small perturbation. Their misinterpretation of this result that triply quasiperiodic motion is not typical in physical systems has stimulated extensive numerical and experimental work on triply quasiperiodic motion,[5, 8] where triply quasiperiodic motion is found with finite probability in the space of relevant control parameters of the system.

3.1   Definitions

Similar to the case of two periods, for three weakly coupled oscillators with dissipation asymptotic dynamics occurs on the invariant torus of T^3. By using the standard method of the Poincaré section, continuous dynamics is reduced to one given by a sequence of two angle coordinates, $\{(x_{1,n}, x_{2,n})\}$, which specifies the time evolution of the state of the system.

The recursion relation for the angles is given by the return map:

$$x_{1,n+1} = f_1(x_{1,n}, x_{2,n}), \quad x_{2,n+1} = f_2(x_{1,n}, x_{2,n}),$$ (4)

where $f = (f_1, f_2)$ is a smooth mapping of a torus onto itself. Let F be the lift of f onto R^2. One of the most extensively studied is the "strongly coupled torus map," which is homotopic to the identity, given by [14]

$$f_1(x_1, x_2) = x_1 + \Omega_1 - \frac{a}{2\pi} \sin(2\pi x_2), \quad f_2(x_1, x_2) = x_2 + \Omega_2 - \frac{a}{2\pi} \sin(2\pi x_1)$$ (5)

where a is the nonlinear parameter and $\Omega = (\Omega_1, \Omega_2)$ the two-dimensional *rotation parameter*. This map represents the limit of the extremely strong coupling between the oscillators. Note that the nonlinearity of an oscillator is entirely generated by the coupling with the phase of the other oscillator.

For three coupled oscillations there are *two* ratios of the periods and the two-dimensional rotation vector ρ is defined by

$$\rho = (\rho_1, \rho_2) = \lim_{n \to \infty} ((x_{1,n}, x_{2,n}) - (x_{1,0}, x_{2,0}))/n.$$ (6)

Unlike the circle maps, the rotation vector, ρ, in general depends on the starting point. When ρ is not well-defined, the definition can be extended to that of the rotation sets.[3]

Our aim is to understand the web-like structure of the bifurcation set in the parameter space that forms the complement of the set of parameter values for quasiperiodic motion. First we need to classify possible dynamics depending on the rotation vector, which is a little bit more involved than the two period case since now two ratios are involved.

In general, the system is said to be *resonant* if the ratios are rationally related, that is, there exists integers l, m, and n (not all zeroes) such that $l\rho_1 + m\rho_2 = n$. If there is only one such relationship, the system is *partially resonant (partially commensurate)*, and phase space dynamics is *doubly quasiperiodic*. If there are two, the system is *fully resonant (fully commensurate)* and phase space dynamics is periodic. Otherwise, the system is said to be *incommensurate* and the phase space dynamics is *triply quasiperiodic*. Note that this provides an elementary classification scheme which need to be refined further in order to describe, for example, the regions of the bistability and the homoclinic chaos, where observed dynamical behaviors may depend on initial conditions.

3.2 Devil's Web

When the map is linear, that is, $a = 0$, we get $\rho = \Omega$. Therefore, the resonance structure is completely determined by the rationality of Ω so that it becomes identical to the structure of rational numbers as shown in Figure 3. The number theoretic structure of the rational numbers is described in detail elsewhere.[15] Note that for a linear case triply quasiperiodic motion occurs with probability one.

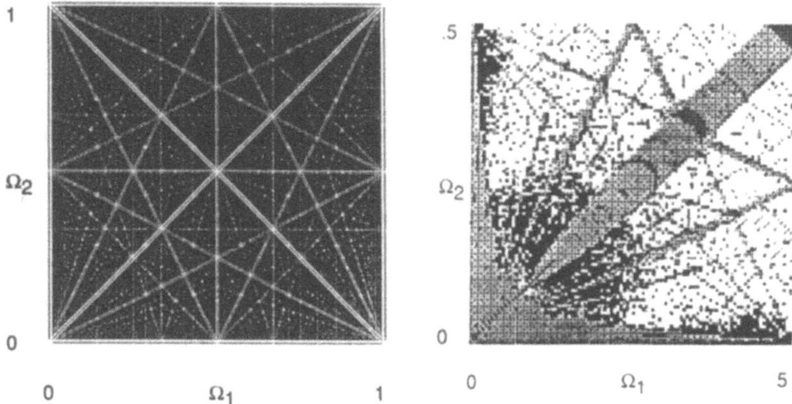

Figure 3. The resonant web: A plot of all rational values of (Ω_1, Ω_2) with period smaller than 100 which represents the phase diagram of the pure translations. Partially incommensurate regions are effectively seen as white strips.

Figure 4. A two-dimensional slice of the perturbed resonant web in the space of Ω_1 and Ω_2 for the strongly coupled torus map with $a = .8$.

When the nonlinearity is turned on, however, a rational point for resonance is perturbed into an open region with a nontrivial shape, called the *resonance region*, where the periodic dynamics with a given period, persists. The lines with rational slopes corresponding to doubly quasiperiodic motion are perturbed into the *(partially) mode-locked strips*. The resulting structure of resonances in the parameter space of (Ω_1, Ω_2, a) is called the *Devil's web of mode-locking*. Figure 4 shows a square-grid parameter scan in the space of (Ω_1, Ω_2) for the family in Equation (5), which gives a two-dimensional slice of the Devil's web with $a = 0.8$ in the parameter space of (Ω_1, Ω_2, a). The regions in white, light gray, and other grays denote regions for triply quasiperiodic, doubly quasiperiodic, and periodic motion, respectively. This cross-section of the phase diagram appears to be very complex with a self-similarity whose scaling aspects have been analyzed in terms of number theory of a pair of rationals[15] and renormalization group theory. A structure similar to this structure has also been observed in the experiments of nonlinear electrical circuits. [5]

The torus topology of the phase space allows different kinds of chaos for coupled oscillators which can be best characterized by the behavior of the limit points of the rotation vector. Let's define the *rotation set*, $\rho^s = (\rho_x^s, \rho_y^s)$ to be the set of all limit points of the rotation vector.[3]

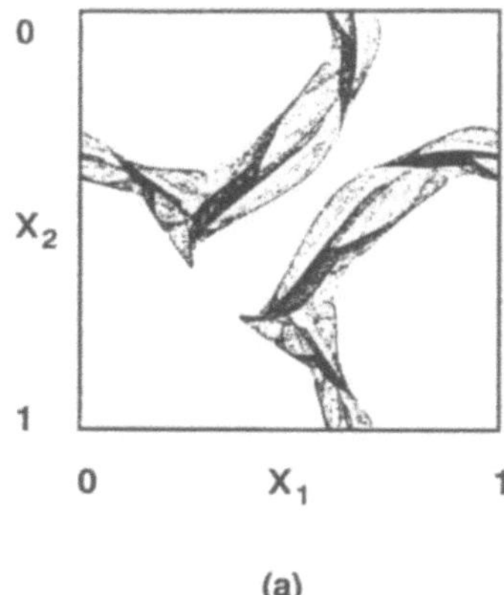
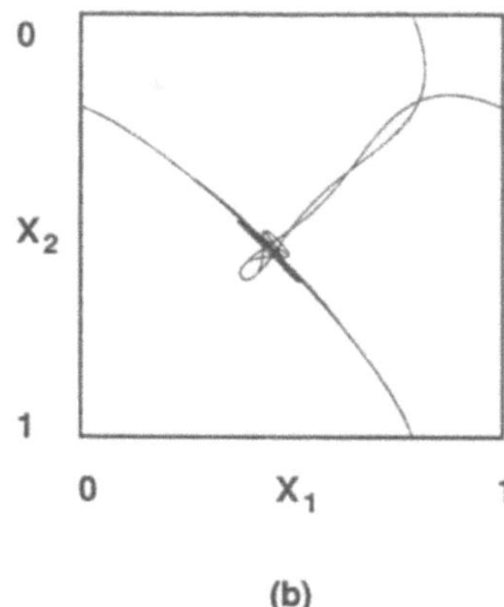

Figure 5. Different forms of chaos: (a)annular chaos and (b) toroidal chaos.

Chaos is said to be *mode-locked.* if ρ^s is a *single* rational point but its phase space dynamics is, though localized, chaotic. Chaos is said to be *annular* if ρ^s is an interval of the rational *line* defined by $l\rho_x^s + m\rho_y^s = n$ with l, m, n integers, not all zeroes. For annular chaos, the fluctuation of orbits grows linearly as a function of the number of iterations but only along the direction of an invariant annular strip which is closed on the torus. Chaos is called *toroidal* if ρ^s occupies an open region, where chaotic oscillation occurs in all directions of the torus. It has been shown that if resonance regions of three different rotation vectors overlap the system exhibits toroidal chaos[17]. Some examples of chaotic orbits are shown in Figure 5, where the square box denotes the phase space of the torus with opposing boundaries identified. More detailed discussion on the rotation vector and the rotation set can be found in Ref. [3].

We have also found that the nontrivial topology of the torus has a significant consequence on dynamics and the bifurcation structure. For example, resonance regions does not have a simple topology since its boundaries are the fold curves of the projection of the torusfull family of periodic solutions in the four-dimensional space of $(\Omega_1, \Omega_2, x_1, x_2)$ onto the parameter space (Ω_1, Ω_2)[13]. One of its consequences is that special points in the parameter space with degenerate eigenvalues of $+1$, called *Takens-Bogdanov points*, exist and persist in the limit of vanishing nonlinearity. It can be shown that the Takens-Bogdanov point always accompanies a small region in the parameter space where chaos of the transversal homoclinic type occurs due to the transversal crossing of the stable and unstable manifolds of periodic saddles. This implies that three coupled oscillators, however small the coupling is, can show chaos. However, as nonlinearity decreases chaos becomes weaker and less probable in the sense that the region in the parameter space showing chaos decreases exponentially in size.

3.3 Strongly Coupled Symmetric Torus Map

We have recently studied dynamics and bifurcations on the diagonal plane of $\Omega_1 = \Omega_2$ of the Devil's web in Figure 4, which corresponds to a two parameter family

$$f_1(x_1, x_2) = x_1 + \Omega - \frac{a}{2\pi} sin2\pi x_2, \quad f_2(x_1, x_2) = x_2 + \Omega - \frac{a}{2\pi} sin2\pi x_1 \tag{7}$$

where a is the nonlinearity, Ω the rotation parameter, and x_1, $x_2 \in S^1$. This map is called as the *strongly coupled symmetric torus map* (SCSTM) since it is invariant with respect to an exchange operation, $S(x_1, x_2) = (x_2, x_1)$. Though there are many interesting questions regarding the role of the symmetry in bifurcations, in this paper we focus on the questions of when and how dynamics can be reduced to that of circle maps and how dynamics deviates away from the results of circle maps as the nonlinearity of the system increases.

The scan of the parameter space of (Ω, a) for the circle map in Equation 2 is shown in Figure 6, which reveals the self-similar structure of the mode-locking. The mode-locked regions (Arnold tongues) are shaded, the quasiperiodic region in white, and the chaotic region in black. Here the autocorrelation algorithm and the phase space partition algorithm by Cumming and Linsay[5] have been used to automatically determine periodic, quasiperiodic, and chaotic dynamics.

For the SCSTM, the diagonal line of $x_1 = x_2$ in the phase space of the torus is an invariant subspace of the exchange operator S. Therefore, orbits with symmetric initial conditions, $x_1 = x_2$, remain on the invariant line so that the oscillators maintain the complete synchrony. Since the SCSTM is equivalent to the circle map on the invariant line, its parameter scan using symmetric initial conditions should be exactly the same as the bifurcation set of circle maps in Figure 6. However, a parameter scan of SCSTM with asymmetric initial conditions in Figure 7 should show some deviations from the circle maps.

Indeed the comparison of Figure 6 and Figure 7 shows differences for large nonlinearity.

Nevertheless the bifurcation set appears very similar to that of circle maps below some critical values of nonlinearity. Numerical evidence is that the invariant line is stable for $a \le a_c$ with some $a_c > 1$ except the tongue with $\rho = (0,0)/1$. Therefore asymptotic dynamics is completely given by one on the invariant line. Note that the region with the stable invariant line includes the critical line of the sine map, $a = 1$. Thus as the nonlinearity increases from zero chaos appears first in the form of annular chaos *on* the invariant line, which, including the scaling structure, can be completely described by the standard circle map theory.

As the nonlinearity is increased further, in particular in the direction of entering the tongue $(0,0)/1$ from below, invariant curves or annular chaos, which are off the invariant line, may occur and dynamics deviates away from that of circle maps. In this case, the oscillators become asynchronous. We present a few preliminary results on off-diagonal dynamics in the followings.

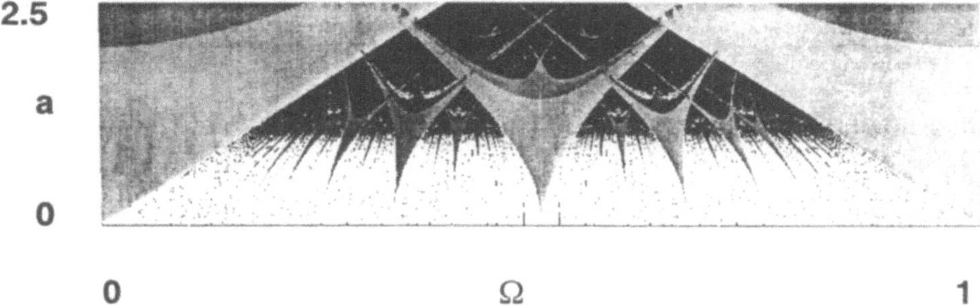

Figure 6. A bifurcation set in the space of (Ω, a) for a circle map showing
an infinity of self-similar mode-locked tongues. Mode-locked tongues are
shaded; the darker is the region, the higher is the periodicity associated
with it. The quasiperiodic regions are in white, and the chaotic regions in
black.

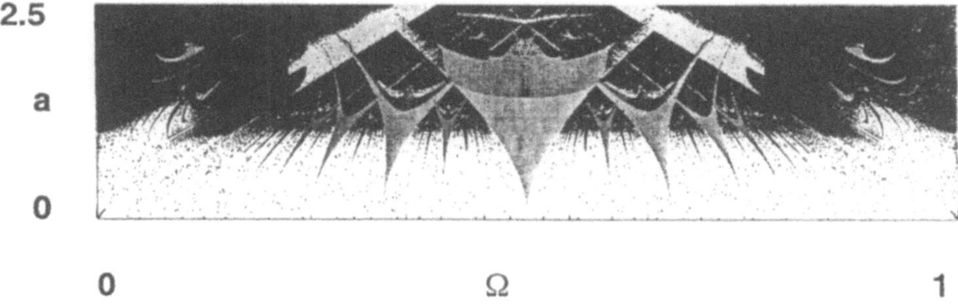

Figure 7. A bifurcation set of SCSTM in the space of (Ω, a) starting from
asymmetric initial conditions. The figure legend is the same as one for Figure
6.

The mechanism for annular chaos just inside the $(0,0)/1$ tongue involves the unstable manifolds of a saddle on the diagonal line. As one enters the $(0,0)/1$ tongue from below, four
symmetrically arranged fixed points of period one appears, two saddles on the diagonal and two
sources off-diagonal. (See Figure 8.)

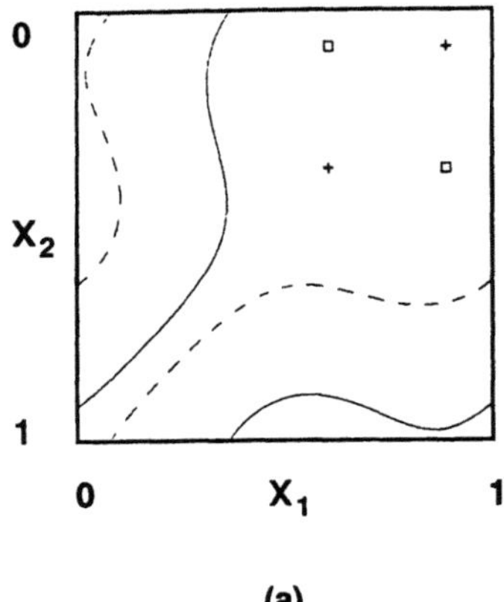
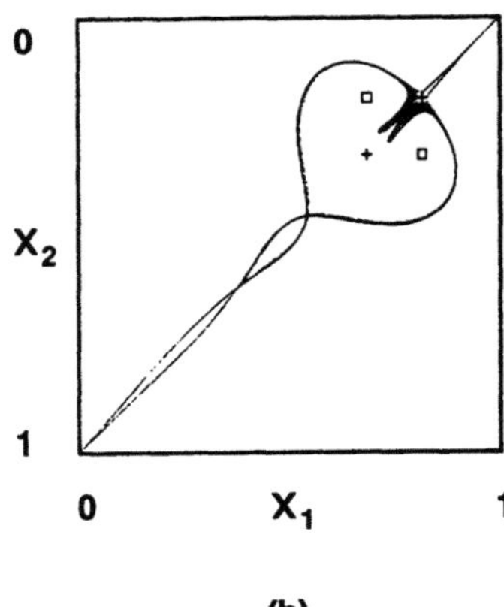

Figure 8. Off-diagonal dynamics: (a) A stable invariant curve. The dotted curve denotes the conjugate attractor. (b) A "strange attractor." The crosses denote the period 1 saddles and the squares the period 1 sources.

The diagonal line is locally repelling but global dynamics is attracted to the neighborhood of the diagonal. One possible dynamics is that the unstable manifolds may accumulate to the asymmetric stable invariant curve as in Figure 8. Note that there exists another stable invariant curve, called the *conjugate attractor*, which can be obtained by a symmetry operation, S.

Another possibility is that unstable manifolds of a saddle cross themselves on the invariant line, and after repeated elongations and reinjections to the neighborhood of a saddle on the invariant line they accumulate onto a "strange attractor" as in Figure 8. This mechanism for annular chaos is apparently due to the noninvertibility of the mapping and the annular nature of the phase space. The attractor with the annular chaos, though chaotic, can be symmetric under S with oscillators, though asynchronous, in phase on the average.

There are many other kinds of dynamics to be explored including the transition to toroidal chaos and the transition between the symmetric and asymmetric orbits, which are the subject of ongoing investigations.

4 Discussion

We have shown that the theory of low-dimensional dynamical models can be and has been successfully applied to study problems of competing periods and that for three coupled oscillators chaos arises spontaneously in different forms of mode-locked, annular, and toroidal chaos as nonlinearity is turned on. The typical mechanism for these types of chaos is provided by transversal homoclinic orbits.

The problem of three coupled oscillations has been challenging because dynamics and the structure of the phase diagram can be very complex due to the nontrivial topology of the torus and also because systematic tools for studying it has been lacking. On the other hand, the large phase and parameter spaces and the topology of the toroidal phase space promise rich dynamical behaviors and exotic phase diagrams of mode-locking, which make an investigation of coupled oscillations very rewarding.

Though many of tools from the two-period problem can be carried over and applied to the study of the three-period problem, we have found that larger phase and parameter spaces require the development of new, more efficient tools. Torus maps have been a fertile ground for testing and investigating bifurcations through an extensive interactive use of analytical and visual-numerical methods, which inspired the development of toolkits for exploring dynamical systems such as "Funtorus," "Kaos," and "DSTools."[9, 12]

In spite of extensive efforts in recent years, many problems in three or more coupled oscillations still remain essentially unsolved, warranting further study. For example, as nonlinearity increases, mode-locked tongues (resonance regions) deform and develop more singularities and bifurcations. It would be interesting to measure the growth exponent of the size of mode-locked tongues as a function of the nonlinearity. It is important to obtain experimental confirmation of analysis and predictions of torus maps. The parameter space scans of torus maps could be compared in detail with those for nonlinear electrical oscillators[2, 5] by checking the topology of the mode-locked regions and searching for transversal homoclinic chaos, and explaining the existence of the chaotic wedge. It will be also interesting to reinterpret the results of classic hydrodynamic experiments involving three-frequency dynamics[16, 7] in the light of recent progresses in understanding torus maps.

While under some assumptions the KAM theory guarantees the persistence of the quasiperiodic orbit for small enough nonlinear perturbations, the question of when and how it breaks down as nonlinearity increases is very important but remains unanswered. To answer this and, therefore, obtain a better description of the quasiperiodic path to chaos, it seems necessary to study the three parameter family of torus maps in a more systematic manner. The connection to the theory of Hamiltonian systems with three degrees of freedom is also an interesting problem which is currently explored through coupled standard maps. Finally, the question of the role of the symmetry in bifurcations of the coupled identical oscillators need to be explored further.

Acknowledgement

S. Kim would like to thank C. Baesens, J. Guckenheimer, S. Ostlund, and R. Mackay, in collaboration with whom much of the results in this paper has been obtained. The work on the SCSTM has been carried out in collaboration with Won Gyu Choe, the detailed description of which will be published elsewhere.

References

[1] V. I. Arnold, AMS Transl. Ser. 2, 46, 213, 1965.

[2] P. Ashwin and G. King, *preprint*, 1991.

[3] C. Baesens, J. Guckenheimer, S. Kim, and R. Mackay, *Physica* **49D**, 387-475, 1991.

[4] P. Bak, *Physics Today* 39-45, December, 1986; see references therein.

[5] A. Cumming and P. S. Linsay, *Phys. Rev. Lett.* **60**, 2719, 1988.

[6] M. J. Feigenbaum, L. P. Kadanoff, and S. J. Shenker, *Physica* **5D**, 370, 1982.

[7] J. P. Gollub and S. V. Benson, *J. Fluid. Mech.* **100**, 449, 1980.

[8] C. Grebogi, E. Ott, J. A. Yorke, *Physica* **15D**, 354, 1985.

[9] J. Guckenheimer and S. Kim, *MSI Tech. Rep.*, **90**, 1990.

[10] M. R. Herman, *Publ. Math. IHES* **49**, 5, 1979; *Commen. Math. Helv.* **58**, 453, 1983.

[11] M. H. Jensen, P. Bak, T. Bohr, *Phys. Rev. Lett.* **50**, 1637, 1983.

[12] S. Kim and J. Guckenheimer, *MSI Tech. Rep.*, **89-48**, 1989.

[13] S. Kim, R. Mackay, and J. Guckenheimer, *Nonlinearity* **2**, 391, 1989.

[14] S. Kim and S. Ostlund, *Phys. Rev. Lett.* **55**, 1165, 1985.

[15] S. Kim and S. Ostlund, *Phys. Rev.* **A 34**, 3426, 1986.

[16] A. Libchaber, S. Fauve, and C. Laroche, *Physica* **7D**, 73, 1983.

[17] J. Llibre and R. Mackay, *Erg. Th. Dyn. Sys.* **11**, 115, 1991.

[18] S. Newhouse, D. Ruelle, and F. Takens, *Commun. Math. Phys.* **64**, 35, 1978.

[19] D. Rand, S. Ostlund, J. Sethna, and E. Siggia, *Phys. Rev. Lett.* **49**, 132, 1982.

[20] D. Ruelle, and F. Takens, *Commun. Math. Phys.* **20**, 167, 1971; **23**, 343, 1971.

Hopf Bifurcation with $Z_4 \times T^2$ Symmetry

E. Knobloch

Department of Physics
University of California
Berkeley, CA 94720 USA

M. Silbert†

Center for Dynamical Systems and Nonlinear Studies
Georgia Institute of Technology
Atlanta, GA 30332 USA

Abstract

Hopf bifurcation on a rotating square lattice is considered. As many as seven primary solution branches are found to exist. Five of these are periodic in time and two are quasiperiodic with two independent frequencies. The stability properties of the solutions are established. A structurally stable attracting heteroclinic cycle connecting four rolls travelling in four orthogonal directions is also found. The relevance of the analysis to overstable convection in a plane rotating layer is discussed.

I. Introduction

In this paper we consider the Hopf bifurcation on a square lattice with the symmetry $Z_4 \times T^2$. Here Z_4 is the symmetry of a square under proper rotations through $\pi/2$ and T^2 is a two–torus. Problems of this type arise in a natural way in pattern selection studies in spatially extended systems that are translation-invariant in two (horizontal) directions but lack reflection symmetry in vertical planes. The reflection symmetry is absent, for example, in all systems that *rotate* about the vertical. In hydrodynamic systems the centrifugal force may often be balanced by an appropriate pressure gradient with the result that no preferred origin is selected in the dynamical equations; the rotation manifests itself only through the Coriolis force. Alternatively, one may suppose that the rotation is sufficiently weak that the centrifugal force can be neglected. In either case the resulting dynamical equations are equivariant under translations and proper rotations about the vertical. If one now restricts attention to solutions that are periodic in two orthogonal directions with the same spatial period, then the resulting field equations have the symmetry $Z_4 \times T^2$. Note that in the absence of rotation an analogous formulation leads to a problem with $D_4 \times T^2$ symmetry,

---

† Present address: Department of Applied Mechanics, Caltech, Pasadena, CA 91125

already discussed by Swift (1984) and Silber & Knobloch (1991). We follow the latter paper and assume the neutrally stable modes at the Hopf bifurcation have the form

$$(v_1(t)e^{ik_c x_1} + v_2(t)e^{ik_c x_2} + w_1(t)e^{-ik_c x_1} + w_2(t)e^{-ik_c x_2})f(y), \tag{1}$$

where (x_1, x_2) are the two orthogonal coordinates in the horizontal, $f(y)$ denotes the vertical structure of the modes and $k_c \neq 0$ is the wavenumber with which the instability sets in. In the following we assume that the linear stability problem takes the form $\dot{z} = \mu(\lambda)z$, where $z \equiv (v_1, v_2, w_1, w_2) \in \mathbf{C}^4$, and $\mu(0) = i\omega_c$, $\mathrm{Re}(\mu'(0)) > 0$. Here λ is the bifurcation parameter. It follows that the quantities $(|v_1|, |w_1|)$ and $(|v_2|, |w_2|)$ represent amplitudes of left- and right-travelling waves in the (x_1, x_2) directions, respectively. The spatial symmetries act on (v_1, v_2, w_1, w_2) as follows:

$$\rho_{\pi/2} : \quad \begin{pmatrix} v_1 \\ v_2 \\ w_1 \\ w_2 \end{pmatrix} \rightarrow \begin{pmatrix} w_2 \\ v_1 \\ v_2 \\ w_1 \end{pmatrix}, \quad \rho_{\pi/2} \in Z_4, \tag{2a}$$

$$(\theta_1, \theta_2) : \quad \begin{pmatrix} v_1 \\ v_2 \\ w_1 \\ w_2 \end{pmatrix} \rightarrow \begin{pmatrix} e^{i\theta_1} v_1 \\ e^{i\theta_2} v_2 \\ e^{-i\theta_1} w_1 \\ e^{-i\theta_2} w_2 \end{pmatrix}, \quad (\theta_1, \theta_2) \in T^2. \tag{2b}$$

In addition, in normal form the dynamical equations will commute with an S^1 phase shift symmetry acting by

$$\phi : \quad z \rightarrow e^{i\phi}z, \quad \phi \in S^1. \tag{2c}$$

In this paper we take the full symmetry group of the dynamical equations near $\lambda = 0$ to be $\Gamma \equiv Z_4 \times T^2 \times S^1$.

It is now simple to show that the most general equations commuting with the above symmetries, truncated at third order, take the form

$$\dot{v}_1 = \mu v_1 + (a|v_1|^2 + b|w_1|^2 + c|v_2|^2 + d|w_2|^2)v_1 + ev_2 w_2 \bar{w}_1 \tag{3a}$$

$$\dot{v}_2 = \mu v_2 + (a|v_2|^2 + b|w_2|^2 + c|w_1|^2 + d|v_1|^2)v_2 + ev_1 w_1 \bar{w}_2 \tag{3b}$$

$$\dot{w}_1 = \mu w_1 + (a|w_1|^2 + b|v_1|^2 + c|w_2|^2 + d|v_2|^2)w_1 + ev_2 w_2 \bar{v}_1 \tag{3c}$$

$$\dot{w}_2 = \mu w_2 + (a|w_2|^2 + b|v_2|^2 + c|v_1|^2 + d|w_1|^2)w_2 + ev_1 w_1 \bar{v}_2. \tag{3d}$$

Note that if the system is not rotating so that the reflection symmetry is present then $c = d$; this bifurcation problem was analyzed by Silber & Knobloch (1991).

In the following section we summarize the information about the possible solutions to equations (3) that can be obtained using group-theoretic arguments alone. We then compute these solutions explicitly and determine their stability properties in terms of the complex coefficients (a, b, c, d, e). In addition we describe some more exotic solutions, before discussing the relevance of the present study to convection in a rotating layer, and in particular to the so-called Küppers-Lortz instability.

II. Group–theoretic Results

Each nontrivial (pattern–forming) solution $\mathbf{z} \neq \mathbf{0}$ breaks the full symmetry Γ, and therefore has symmetry characterized by an isotropy subgroup $\Sigma_{\mathbf{z}}$ of Γ:

$$\Sigma_{\mathbf{z}} = \{\sigma \in \Gamma : \sigma \mathbf{z} = \mathbf{z}\}. \tag{4}$$

The successive breaking of the full symmetry is summarized by the lattice of isotropy subgroups of Γ shown in fig. 1. Associated with each isotropy subgroup is a linear subspace $Fix(\Sigma)$, which is invariant under the dynamics (3):

$$Fix(\Sigma) = \{\mathbf{z} \in \mathbf{C}^4 : \sigma \mathbf{z} = \mathbf{z}, \forall \sigma \in \Sigma\}. \tag{5}$$

Hence all solutions with symmetry $\Sigma \subset \Gamma$ evolve under (3) restricted to the subspace $Fix(\Sigma)$. In table 1 we list all the isotropy subgroups of Γ (up to conjugacy) together with their fixed point subspaces. We identify solutions in $Fix(\Sigma)$ with spatio-temporal patterns using the expression (1). The equivariant Hopf theorem guarantees the existence of primary solution branches with two-dimensional fixed point subspaces (Golubitsky & Stewart 1985). In the present case these are the travelling rolls (TR), the standing rolls (SR), the standing squares (SS) and the alternating rolls (AR). It should be noted that in the corresponding $D_4 \times T^2 \times S^1$ equivariant problem an *additional* primary branch, called travelling squares (TS), is present. All of these solutions are periodic in time.

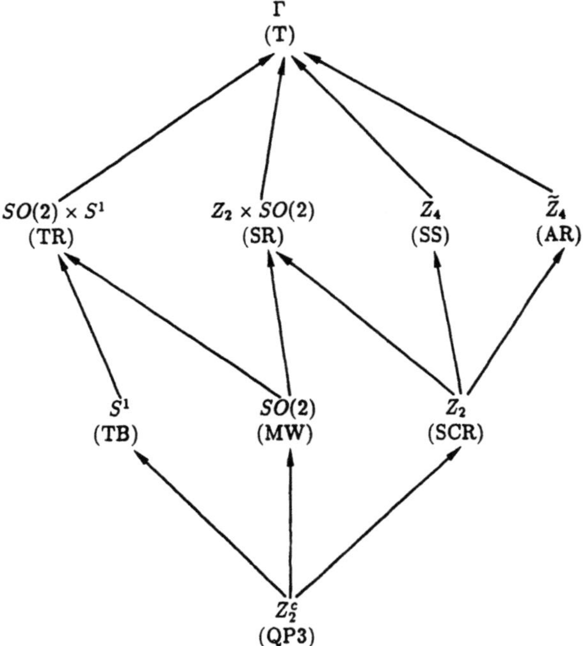

Fig. 1. Lattice of isotropy subgroups. Arrows indicate inclusion.

Name	$Fix(\Sigma)$	Σ
I. Trivial Solution (T)	$(0,0,0,0)$	Γ
II. Travelling Rolls (TR)	$(z,0,0,0)$	$SO(2) \times S^1$
III. Standing Rolls (SR)	$(z,0,z,0)$	$Z_2 \times SO(2)$
IV. Standing Squares (SS)	(z,z,z,z)	Z_4
V. Alternating Rolls (AR)	(z,iz,z,iz)	$\widetilde{Z}_4$
VI. Standing Cross–Rolls (SCR)	(z_1,z_2,z_1,z_2)	Z_2
VII. Travelling Bimodal (TB)	$(z_1,z_2,0,0)$	S^1
VIII. Modulated Wave (MW)	$(z_1,0,z_2,0)$	$SO(2)$
IX. Triply Periodic (QP3)	(z_1,z_2,z_3,z_4)	Z_2^c

Table 1: The solutions corresponding to the different isotropy subgroups of Γ. The coordinates specifying each $Fix(\Sigma)$ are (v_1, v_2, w_1, w_2). For a description of the symmetries Σ see table 2. The trivial symmetry Z_2^c is contained in every Σ.

Σ	Generators of Σ
$SO(2)$	$[(0,\theta),0], \quad \theta \in SO(2)$
S^1	$[(-\phi,-\phi),\phi], \quad \phi \in S^1$
Z_2	$\rho_{\pi/2}^2 \in Z_2$
Z_4	$\rho_{\pi/2} \in Z_4$
$\widetilde{Z}_4$	$[\rho_{\pi/2}(0,\pi),\pi/2] \in \widetilde{Z}_4$
Z_2^c	$[(\pi,\pi),\pi] \in Z_2^c$

Table 2: Generators of isotropy subgroups in table 1 in terms of the action (2) of the group Γ on $\mathbf{C}^4$. A group element is specified by $[\rho_{\pi/2}^n(\theta_1,\theta_2),\phi] \in Z_4 \times T^2 \times S^1$, where $\rho_{\pi/2}^n \in Z_4$ ($n \in \{0,1,2,3\}$), $(\theta_1,\theta_2) \in T^2$, $\phi \in S^1$.

III. Stability Results

Since the four patterns (TR, SR, SS, AR) are periodic in time (with frequencies ω near ω_c) we substitute $\mathbf{z} = e^{i\omega t}\tilde{\mathbf{z}}$ into equations (3) and drop the tildes:

$$
\begin{aligned}
\dot{v}_1 &= \nu v_1 + (a|v_1|^2 + b|w_1|^2 + c|v_2|^2 + d|w_2|^2)v_1 + ev_2 w_2 \bar{w}_1 \\
\dot{v}_2 &= \nu v_2 + (a|v_2|^2 + b|w_2|^2 + c|w_1|^2 + d|v_1|^2)v_2 + ev_1 w_1 \bar{w}_2 \\
\dot{w}_1 &= \nu w_1 + (a|w_1|^2 + b|v_1|^2 + c|w_2|^2 + d|v_2|^2)w_1 + ev_2 w_2 \bar{v}_1 \\
\dot{w}_2 &= \nu w_2 + (a|w_2|^2 + b|v_2|^2 + c|v_1|^2 + d|w_1|^2)w_2 + ev_1 w_1 \bar{v}_2 \, .
\end{aligned}
\tag{6}
$$

Here $\nu = \mu(\lambda) - i\omega$. In the following we write (without loss of generality) $\nu = \lambda + i\sigma$, with λ, σ both real. The four patterns now correspond to fixed points of the system (6) for appropriate choice of σ, where σ determines the frequency ω. The (linear orbital) stability of these solutions is conveniently determined by forming the isotypic decomposition of $\mathbf{C}^4$ for each isotropy subgroup. This determines the coordinates that block-diagonalize the 8×8 Jacobian matrix for each solution. The details are similar to the $D_4 \times T^2 \times S^1$ case (Silber & Knobloch 1991) and are omitted. The results are summarized in table 3. Note that the results for SS and AR are related by the parameter symmetry $e \to -e$.

IV. Other Primary Branches

As in the $D_4 \times T^2 \times S^1$ problem the Hopf bifurcation with $Z_4 \times T^2 \times S^1$ symmetry can produce (generically) primary branches with submaximal isotropy. These reside in fixed point subspaces with dimension ≥ 4 and are not always present. The simplest example is provided by the temporally periodic standing cross-rolls $(SCR1)$. Restricting (3) to the SCR subspace of table 1 we note that the resulting vector field on $\mathbf{C}^2$ has $D_4 \times S^1$ symmetry:

$$
\begin{aligned}
\dot{z}_1 &= \mu z_1 + (a+b)|z_1|^2 z_1 + (c+d)|z_2|^2 z_1 + ez_2^2 \bar{z}_1 \\
\dot{z}_2 &= \mu z_2 + (a+b)|z_2|^2 z_2 + (c+d)|z_1|^2 z_2 + ez_1^2 \bar{z}_2 \, .
\end{aligned}
\tag{7}
$$

This equivariant bifurcation problem was analyzed by Swift (1988). He observed that there is an *unstable* branch of periodic solutions with $z_1 z_2 \neq 0$, $|z_1| \neq |z_2|$. We make use of the symmetry to select the representative $(v_1, v_2, w_1, w_2) = (r_1 e^{i\omega t}, r_2 e^{i(\omega t + \psi)}, r_1 e^{i\omega t}, r_2 e^{i(\omega t + \psi)})$ from the group orbit of $SCR1$, where $\psi \in (0, \pi)$ and $r_1 > r_2 > 0$. Setting $z_1 = r_1 e^{i\omega t}$, $z_2 = r_2 e^{i(\psi + \omega t)}$ in equations (7) we determine the fixed amplitudes (r_1, r_2) and phase ψ:

$$
r_1^2 + r_2^2 = -2\lambda \, \mathrm{Im}(f\bar{e})\{2(a_r + b_r)\mathrm{Im}(f\bar{e}) + e_i(|f|^2 - |e|^2)\}^{-1}
\tag{8a}
$$

$$
\frac{r_1^2}{r_2^2} = \frac{\mathrm{Im}(f\bar{e}) - |e|^2 \sin 2\psi}{\mathrm{Im}(f\bar{e}) + |e|^2 \sin 2\psi}
\tag{8b}
$$

$$
\cos 2\psi = \mathrm{Re}(f\bar{e})/|e|^2,
\tag{8c}
$$

where $f \equiv a + b - c - d$, and $a_r \equiv \mathrm{Re}(a)$, $e_i \equiv \mathrm{Im}(e)$, *etc.* Note that the $SCR1$ exist in the open regions of the coefficient space satisfying $|\mathrm{Re}(f\bar{e})| < |e|^2 < |f|^2$. In addition Swift

Pattern	Branching Equation	Stability
TR	$\lambda + a_r|z|^2 = 0$	$sgn(a_r),\ sgn(b_r - a_r)^*,$ $sgn(c_r - a_r)^*,\ sgn(d_r - a_r)^*$
SR	$\lambda + (a_r + b_r)|z|^2 = 0$	$sgn(a_r + b_r),\ sgn(a_r - b_r),$ $sgn(s_1 + s_2)^{**} = -sgn(f_r),$ $sgn(-s_1 s_2) = sgn(|e|^2 - |f|^2)$
SS	$\lambda + (a_r + b_r + c_r + d_r + e_r)|z|^2 = 0$	$sgn(a_r + b_r + c_r + d_r + e_r),$ $sgn(a_r - b_r - e_r)^*,$ $sgn(s_1 + s_2) = sgn(f_r - 3e_r),$ $sgn(-s_1 s_2) = sgn(\mathrm{Re}(f\bar{e}) - |e|^2)$
AR	$\lambda + (a_r + b_r + c_r + d_r - e_r)|z|^2 = 0$	$sgn(a_r + b_r + c_r + d_r - e_r),$ $sgn(a_r - b_r + e_r)^*,$ $sgn(s_1 + s_2) = sgn(f_r + 3e_r),$ $sgn(-s_1 s_2) = sgn(-\mathrm{Re}(f\bar{e}) - |e|^2)$

Table 3: A solution is stable if the signed quantities are all negative. The quantity $f \equiv a+b-c-d$ and the subscript r denotes the real part. s_1 and s_2 are two of the eigenvalues of the Jacobian matrix. An asterisk indicates the presence of a complex conjugate pair of eigenvalues. A double asterisk indicates that the eigenvalues s_1 and s_2 are repeated.

showed, using topological arguments, that there exists, in an open region of the coefficient space, a branch of *quasiperiodic* solutions to (7) which we denote by $SCR2$. In the case where the SR, SS and AR solutions all bifurcate supercritically *sufficient* conditions for the quasiperiodic solution to be stable in the SCR-subspace are:

$$-3e_r < f_r < 0 \quad \text{and} \quad |e|^2 < \mathrm{Re}(f\bar{e}),$$
$$\text{or} \quad 3e_r < f_r < 0 \quad \text{and} \quad |e|^2 < -\mathrm{Re}(f\bar{e}), \tag{9}$$
$$\text{or} \quad f_r > 3|e_r| \quad \text{and} \quad |e|^2 > |f|^2.$$

When one of the above conditions is satisfied the $SCR1$ solution does not exist, although in other cases the quasiperiodic $SCR2$ solution can co-exist stably with the $SCR1$ solution. The stability of $SCR2$ to perturbations out of the SCR subspace has not been determined. However, we expect that it can be stable since, in a particular degenerate limit, it arises

from a supercritical Hopf bifurcation from one of the SS, AR or SR solutions as one of
the cubic coefficients f or e is varied (Swift 1988).

Additional quasiperiodic solutions with submaximal symmetry may be investigated
explicitly as fixed points of appropriately transformed equations using the $T^2 \times S^1$ symmetry, *i.e.*, by going to a moving frame in which these are single–frequency solutions. Unlike
$SCR2$, we do not expect these quasiperiodic solutions to phase–lock if the S^1 normal form
symmetry is broken. Consider first the travelling bimodal pattern which we denote by TB.
This is the analogue of the periodic travelling squares (TS) in the $D_4 \times T^2 \times S^1$–equivariant
problem. The TB take the form $(v_1, v_2, w_1, w_2) = (r_1 e^{i\omega_1 t}, r_2 e^{i\omega_2 t}, 0, 0)$, where generically
$\omega_1 \neq \omega_2$. From (3) it follows that

$$\lambda + i\sigma_1 + ar_1^2 + cr_2^2 = 0$$
$$\lambda + i\sigma_2 + ar_2^2 + dr_1^2 = 0 . \tag{10}$$

From the real parts of these equations we find that

$$\begin{pmatrix} r_1^2 \\ r_2^2 \end{pmatrix} = \frac{-\lambda}{a_r^2 - c_r d_r} \begin{pmatrix} a_r - c_r \\ a_r - d_r \end{pmatrix} . \tag{11}$$

Hence TB exist only when $a_r - c_r$ and $a_r - d_r$ have the *same* sign. The corresponding frequencies ω_1, ω_2 are then readily obtained from the imaginary parts of (10). Both
frequencies are near ω_c for $|\lambda| \ll 1$.

The calculation of the stability of the TB pattern can be divided into two parts, by
considering separately perturbations within the TB fixed point subspace and out of it.
The two nonzero eigenvalues s_1, s_2 describing stability within the TB fixed point subspace
satisfy

$$s_1 + s_2 = 2a_r(r_1^2 + r_2^2), \quad s_1 s_2 = 4(a_r^2 - c_r d_r)r_1^2 r_2^2, \tag{12}$$

so that the requirement for stability in this subspace is that $a_r < 0$, $a_r^2 - c_r d_r > 0$. The
TB is then supercritical if $a_r - c_r < 0$, $a_r - d_r < 0$. Note that when this is the case both
TR_1 and TR_2 (travelling rolls in the x_1, x_2 directions) are unstable in this subspace (see
fig. 2a). If d_r is now changed so that $a_r - d_r \to 0$, the TB fixed point coalesces with
the TR_1 fixed point, and a heteroclinic orbit will form connecting TR_2 to TR_1 provided
all orbits within the TB invariant subspace remain bounded (fig. 2b). Also of interest is
the opposite case, when the TB is unstable, but the TR_1, TR_2 are stable (fig. 2c). Then
as $a_r - d_r \to 0$ (for example) the TR_1 solution becomes unstable to TR_2, to which it is
now connected by a heteroclinic orbit (fig. 2d). Thus the disappearance of the TB state
is intimately related to the loss of stability of the travelling roll state to travelling rolls
oriented at 90° to the original ones. This is the travelling wave analogue of the so-called
Küppers-Lortz instability of stationary rolls (Küppers & Lortz 1969; Knobloch & Silber
1990).

To determine the stability of the TB with respect to perturbations out of the TB
invariant subspace let $(w_1, w_2) = (\tilde{w}_1 e^{i\alpha_1 t}, \tilde{w}_2 e^{i\alpha_2 t})$, where the frequencies satisfy the relation $\alpha_1 - \alpha_2 + \omega_1 - \omega_2 = 0$. Linearizing equations (3c,d) about $\tilde{w}_1 = \tilde{w}_2 = 0$ and dropping
tildes one finds

$$\dot{w}_1 = (\mu - i\alpha_1)w_1 + (br_1^2 + dr_2^2)w_1 + er_1 r_2 w_2$$
$$\dot{w}_2 = (\mu - i\alpha_2)w_2 + (br_2^2 + cr_1^2)w_2 + er_1 r_2 w_1 . \tag{13}$$

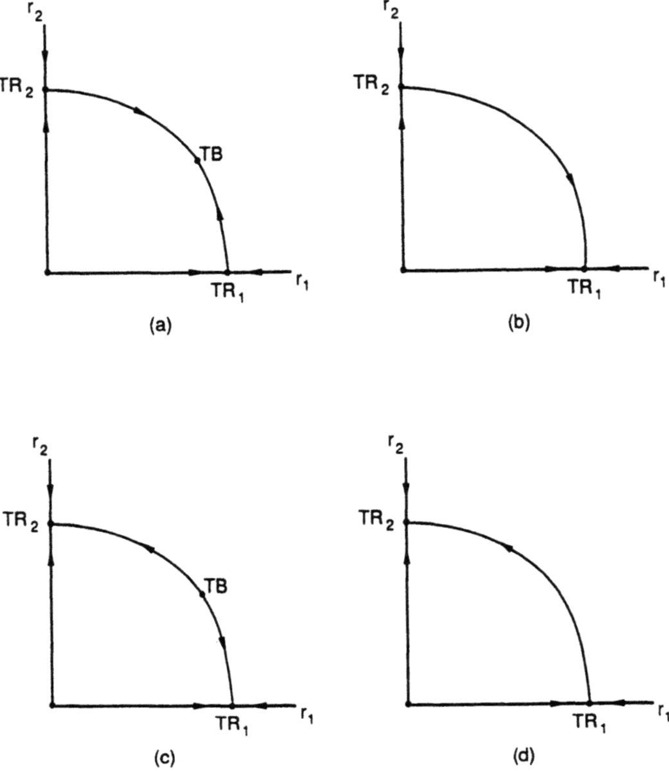

Fig. 2. Typical phase portraits in the TB invariant subspace of equations (3) showing the formation of a structurally stable heteroclinic orbit connecting two orthogonal travelling roll states when the TB state ceases to exist. The portraits are drawn for $a_r < 0$ and (a) $a_r - c_r < 0$, $a_r - d_r < 0$, $a_r^2 - c_r d_r > 0$, (b) $a_r - c_r < 0$, $a_r - d_r > 0$, (c) $a_r - c_r > 0$, $a_r - d_r > 0$, (d) $a_r - c_r > 0$, $a_r - d_r < 0$.

The four eigenvalues $s_\pm$, $\bar{s}_\pm$ are given by

$$s_\pm = \frac{1}{2}\left(\frac{\lambda}{a_r^2 - c_r d_r}\right)[(c_r - d_r)^2 + (a_r - b_r)(2a_r - c_r - d_r) + i\Omega \\ \pm \{(c_r - d_r)^2 f^2 + 4e^2(a_r - c_r)(a_r - d_r)\}^{1/2}], \tag{14}$$

where Ω is real and so does not affect the stability properties. Note that the argument of the square root is complex so that the stability requirement, $\mathrm{Re}(s_\pm) < 0$, takes a somewhat unwieldy form which we omit. However, it is known that the solution can be stable when $c_r = d_r$ (Silber & Knobloch 1991) and hence, by continuity, TB can be stable for $|c_r - d_r|$ sufficiently small.

The vector field (3) restricted to the fixed–point subspace MW in table 1 has $O(2)$ symmetry. It is easy to verify that generically there are no primary branches of the form $(v_1, v_2, w_1, w_2) = (z_1, 0, z_2, 0)$, $|z_1| \neq |z_2|$ and $z_1 z_2 \neq 0$, $i.e.$, quasiperiodic (modulated wave) solutions with $SO(2)$ symmetry. The possible existence of a triply periodic state with the trivial isotropy Z_2^c (state $QP3$ in table 1) has not been established. Such a state would satisfy the fixed point equations

$$\nu_1 + ar_1^2 + br_3^2 + cr_2^2 + dr_4^2 + e\left(\frac{r_2 r_3 r_4}{r_1}\right) e^{2i\psi} = 0$$

$$\nu_2 + ar_2^2 + br_4^2 + cr_3^2 + dr_1^2 + e\left(\frac{r_1 r_3 r_4}{r_2}\right) e^{-2i\psi} = 0$$

$$\nu_3 + ar_3^2 + br_1^2 + cr_4^2 + dr_2^2 + e\left(\frac{r_1 r_2 r_4}{r_3}\right) e^{2i\psi} = 0 \tag{15}$$

$$\nu_4 + ar_4^2 + br_2^2 + cr_1^2 + dr_3^2 + e\left(\frac{r_1 r_2 r_3}{r_4}\right) e^{-2i\psi} = 0 \; ,$$

where $\nu_j \equiv \lambda + i\sigma_j$ $(j = 1, 2, 3, 4)$ and the σ_j satisfy $\sigma_1 - \sigma_2 + \sigma_3 - \sigma_4 = 0$. Here we have used the $T^2 \times S^1$ symmetry to select the representative $(v_1, v_2, w_1, w_2) = (r_1 e^{i\omega_1 t}, r_2 e^{i(\omega_2 t + \psi)},$ $r_3 e^{i\omega_3 t}, r_4 e^{i(\omega_4 t + \psi)})$, where $r_j > 0$ $(j = 1, 2, 3, 4)$, $\psi \in [0, \pi)$, and $\omega_1 - \omega_2 + \omega_3 - \omega_4 = 0$. Equations (15) constitute a set of eight nonlinear equations for the amplitudes and phase $(r_1, r_2, r_3, r_4, \psi)$ and the frequencies $(\omega_1, \omega_2, \omega_3)$ (or, equivalently, $\sigma_1, \sigma_2, \sigma_3$). We hope to report on existence and stability of solutions of (15) in a future publication.

In the above discussion we have established the existence of up to seven primary branches in the Hopf bifurcation with $Z_4 \times T^2$ symmetry. From table 3 and the subsequent discussion of the existence and stability properties of the SCR and TB states it is possible to show that all may be supercritical but unstable. When this is the case other types of solutions may be present. For $a_r < 0$, $(a_r - c_r)(a_r - d_r) < 0$ and $c_r + d_r + 2a_r < 0$, there exists a structurally stable heteroclinic cycle connecting four travelling roll states as shown in fig. 3. Sufficient conditions for the cycle to be asymptotically stable are that $b_r < a_r$, and either $min(-2a_r, a_r - d_r, c_r - b_r) > c_r - a_r > 0$ or $min(-2a_r, a_r - c_r, d_r - b_r) > d_r - a_r > 0$. These existence and stability results follow from propositions 2.4 and 2.6 and theorem 2.10 of Melbourne et $al.$ (1989). Fig. 4 illustrates the approach to such a cycle for $\lambda + i\omega = 0.2 + i$, $a = -1.0 + 1.5i$, $b = -1.6 - 2.5i$, $c = -0.6 - 0.2i$, $d = -1.8 + 3.2i$, and $e = 0.3 + 1.4i$. The four panels show $|v_1(t)|$, $|w_2(t)|$, $|w_1(t)|$ and $|v_2(t)|$ and demonstrate clearly the successive switching between the four travelling roll states, and the increasing length of time spent in each state, as the solution approaches the heteroclinic cycle. Note that for this choice of coefficients $(r, 0, 0, 0)$ is stable with respect to all infinitesimal perturbations except those in the $(0, 0, 0, w_2)$ direction, thereby accounting for the switching sequence. In particular it is the absence of the reflection symmetry in the group Z_4 that is responsible for the preference for the clockwise switching indicated in fig. 3 over counterclockwise switching. We note that the structural stability of the heteroclinic cycle is due, in part, to the S^1 normal form symmetry. If this symmetry is weakly broken a remnant cycle is expected to persist. Such a cycle is, however, no longer $heteroclinic$ – the length of time spent near each travelling wave state remains finite (Melbourne 1989).

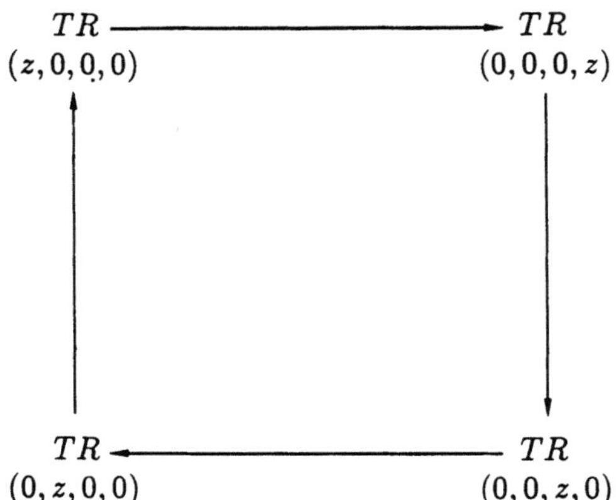

Fig. 3. The structurally stable attracting heteroclinic cycle shown in fig. 4 connects four travelling roll solutions as shown above. The TR state is labelled by its (v_1, v_2, w_1, w_2) coordinates.

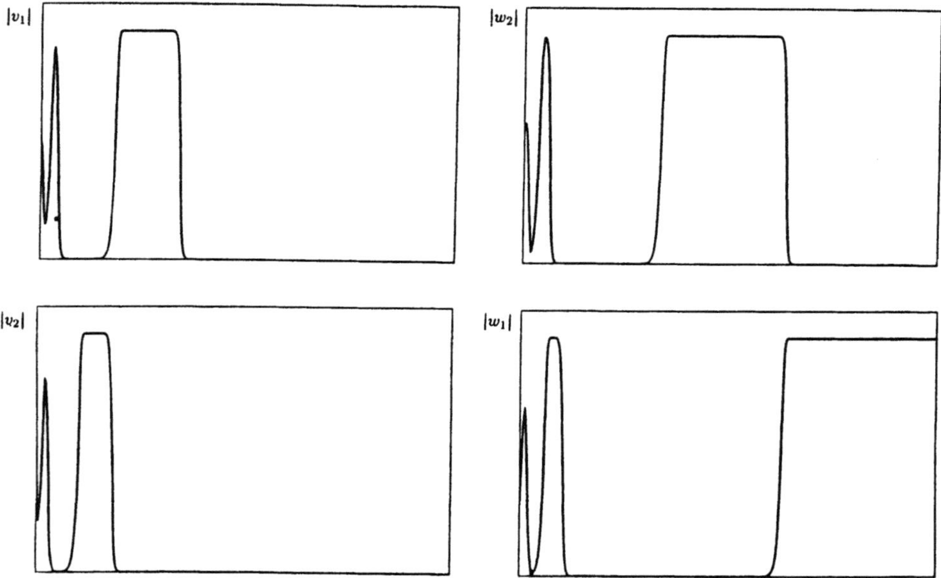

Fig. 4. Solution of equations (3) with a particular choice of coefficients (see text). Clockwise, the four panels show $|v_1(t)|$, $|w_2(t)|$, $|w_1(t)|$ and $|v_2(t)|$ and illustrate the approach to an attracting heteroclinic cycle from arbitrarily chosen initial conditions.

V. Discussion and Conclusion

In this report we have considered the Hopf bifurcation on a square lattice without reflection in vertical planes. We have demonstrated that this bifurcation can give rise to as many as seven distinct primary branches, four of which are always present. Five of the seven branches correspond to singly periodic solutions in time. These we have called travelling rolls (TR), standing rolls (SR), standing squares (SS), alternating rolls (AR) and periodic standing cross-rolls $(SCR1)$, by analogy with the corresponding patterns in the Hopf bifurcation with $D_4 \times T^2$ symmetry. It should be noted, however, that the presence of rotation will break the reflection symmetry in vertical planes present in the planforms presented by Silber & Knobloch (1991). See Goldstein et $al.$ (1991) for a discussion. In addition to these, there are, in open regions in coefficient space, a quasiperiodic solution in the SCR subspace $(SCR2)$ and a two–frequency travelling bimodal state (TB). We have determined the stability of all but the quasiperiodic $SCR2$ state. We have also demonstrated that when all maximal isotropy solutions are supercritical but unstable, a structurally stable, asymptotically stable heteroclinic cycle may be present ($cf.$ Melbourne et $al.$ 1989). This cycle connects rolls travelling in four orthogonal directions. Note that Melbourne et $al.$ call such a cycle $homoclinic$ since it connects identical states related by symmetry.

The possibility of finding quasiperiodic states as primary branches in problems of this type appears not to have been noticed before. We have checked that such solutions are also present in the $Z_2 \times T^2$ and $Z_6 \times T^2$–equivariant Hopf bifurcations. We will report on these solutions in a future publication.

The present study is motivated by the problem of pattern formation in a plane layer of fluid heated uniformly from below and rotated with a constant angular velocity about the vertical. When the Prandtl number of the fluid is large enough the initial instability is a steady state one. On the rhombic lattice the resulting problem is described by a pair of real equations of the form (Knobloch & Silber 1990):

$$\dot{r}_1 = (\lambda + ar_1^2 + cr_2^2)r_1$$
$$\dot{r}_2 = (\lambda + ar_2^2 + dr_1^2)r_2 \tag{16}$$

($cf.$ equation (10)). When the rotation rate is small, the two sets of rolls (denoted R_1 and R_2) oriented at an angle θ to one another are both stable. The mixed mode state (r_1, r_2), $r_1 \neq r_2$ and $r_1 r_2 \neq 0$, is called bimodal and separates the two solutions as in fig. 2c. With increasing rotation rate the bimodal pattern approaches the rolls R_1 and coalesces with them as in fig. 2d. The pattern R_1 is then unstable with respect to R_2, which in turn is unstable to another set of rolls also at angle θ, $etc.$ The resulting instability is called the Küppers-Lortz instability and has been observed experimentally (Busse & Heikes 1980). The instability is made possible by the lack of reflection symmetry in vertical planes.

The analysis of the present report indicates the existence of analogous instabilities for oscillatory patterns. Such patterns arise in fluids with sufficiently small Prandtl number (where they are difficult to access), or more accessibly in rotating binary fluid mixtures. In particular TR_1 lose stability to TR_2 when the pattern TB ceases to exist. If this comes about when $a_r = d_r$ then the instability is in the counterclockwise (corotating) direction;

if instead $a_r = c_r$ then it is in the clockwise direction. There is no analogous instability for standing rolls on the rotating square lattice since the stability properties of the SR solution are the same for $c \neq d$ as they are when c is forced by reflection symmetry to equal d. In contrast on the rotating hexagonal lattice heteroclinic orbits connecting standing rolls do exist (Swift & Barany 1991) and this is also the case for the rotating rhombic lattice.

Acknowledgements

We have benefitted from discussions with Ian Melbourne. The work of E.K. was supported by NSF grant DMS-8814702. M.S. acknowledges support from ONR grant N00014-91-J-1257.

References

Busse, F.H. and Heikes, K.E. 1980 Science **208**, 173-175.

Goldstein, H.F., Knobloch, E. and Silber, M. 1991 preprint.

Golubitsky, M. and Stewart, I.N. 1985 Arch. Rat. Mech. Anal. **87**, 107-165.

Knobloch, E. and Silber, M. 1990 in *Nonlinear Structures in Physical Systems - Pattern Formation, Chaos and Waves*, edited by L. Lam and H.C. Morris, Springer-Verlag, New York.

Küppers, G. and Lortz, D. 1969 J. Fluid Mech. **35**, 609-620.

Melbourne, I. 1989 Dyn. Diff. Eqn. **1**, 347-367.

Melbourne, I., Chossat, P. and Golubitsky, M. 1989 Proc. Roy. Soc. Edinburgh **113A**, 315-345.

Silber, M. and Knobloch, E. 1991 Nonlinearity 4, 1063-1107.

Swift, J.W. 1984 *Bifurcation and Symmetry in Convection*, Ph. D. Thesis, University of California at Berkeley.

Swift, J.W. 1988 Nonlinearity 1, 333-377.

Swift, J.W. and Barany, E. 1991 Euro. J. Mech B **10**, no. 2-Suppl., 99-104.

Forced Symmetry Breaking from $\mathbf{O(3)}$

Reiner Lauterbach
Universität Augsburg, FRG

December 28, 1991

Abstract

Here we consider the flow near a manifold of equilibria of a specific orbit type of an $\mathbf{O(3)}$-equivariant ordinary differential equation if we break the symmetry of the underlying system. This continues joint work with Mark Roberts [5] on cases of breaking $\mathbf{SO(3)}$-equivariance. Applications to PDE's are given in [6].

1 Introduction

Suppose

$$\dot{x} = f(x) \tag{1}$$

is an ODE in $\mathbb{R}^n$, where $f : \mathbb{R}^n \to \mathbb{R}^n$ is C^1 and $\mathbf{O(3)}$-equivariant. In the context of *spontaneous symmetry breaking* (see [2]) one is interested in the question of bifurcation of less symmetric states from a state having the symmetry of the equation (here $\mathbf{O(3)}$). We want to study questions related to breaking the symmetry of the system[1]. More precisely we assume that M is a manifold of equilibria of (1) and ask what kind of flow can be expected near M if we perturb the equation with a term having less symmetry, i.e. we look at

$$\dot{x} = f(x) + \varepsilon h(x), \tag{2}$$

---

[1] J. Marsden has suggested calling this phenomenon "System symmetry breaking"

where $h : \mathbb{R}^n \to \mathbb{R}^n$ is equivariant with respect to a subgroup K of $O(3)$. In the case of breaking $SO(3)$-equivariance this question was adressed in [5]. In [6] the case of partial differential equations with $O(3)$ will be considered. Here, as well as in [5], [6] we assume that M is a single group orbit. The main difference to the $SO(3)$ case is the fact that the containment relations of subgroups are more subtle. This leads to a slightly different structure intypical $SO(3)$-cases. Another difference to the work in [5] is the following:In principle the results presented here and in [5] just depend on the isotropy type of the points on M and the subgroup K. There is no dependence on representations of the underlying group. However one way of producing the situation is through a spontaneous symmetry breaking bifurcation. However, in one parameter bifurcation problems with $SO(3)$ or $O(3)$-equivariance the tetrahedral group T never occurs as an isotropy subgroup of any solution which can be obtained with the so called 'Equivariant Branching Lemma' (compare [4]). This group played an important rôle for the heteroclinic cycles in [5]. The groups O^- (this group is the group of all motions of a tetrahedron, including reflections) and $O(2)^-$, which are considered here are isotropy subgroups in the seven dimensional representation of $O(3)$. In [6] similar results will be applied to discuss the flow described by some partial differential equations near a manifold of equilibria.
In the following we assume

- M is a single group orbit, and

- M is normally hyperbolic (for a definition see [3]).

As a consequence there exists a number $\varepsilon_0 > 0$ such that for all $0 < \varepsilon < \varepsilon_0$ there exists a unique manifold M_ε near M which is invariant under the flow of (2) and diffeomorphic to M. We are interested in features of the flow restricted to this manifold. In [5] it is shown that this flow is K-equivariantly diffeomorphic to a K-equivariant flow on $O(3)/H$ where H is the isotropy subgroup of some point on M. We distinguish two cases

- $K = O^-$ and $H = O(2)^-$

- $H = O^-$ and $K = O(2)^-$ (the dual problem, see [5]).

In the sequel we study K-equivariant flows on $M = O(3)/H$ for these two choices.

2   Subconjugacy

It is wellknown that for a subgroup $K_1 \subset K$ the set $\mathrm{Fix}(K_1) \subset M$ is flow invariant. In order to describe this fixed point subspace it is necessary to introduce the set $N(K_1, H)$, which was first used in [4] in an entirely different context.

Definition 2.1 (a) *Set*

$$N(K_1, H) = \{g \in \mathbf{O}(3) \mid g^{-1}K_1 g \subset H\}. \tag{3}$$

(b) If $N(K_1, H) \neq \emptyset$ we say that K_1 is subconjugate to H.

Lemma 2.1 ([4]) *The set $N(K_1, H)$ is invariant under multiplication with elements from K_1 (even $N(K_1)$) from the left and elements from H from the right.*

In view of this lemma one may factor $N(K_1, H)$ by this action of H. Then we have

Lemma 2.2 ([5])
$$\mathrm{Fix}(K_1) = N(K_1, H)/H.$$

In order to find the fixed point subspaces of subgroups of K on $\mathbf{O}(3)/H$ one determines those subgroups K_1 of K which are subconjugate to H and the set of elements in $N(K_1, H)$. In order to discuss the examples we look at their poset of isotropy subgroups. Let us first recall a lemma due to IHRIG & GOLUBITSKY [4] concerning class III subgroups of $\mathbf{O}(3)$. Those subgroups of $\mathbf{O}(3)$ which are contained in $\mathbf{SO}(3)$ are called class I. Those subgroups of $\mathbf{O}(3)$ which are not of class I and contain $-\mathbb{1}$ are said to be of class II. The remaining ones are called class III subgroups. Any class III subgroup K can be characterized in one-to-one fashion by a pair (L, J) of class I subgroups, where $J = K \cap \mathbf{SO}(3)$ and $L = \pi(K)$, and π denotes the canonical projection. For more details see [4] or [2]. We write $K = (L, J)$. Let $\mathbf{O}$ be the group of all rotations of an octahedron. Then $\mathbf{O}^- = (\mathbf{O}, \mathbf{T})$.

Lemma 2.3 *Let $K_1 = (L_1, J_1)$ and $K_2 = (L_2, J_2)$. Then $K_1 \subset K_2$ if and only if $L_1 \subset L_2$, $J_1 \subset J_2$ and $L_1 \not\subset J_2$.*

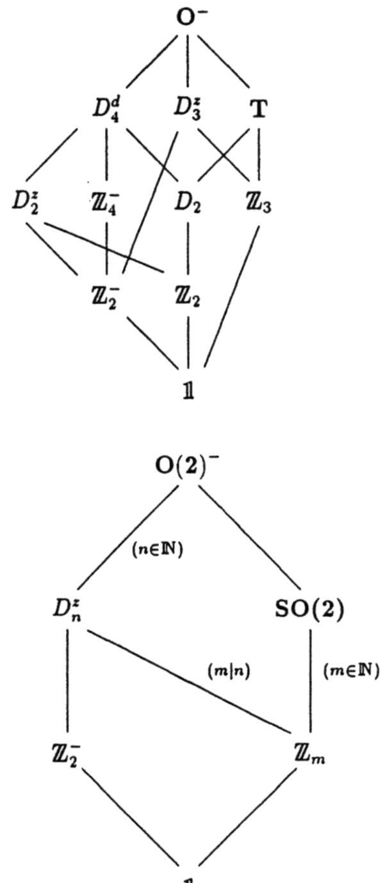

Figure 1: The lattices of subgroups of O^- and $O(2)^-$.

The lattices of (closed) subgroups of $\mathbf{O}^-$ and $\mathbf{O}(2)^- = (\mathbf{O}(2), \mathbf{SO}(2))$ have the form given in figure 1.

Therefore we conclude that the only subgroups of either group subconjugate to the other group is (conjugate to) an element of the following list

$$S = \{\{\mathbb{1}\},\ \mathbb{Z}_2,\ \mathbb{Z}_3,\ \mathbb{Z}_2^-,\ D_2^z,\ D_3^z\}, \tag{4}$$

with $D_k^z = (D_k, \mathbb{Z}_k)$ for $k = 1, 2$.

3   The axisymmetric case

Here we investigate the case, that the remaining symmetry in the equation is $\mathbf{O}(2)^-$, i.e. $K = \mathbf{O}(2)^-$. First we have to give the list of sets $N(K_1, \mathbf{O}^-)$ for $K_1 \in S$.

Lemma 3.1 *For the nontrivial subgroups $K_1 \in S$ we have*

$$
\begin{aligned}
N(\mathbb{Z}_2, \mathbf{O}^-) &= N(\mathbb{Z}_2, \mathbf{T}) \\
N(\mathbb{Z}_3, \mathbf{O}^-) &= N(\mathbb{Z}_3, \mathbf{T}) \\
N(\mathbb{Z}_2^-, \mathbf{O}^-) &= N(\{\mathbb{1}\}, \mathbf{T}) \cap N(\mathbb{Z}_2, \mathbf{O}) \\
N(D_2^z, \mathbf{O}^-) &= \{g \in \mathbf{O}(3) \mid g^{-1}D_2 g \subset \mathbf{O} \wedge g^{-1}\mathbb{Z}_2 g \subset \mathbf{T} \wedge g^{-1}D_2 g \not\subset \mathbf{T}\} \\
N(D_3^z, \mathbf{O}^-) &= N(\mathbb{Z}_3, \mathbf{T}) \cap N(D_3, \mathbf{O})
\end{aligned}
$$

<u>Proof:</u> The first two assertions follow from the fact that a subgroup of $\mathbf{SO}(3)$ is subconjugate to a class III subgroup K if and only if it is subconjugate to the intersection of $\mathbf{SO}(3) \cap K$, since $\mathbf{SO}(3)$ is normal in $\mathbf{O}(3)$. The last two assertions are consequences of Lemma 2.3. The third assumption follows directly from the definitions. $\square$

Since the sets on the right hand side are known from [5] and [1] we have the description of $N(K_1, \mathbf{O})$. Observe however that in [5] the sets $N(K_1, H)$ are given for pairs of subgroups of $\mathbf{SO}(3)$ within $\mathbf{SO}(3)$. If we look at the same situation in $\mathbf{O}(3)$ we have $N_{\mathbf{O}(3)}(K_1, H) = N_{\mathbf{SO}(3)}(K_1, H) \cup (-N_{\mathbf{SO}(3)}(K_1, H))$.

Lemma 3.2

$$N(\mathbb{Z}_2, \mathbf{O}^-) = (\mathbf{O}(2)\mathbf{O}) \cup (-\mathbf{O}(2)\mathbf{O})$$

$$
\begin{aligned}
N(\mathbb{Z}_3, \mathbf{O}^-) &= (\mathbf{O}(2)\mathbf{O}) \cup (-\mathbf{O}(2)\mathbf{O}) \\
N(\mathbb{Z}_2^-, \mathbf{O}^-) &= (\mathbf{O}(2)\mathbf{O}) \cup (-\mathbf{O}(2)\mathbf{O}) \\
N(D_2^z, \mathbf{O}^-) &= N(\mathbf{O}) \\
N(D_3^z, \mathbf{O}^-) &= (\mathbf{O}(2)\mathbf{O} \cap D_6\mathbf{O}) \cup (-(\mathbf{O}(2)\mathbf{O} \cap D_6)) \\
&= (D_6\mathbf{O}) \cup (-D_6\mathbf{O}).
\end{aligned}
$$

<u>Proof:</u> Compare [5] and use the remark just made. The set $N(D_2^z, \mathbf{O}^-)$ is a group by theorem A.2 in [1]. Therefore it contains $N(\mathbf{O}^-)$ which is a maximal subgroup of $\mathbf{O}(3)$ and equals $N(\mathbf{O})$. $\qquad\Box$
Observe that the meaning of $\mathbf{O}(2)\mathbf{O}$ is ambiguous. In all the cases $\mathbf{O}(2)$ stands for the normalizer of the smaller group on the left side (in the last case it means the normalizer of $\mathbb{Z}_3 \subset D_3$ in $\mathbf{SO}(3)$). Factoring out the action of $\mathbf{O}^-$ yields the structure for the fixed set of the various groups. If this set is a circle we denote it by S^1, finite sets are described by the number of elements. Observe that $\mathbf{O}(2)^-$ contains infinitely many subgroups conjugate to $\mathbb{Z}_2^-$, D_2^z and D_3^z respectively, but only one group conjugate to $\mathbb{Z}_2$ and $\mathbb{Z}_3$.

Lemma 3.3

$$
\begin{aligned}
\operatorname{Fix}(\mathbb{Z}_2) &= S^1 \\
\operatorname{Fix}(\mathbb{Z}_3) &= S^1 \cup S^1 \\
\operatorname{Fix}(\mathbb{Z}_2^-) &= S^1 \\
\operatorname{Fix}(D_2^z) &= 2pts \\
\operatorname{Fix}(D_3^z) &= 4pts.
\end{aligned}
$$

<u>Proof:</u> We know that

$$
\operatorname{Fix}(\mathbb{Z}_2) = N_{\mathbf{O}(3)}(\mathbb{Z}_2, \mathbf{O}^-)/\mathbf{O}^- = N_{\mathbf{SO}(3)}(\mathbb{Z}_2, \mathbf{T})/\mathbf{T} = S^1.
$$

The last equality is proved in [5]. Note, that in order to prove the second equality in this line a subtlety is involved. Two copies of $\mathbb{Z}_2$ in $\mathbf{O}$ are conjugate under elements from $\mathbf{O}$ iff they are conjugate under elements from $\mathbf{T}$. This is true although there are two conjugacy classes of $\mathbb{Z}_2$ in $\mathbf{O}$.
The second assertion is derived in a similar fashion [5]:

$$
\operatorname{Fix}(\mathbb{Z}_3) = N_{\mathbf{O}(3)}(\mathbb{Z}_3, \mathbf{O}^-)/\mathbf{O}^- = N_{\mathbf{SO}(3)}(\mathbb{Z}_3, \mathbf{T})/\mathbf{T} = S^1 \cup S^1.
$$

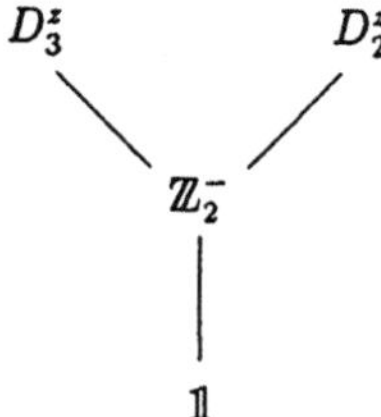

Figure 2: The poset of isotropy subgroups in the axisymmetric case

The third assertion can be seen from this computation

$$\mathrm{Fix}(\mathbb{Z}_2^-) = N_{\mathbf{O}(3)}(\mathbb{Z}_2^-, \mathbf{O}^-)/\mathbf{O}^- = ((\mathbf{O}(2)\mathbf{O}) \cup (-\mathbf{O}(2)\mathbf{O}))/\mathbf{O}^-.$$

A consideration as in the first assertion leads to

$$\mathrm{Fix}(\mathbb{Z}_2^-) = S^1.$$

The fourth assertion comes from the fact that $N(\mathbf{O})/\mathbf{O}^-$ has two elements. Finally the last statement can be seen in the following way.

$$\mathrm{Fix}(D_3^z) = ((D_6\mathbf{O} \cup (-D_6\mathbf{O}))/\mathbf{O}^-) = D_6\mathbf{O}/\mathbf{T}.$$

This last expression gives 4 points.   □

<u>Remark:</u> It turns out that the isotropy subgroups are D_3^z, D_2^z, $\mathbb{Z}_2^-$. The fixed point sets of $\mathbb{Z}_3$, $\mathbb{Z}_2$ are filled with points of isotropy D_3^z and D_2^z respectively. The partially ordered set of conjugacy classes of isotropy subgroups for this action is given in figure 2.
The main result of this section describes some features of a $\mathbf{O}(2)^-$-equivariant flow on $\mathbf{O}(3)/\mathbf{O}^-$.

Theorem 3.1 *For sufficiently small $\varepsilon > 0$ the flow on M_ε has infinitely many invariant circles of points with $\mathbb{Z}_2^-$ symmetry and one circle of equilibria with D_2^z symmetry and two circles of equilibria with D_3^z symmetry. On each invariant $\mathbb{Z}_2^-$ circle there are two points with D_2^z symmetry and four points with D_3^z symmetry.*

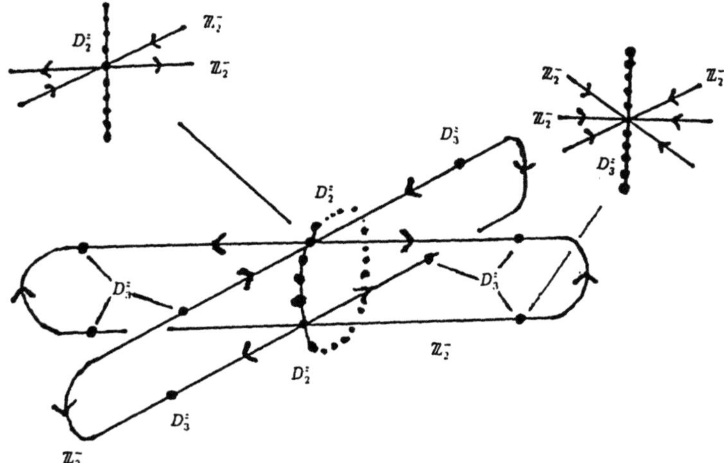

Figure 3: A possibility for the flow on the invariant complex in the axisymmetric case

The proof follows from the considerations given above. $\qquad\qquad$ $\Box$
A possible flow for this case is indicated in figure 3.
In [6] we decide for some PDE's which of these alternatives occur.

4   The dual case

In this section we briefly describe flows which are equivariant with respect to the action of O^- on $O(3)/O(2)^-$. This homogeneous space is a connected two manifold. We are not going to explore this fact. The same procedure as before leads to:

Lemma 4.1

$$
\begin{aligned}
N(\mathbb{Z}_2^-, O(2)^-) &= O(2)O(2) \cup -O(2)O(2) \\
N(\mathbb{Z}_2, O(2)^-) &= O(2) \cup -O(2) \\
N(\mathbb{Z}_3, O(2)^-) &= O(2) \cup -O(2) \\
N(D_2^z, O(2)^-) &= N(\mathbb{Z}_2, O(2)^-) \\
N(D_3^z, O(2)^-) &= O(2) \cup -O(2)
\end{aligned}
$$

<u>Proof:</u> Follows directly from the definitions. The fourth assertion comes from the obvious relation

$$N(D_2^z, \mathbf{O}(2)^-) \subset N(\mathbb{Z}_2, \mathbf{O}(2)^-)$$

and the implication $\mathbb{Z}_2 \subset D_2$ and $\mathbb{Z}_2 \subset \mathbf{SO}(2)$ then $D_2 \subset \mathbf{O}(2)$. $\quad\square$

<u>Remark:</u> Observe that the two copies of $\mathbf{O}(2)$ in the first assertion are different. One of them is the normalizer of $\mathbb{Z}_2$, the other one is the normalizer of $\mathbf{O}(2)$. Here $\mathbb{Z}_2$ is one of the reflections on $\mathbf{O}(2)$. The next lemma describes the fixed sets of these subgroups.

Lemma 4.2

$$
\begin{aligned}
\mathrm{Fix}(\mathbb{Z}_2^-) &= S^1 \cup S^1 \\
\mathrm{Fix}(\mathbb{Z}_2) &= 2pts \\
\mathrm{Fix}(\mathbb{Z}_3) &= 2pts \\
\mathrm{Fix}(D_2^z) &= 2pts. \\
\mathrm{Fix}(D_3^z) &= 2pts.
\end{aligned}
$$

<u>Proof:</u> The last four statements follow from the fact that the set $N(K_1, \mathbf{O}(2)^-)$ is a group and the subgroup which has to be factored out is a subgroup of index 2. To prove the first assertion we observe that the action of $\mathbf{O}(2)^-$ which has to be factored leads to sets $\mathbf{O}(2)$ and $\mathbf{O}(2)\gamma$, where γ is in the nontrivial coset of $\mathbf{O}(2)^-$ in $\mathbf{O}(2)\cup(-\mathbf{O}(2))$. The set $\{\mathbf{O}(2), \mathbf{O}(2)\gamma\}$ consists of two disjoint circles. $\quad\square$

Therefore we find that the isotropy subgroups in this case are: $\mathbb{Z}_2^-$, D_2^z, D_3^z. The poset of isotropy subgroups is the same as in the other case. If we look at the flow we see that the points with isotropy D_2^z, D_3^z have to be equilibria. On the circles with isotropy $\mathbb{Z}_2^-$ the flow is more interesting. Each group $\mathbb{Z}_2^-$ is contained in two of the four groups of conjugacy class D_3^z (see [1], theorem A.2). On each circle which is fixed under $\mathbb{Z}_2^-$ we find one equilibrium with isotropy $D_3^{z(1)}$ and $D_3^{z(2)}$ respectively. There is a group element in the normalizer of $\mathbb{Z}_2^-$ interchanging the rôle of $D_3^{z(1)}$ and $D_3^{z(2)}$. Furthermore each circle contains two points with isotropy D_2^z. Therefore the two types of equilibria of D_3^z type have the same stability properties. Group theory does not suffice to give more details on these flows. In finite dimensions it is possible to construct flows with all possible flows, which are not excluded by group theory (see [5]). Again for some PDE's it is possible to discuss these invariant complexes and their flows (see [6]). We have proved

Theorem 4.1 O^- *equivariant flows on* $O(3)/O(2)^-$ *have always equilibria with isotropy* D_2^z *and* D_3^z. *There exist connecting orbits having symmetry* $\mathbb{Z}_2^-$.

References

[1] P. CHOSSAT, R. LAUTERBACH & I. MELBOURNE, Steady-state bifurcation with O(3)-symmetry, *Arch. Rat. Mech. Anal.*, **113**(4), 313-376, (1991)

[2] M. GOLUBITSKY, I. STEWART & D. SCHAEFFER, *Singularities and Groups in Bifurcation Theory*, Vol. II Springer Verlag, (1988)

[3] M. W. HIRSCH, C. C. PUGH & M. SHUB, *Invariant Manifolds, Lecture Notes in Mathematics* **583**. Springer Verlag, (1977)

[4] E. IHRIG & M. GOLUBITSKY, Pattern selection with O(3)-symmetry, *Physica 13D*, pages 1-13, (1984)

[5] R. LAUTERBACH & M. ROBERTS, Heteroclinic cycles in dynamical systems with broken spherical symmetry, *J. Diff. Equat.*, (to appear)

[6] R. LAUTERBACH & M. ROBERTS, (in preparation)

International Series of Numerical Mathematics, Vol. 104, © 1992 Birkhäuser Verlag Basel

Utilization of Scaling Laws and Symmetries in the Path Following of a Semilinear Elliptic Problem*

Z. Mei

Fachbereich Mathematik, Universität Marburg, 3550 Marburg/Lahn, FRG and
Department of Mathematics, Xi'an Jiaotong University, Xi'an 710049, PRC

Abstract

Path following of a semilinear elliptic problem is considered at its simple and corank-2 bifurcation points. Exploiting the scaling laws and symmetries of the problem, we show that bifurcation points can be divided into equivalent classes and path following of solution branches can be confined to the fundamental domains, which reduce the computational work and improve the numerical conditioning of the problem.

1. Introduction

Consider the semilinear elliptic problem on the unit square $\Omega := [0,1] \times [0,1]$ with Dirichlet boundary conditions:

$$Find \quad (u,\lambda) \in X_0 \times \mathbf{R}, \quad such\ that \quad G(u,\lambda) := \Delta u + \lambda f(u) = 0, \qquad (1.1)$$

where $f : \mathbf{R} \mapsto \mathbf{R}$ is a smooth odd function, $f'(0) \cdot f'''(0) \neq 0$ and

$$X_0 := \left\{ u \in C^{2,s}(\Omega) \mid u|_{\partial\Omega} = 0 \right\}, \quad Y_0 := C^{0,s}(\Omega), \quad G : X_0 \times \mathbf{R} \mapsto Y_0. \qquad (1.2)$$

For simplicity, we take the normalization $f'(0) = 1$ and assume

$$f'''(0) < 0. \qquad (1.3)$$

The case $f'''(0) > 0$ can be treated similarly. Due to $f(0) = 0$, the problem (1.1) has a trivial solution curve $\{(0,\lambda); \lambda \in \mathbf{R}\}$. The set of bifurcation points of (1.1) on $\{(0,\lambda); \lambda \in \mathbf{R}\}$ consists of $(0,\lambda_0)$ with $\lambda_0 = (m^2 + n^2)\pi^2, m,n \in \mathbf{N}$ as eigenvalues of the Laplacian $-\Delta$ on Ω with the Dirichlet boundary conditions. At simple and corank-2 bifurcation points, it is known by a modified Lyapunov-Schmidt method that (1.1) has one and four nontrivial solution branches respectively (cf. Allgower/Böhmer/Mei [1], Mei [17, 18]). In this paper we consider a path following of these solution branches with considerations of the symmetries of (1.1). In particular, we show that scaling laws hold for (1.1), i.e. solution branches at some bifurcation points can be shifted to others by a simple rescaling. Thus path following may be confined to the selected bifurcation points.

To study the (hidden) symmetries and the scaling laws of the solution branches, we follow Healey/Kielhöfer [15] and embed (1.1) into a periodic problem with a larger symmetry group. Let

$$C_T^{k,s}(\mathbf{R}^2) := \{u \in C^{k,s}(\mathbf{R}^2), \quad u \text{ has the period 2 in } x,y\}, \quad k = 0,1,2 \qquad (1.4)$$

* The work was supported by the Deutsche Forschungsgemeinschaft, F. R. Germany.

and

$$X := C_T^{2,s}(\mathbf{R}^2), \qquad Y := C_T^{0,s}(\mathbf{R}^2), \tag{1.5}$$

endowed with the Hölder norms $\| \cdot \|_{2,s}$ and $\| \cdot \|_{0,s}$ respectively. We consider the group $\Gamma := Z_2 \times D_4 \times T_2$, where T_2 is a translation group in the (x,y)-plane:

$$T_2 := \{(\theta_1, \theta_2) \mid \theta_i \in \mathbf{R} \,(mod\ 2)\}. \tag{1.6}$$

The action of T_2 on Y is defined by

$$(\theta_1, \theta_2)u(x,y) := u(x - \theta_1, y - \theta_2) \qquad \text{for all } (\theta_1, \theta_2) \in T_2,\ u \in Y. \tag{1.7}$$

D_4 is the symmetry group of the square $[-1,1] \times [-1,1]$, see Fig. 1.1, and the action of D_4 on Y is generated by

$$S_1 u(x,y) = u(-x,y),\ S_2 u(x,y) = u(y,x) \text{for all } u \in Y. \tag{1.8}$$

The action of $Z_2 := \{I, -I\}$ is defined as

$$(\pm I)u(x,y) = \pm u(x,y) \qquad \text{for all } u \in Y. \tag{1.9}$$

It is easy to verify that the operator G in (1.1) maps X into Y and is Γ-equivariant, i.e.

$$G(\gamma u, \lambda) = \gamma G(u, \lambda) \quad \text{for all } \gamma \in \Gamma,\ u \in X,\ \lambda \in \mathbf{R}. \tag{1.10}$$

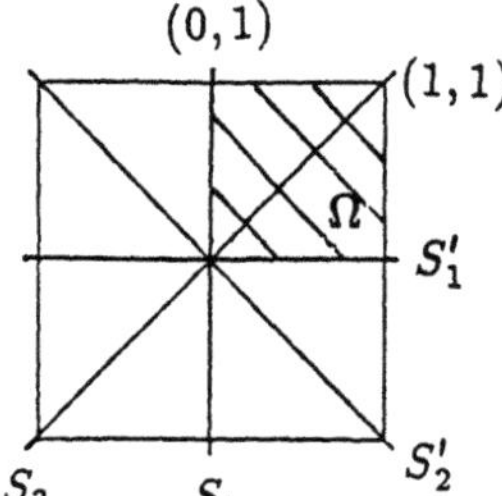

Fig. 1.1: D_4-group.

Now let us consider the periodic problem:

$$\textit{Find} \quad (u, \lambda) \in X \times \mathbf{R}, \qquad \textit{such that} \quad G(u, \lambda) = 0. \tag{1.11}$$

The restriction of (1.11) to $X_0 \times \mathbf{R}$ coincides obviously with the original problem (1.1). The regularity of solutions of (1.1) follows from the smoothness of f and the homogeneous Dirichlet boundary conditions.

For any subgroups Σ of Γ, Σ-symmetric solution branches of (1.1) are determined by the reduced problem:

$$\textit{Find} \quad (u, \lambda) \in X^{\Sigma} \times \mathbf{R}, \qquad \textit{such that} \quad G^{\Sigma}(u, \lambda) = 0, \tag{1.12}$$

where $G^{\Sigma} := G|_{X^{\Sigma} \times \mathbf{R}}$ is the restriction of G to $X^{\Sigma} \times \mathbf{R}$. The problems with Dirichlet or Neumann boundary conditions correspond to (1.12) with special subgroups $\Sigma \preceq \Gamma$, see e.g. Lemma 2.1 below. This embedding of (1.1) into (1.11) and the group Γ allow proper descriptions of the so-called hidden symmetries of solution branches of (1.1), as well as of Neumann boundary value problems. The reduced problem (1.12) is very useful in the analysis of the existence of bifurcating solution branches as well as in their parameterizations, see e.g. Golubitsky/Stewart/Schaefer [14], Vanderbauwhede [21], Allgower/Böhmer/Mei [4, 5]. Nevertheless, in the numerical approximations, discretizing (1.12) in the fixed point space X^{Σ} requires special schemes, and path following with (1.12) is often cubmersome.

For these reasons, we want to transform (1.12) into the forms of classical PDEs on fundamental domains of Ω (cf. Bossavit [10] for linear problems, and Allgower/Böhmer/Mei [2]). Thereafter, it may be discretized by the standard numerical methods and the solution branches can be traced with continuation methods.

2. Reduced Problems and Scaling Laws

We consider in this section an equivalent form of (1.1) and the scaling properties of bifurcating solution branches mentioned in Budden/Norbury [11, 12] for (1.1) at simple and double bifurcation points. Scaling laws have also been observed in Kuramoto-Sivashinsky equations and porous medium convection, see e.g. Scovel/Kevrekids/Nicolaenko [20], and for a general setting, Aston [7].

Let, e.g.,

$$\tilde{\Sigma}_M := \{-S_1, -S_1'; I, R^2\} \tag{2.1}$$

be a subgroup of Γ, generated by $-S_1, -S_1'$. Here and in the sequel we will separate the generators of a group from the other elements with the semicolon ";".

Lemma 2.1: *The space X_0 in (1.2) is isomophic to $X^{\tilde{\Sigma}_M}$ and the problem (1.1) is equivalent to*

$$Find \quad (u, \lambda) \in X^{\tilde{\Sigma}_M} \times \mathbf{R}, \qquad such\ that \quad G^{\tilde{\Sigma}_M}(u, \lambda) = 0. \tag{2.2}$$

Proof: Let $\rho : u \in Y \mapsto \rho(u) := u|_\Omega \in Y_0$ be the restriction mapping. For any $u \in X^{\tilde{\Sigma}_M}$, the anti-symmetries $-S_1, -S_1'$ of u and its 2-periodicity in the x, y directions imply that $\rho(u) \in X_0$. Moreover, ρ is linear and obviously maps $X^{\tilde{\Sigma}_M}$ into X_0 injectively. On the other hand, for any $u \in X_0$, let $\tilde{u} \in Y^{\tilde{\Sigma}_M}$ be the extension of u to the whole space $\mathbf{R}^2$ by the reflections $-S_1, -S_1'$ and the 2-periodicity in the x, y directions. Since u is $C^{2,s}$-continuous in Ω, the function $\tilde{u}$ is also $C^{2,s}$-continuous inside of $S_i(\Omega)$ and $(\theta_1, \theta_2)(\Omega)$ for θ_i (mod 1) $\neq 0$, $i = 1, 2$. We claim that $\tilde{u}$ is also $C^{2,s}$-continuous across the lines $x = i$, $y = j$ for all $i, j \in \mathbf{N}$, in other words, $\tilde{u} \in X^{\tilde{\Sigma}_M}$. As an example, we consider the $C^{2,s}$-continuity of $\tilde{u}$ at $x = 0$. The extension via the reflection $-S_1$ shows

$$\tilde{u}(-x, y) = -\tilde{u}(x, y) = -u(x, y), \quad \frac{\partial^i \tilde{u}}{\partial x^i}(-x, y) = (-1)^{i+1} \frac{\partial^i \tilde{u}}{\partial x^i}(x, y), \ \forall\, x \in (0, 1), i = 1, 2.$$

Hence,

$$\frac{\partial \tilde{u}}{\partial x}\Big|_{x=0+} = \lim_{\epsilon \to 0+} \frac{1}{\epsilon}[u(\epsilon, y) - u(0, y)] = \lim_{\epsilon \to 0+} \frac{1}{-\epsilon}[u(-\epsilon, y) - u(0, y)] = \frac{\partial \tilde{u}}{\partial x}\Big|_{x=0-},$$

$$\frac{\partial^2 \tilde{u}}{\partial x^2}\Big|_{x=0+} = \lim_{\epsilon \to 0+} \frac{1}{\epsilon}[\frac{\partial u}{\partial x}(\epsilon, y) - \frac{\partial u}{\partial x}(0, y)] = -\lim_{\epsilon \to 0+} \frac{1}{-\epsilon}[\frac{\partial u}{\partial x}(-\epsilon, y) - \frac{\partial u}{\partial x}(0, y)] = -\frac{\partial^2 \tilde{u}}{\partial x^2}\Big|_{x=0-}$$

and

$$\frac{\partial^2 \tilde{u}}{\partial x^2}\Big|_{x=0} = \lim_{\epsilon \to 0} \frac{1}{2\epsilon}[\frac{\partial u}{\partial x}(\epsilon, y) - \frac{\partial u}{\partial x}(-\epsilon, y)] = 0.$$

Therefore, $\frac{\partial^i \tilde{u}}{\partial x^i}$ is continuous at $x = 0, i = 1, 2$. The Hölder continuity of $\frac{\partial^i \tilde{u}}{\partial x^i}$ can be proved similarly. Thus $\tilde{u} \in X \cap Y^{\tilde{\Sigma}_M} = X^{\tilde{\Sigma}_M}$ and ρ maps $X^{\tilde{\Sigma}_M}$ onto X_0. Consequently, ρ is an isomorphism from $X^{\tilde{\Sigma}_M}$ onto X_0.

If $(u(t), \lambda(t))$ is a $\bar{\Sigma}_M$-symmetric solution branch of (1.11), then $\rho(u(t)) \in X_0$ and

$$G(\rho(u(t)), \lambda(t)) = G(u(t), \lambda(t))|_\Omega = \rho(G(u(t), \lambda(t))) = 0.$$

Hence $(\rho(u(t)), \lambda(t))$ is a solution of (1.1). Conversely, if $(u(t), \lambda(t)) \in X_0 \times \mathbf{R}$ satisfies (1.1), extending $u(t)$ to the whole domain space $\mathbf{R}^2$ by the reflections $-S_1, -S_1'$ and the periodicities, we obtain $\rho^{-1}(u) \in X^{\Sigma_M}$ and

$$G(\rho^{-1}(u(t)), \lambda(t)) = \rho^{-1}\rho G(\rho^{-1}(u(t)), \lambda(t)) = \rho^{-1}G(u(t), \lambda(t)) = 0,$$

i.e. $(\rho^{-1}(u(t)), \lambda(t))$ is a $\bar{\Sigma}_M$-branch of (1.1).   ∎

Remark 2.1: Lemma 2.1 shows also that if (1.11) has a Σ-solution branch and $\bar{\Sigma}_M \preceq \Sigma$, then this Σ-branch also satisfies (1.1). In particular, all isotropy groups in the following sections contain $\bar{\Sigma}_M$ as proper subgroups.

For any integer $k \in \mathbf{N}$, we define an epimorphism $\beta_k : \Gamma \mapsto \Gamma$ by

$$\beta_k(S_i) = S_i, \ i = 1, 2, \quad \beta_k(\pm I) = \pm I, \quad \beta_k((\theta_1, \theta_2)) = (k\theta_1, k\theta_2). \tag{2.3}$$

The kernel of β_k is the finite cyclic group:

$$\mathrm{kern}(\beta_k) = Z_k := \{\overbrace{(\tfrac{2}{k}, \tfrac{2}{k}) \circ \cdots \circ (\tfrac{2}{k}, \tfrac{2}{k})}^{i \ \text{times}}, \ i = 1, \ldots, k\}. \tag{2.4}$$

Evidently,

$$Y^{Z_k} = \{u \in Y \mid u(x - \tfrac{2}{k}, y - \tfrac{2}{k}) = u(x, y)\}, \qquad X^{Z_k} = X \cap Y^{Z_k}.$$

Define a mapping $h_k : u \in Y \mapsto h_k u \in Y^{Z_k}$ by

$$h_k u(x, y) = u(kx, ky) \qquad \text{for all } u \in Y. \tag{2.5}$$

Correspondingly, h_k maps X into X^{Z_k} and

$$h_k(\theta_1, \theta_2) = (k\theta_1, k\theta_2)h_k, \qquad h_k S_i = S_i h_k, \quad i = 1, 2. \tag{2.6}$$

Moreover, the following scaling laws hold:

$$k^2 h_k G(u, \lambda) = G(h_k u, k^2 \lambda) \qquad \text{for all } (u, \lambda) \in X \times \mathbf{R}, \ k \in \mathbf{N}. \tag{2.7}$$

Hence, solutions of the equation (1.11) have the scaling properties (cf. Aston [7]):

Theorem 2.1: *Let $(0, \lambda_0)$ be a bifurcation point of (1.11). Then $(u(t), \lambda(t))$ is a solution branch of (1.11) passing through $(0, \lambda_0)$ if and only if $(h_k u(t), k^2 \lambda(t))$ is a solution branch of (1.11) passing through $(0, k^2 \lambda_0)$.*

Since $X^{\tilde{\Sigma}_M}$ is $\beta_k(\Gamma)$-invariant and h_k maps $X^{\tilde{\Sigma}_M}$ into itself, the above scaling laws also hold for the reduced problem (2.2). In other words, solution branches of the Dirichlet boundary value problem (1.1) have also the scaling properties. In particular, if two bifurcation points $(0, k^2\lambda_0)$ and $(0, \lambda_0)$ have the same corank, the numbers of bifurcating solution branches of (1.1) at those points are identical, see e.g. Allgower/Böhmer/Mei [1], Mei [17, 18]. The scaling laws imply that all solution branches at $(0, \lambda_0)$ can be shifted to those at $(0, k^2\lambda_0)$ and vice versa. In this way, the scaling laws of (1.11) divide the set of bifurcation points into "equivalent" classes, i.e. if for any $\lambda_0 = (m^2 + n^2)\pi^2$ in which m, n have no common factors, then solution branches at all points in the set

$$S_{\lambda_0} := \{(0, k^2\lambda_0), k \in \mathbf{N} \mid k^2\lambda_0 \text{ has the same multiplicity as } \lambda_0\}.$$

can be generated from those at a single point. Hence, solution branches at different points in S_{λ_0} may be treated as "equivalent". Numerical path following across the bifurcation points in S_{λ_0} should be done at one of these points. Usually, we choose $(0, \lambda_0)$ with the smallest λ_0 to achieve the best numerical conditioning in the path following process, i.e. we will consider only the case $\lambda_0 = (m^2 + n^2)\pi^2$ and m, n have no common factor.

Remark 2.2: For all $k \in N$, the corank of $(0, k^2\lambda_0)$, i.e. the multiplicity of $k^2\lambda_0$ in our case, is always greater than or equal to that of $(0, \lambda_0)$. If the corank of $(0, \lambda_0)$ is strictly less than that of some $(0, k_0^2\lambda_0)$, then the numbers of bifurcating solution branches at these two points are different, e.g. at the simple and corank-3 bifurcation points $(0, 2\pi^2)$ and $(0, 5^2 \cdot 2\pi^2)$, the problem (1.1) has one and thirteen different nontrivial solution branches, respectively, see Allgower/Böhmer/Mei [1], Mei [18]. In this case, the scaling laws hold only among the solution branches at $(0, \lambda_0)$ and a small part of the solution branches at $(0, k_0^2\lambda_0)$.

3. Path Following beyond Simple Bifurcation Points

Simple bifurcation points of (1.1) are in the form $(0, \lambda_0) = (0, 2k^2\pi^2), k \in \mathbf{N}$ and λ_0 is a simple eigenvalue of the Laplacian $-\Delta$ on Ω with Dirichlet boundary conditions. The Crandall/Rabinowitz [13] theory shows that (1.1) has exactly one nontrivial solution branch passing through a simple bifurcation point $(0, \lambda_0)$ (cf. also Allgower/Böhmer/Mei [1, 5], Mei [16, 18]). Let $(u(t), \lambda(t))$ be the nontrivial solution branch across $(0, 2\pi^2)$. According to the scaling laws above, the nontrivial solution branch at a simple bifurcation point $(0, 2k^2\pi^2)$ is given by $(h_k u(t), k^2\lambda(t))$, where h_k is defined in (2.5). Thus path following across all simple bifurcation points of (1.1) can be confined to the single point $(0, 2\pi^2)$.

Let $\Sigma_1 := \{-S_1, S_2, (1,0)S_1, (0,1)S_1'; \cdots\}$ be a subgroup of Γ with the generators $-S_1$, S_2 and $(1,0)S_1$, $(0,1)S_1'$ which represent compositions of the actions in (1.7) and (1.8). Since $\tilde{\Sigma}_M$ is a proper subgroup of Σ_1, any Σ_1-solution branch of (1.11) satisfies also (2.2), i.e. (1.1), see Lemma 2.1. The Γ-equivariance of G, the condition (1.3) and

$$N(D_u G^{\Sigma_1}(0, 2\pi^2)) = \text{span}[\phi] \subset X^{\Sigma_1}, \qquad \phi := 2\sin \pi x \sin \pi y, \qquad (3.1)$$

imply that $(0, 2\pi^2)$ is also a simple bifurcation point of the reduced problem $G^{\Sigma_1}(u, \lambda) = 0$. Thus it has a nontrivial solution branch at $(0, 2\pi^2)$, which is a Σ_1-symmetric solution branch of (1.11) and (1.1). To trace this Σ_1-branch, we transform $G^{\Sigma_1}(u, \lambda) = 0$ into a classical PDE so that we can discretize it with the standard numerical methods.

Choosing a fundamental domain Ω_1 of $\mathbf{R}^2$ with respect to Σ_1 (cf. Bossavit [10]) as

$$\Omega_1 := \{(x,y) \mid 0 \le x \le y \le \tfrac{1}{2}\}, \tag{3.2}$$

we define

$$X_1 := \{u \in C^{2,\bullet}(\bar{\Omega}_1) \,\big|\, u\big|_{x=0} = \frac{\partial u}{\partial x}\big|_{x=0.5} = 0, \, (\frac{\partial u}{\partial x} - \frac{\partial u}{\partial y})\big|_{x=y} = 0\}, \; Y_1 := C^{0,\bullet}(\bar{\Omega}_1). \tag{3.3}$$

For any $u \in X^{\Sigma_1}$, its anti-symmetries $-S_1$, $-S_1'$ and the symmetries $(1,0)S_1$, $(0,1)S_1'$ imply the symmetries with respect to $x = 1/2$, $y = 1/2$, respectively. Together with the symmetry S_2, one sees that the restriction of u to Ω_1 belongs to X_1. Moreover,

Theorem 3.1: *The space X^{Σ_1} is isomorphic to X_1 and the reduced problem $G^{\Sigma_1}(u,\lambda) = 0$ is equivalent to the following PDE on Ω_1:*

$$\text{Find} \quad (u,\lambda) \in X_1 \times \mathbf{R}, \qquad \text{such that} \quad G_1(u,\lambda) := \Delta u + \lambda f(u) = 0. \tag{3.4}$$

The problem (3.4) is in fact a simple restriction of (1.1) to Ω_1 with the corresponding boundary conditions. Nevertheless, it is the Γ-equivariance of G which justify the validity of this restriction.

Due to (3.1), $(0, 2\pi^2)$ is still a simple bifurcation point of (3.4). The nontrivial solution branch of (3.4) at $(0, 2\pi^2)$ is in the form:

$$(u(t), \lambda(t)) = (t\alpha(t)\phi + t^3 v(t), 2\pi^2 + t^2), \quad t \in \mathbf{R}, \; v(t) \in R(D_u G_1(0, 2\pi^2)), \tag{3.5}$$

see Mei [16, 18], Allgower/Böhmer/Mei [1, 5]. If $f'''(0) > 0$, then $\lambda(t)$ should be replaced by $2\pi^2 - t^2$. The functions $\alpha(t), v(t)$ are determined as nonsingular solutions of the enlarged system:

$$F(v, \alpha, t) := \begin{pmatrix} G_1(t\alpha\phi + t^3 v, 2\pi^2 + t^2)/t^3 \\ (\phi, v) \end{pmatrix} = 0, \tag{3.6}$$

where $(\cdot, \cdot)$ represents the $L^2(\Omega_1)$-product in X_1. At $t = 0$, F is defined by its limit and

$$F(v, \alpha, 0) := \begin{pmatrix} \Delta v + 2\pi^2 v + \alpha\phi + \pi^2 f'''(0)\alpha^3 \phi^3/3 \\ (\phi, v) \end{pmatrix} = 0. \tag{3.7}$$

In fact, taking the $L^2(\Omega_1)$-product of ϕ with the first component in (3.7) yields an equation for α with the solutions $\alpha^0 = \pm\frac{2}{\pi}\sqrt{1/3\,f'''(0)}$ or $\alpha^0 = 0$ which corresponds to the trivial solution curve of (3.4). Substituting $\alpha^0 = \pm\frac{2}{\pi}\sqrt{1/3\,f'''(0)}$ into (3.7), one gets a linear system for v, which has a unique solution $v^0 \in R(D_u G_1(0, 2\pi^2))$. Then $(v^0, \alpha^0, 0)$ is a nonsingular solution of (3.7). Applying the implicit function theorem to F at $(v^0, \alpha^0, 0)$ yields a unique solution curve $(v(t), \alpha(t), t)$ of (3.6) passing through $(v^0, \alpha^0, 0)$. Substituting $v(t), \alpha(t)$ into (3.5), one gets a nontrivial solution branch of (3.4) at $(0, 2\pi^2)$. To approximate this solution branch, let h be the parameter of discretization and let Δ_h be the discretizations of the Laplacian Δ in the subdomain Ω_1 with the finite difference, or finite element method or any other numerical method, and ϕ_h be the discrete projection function of ϕ. Substituting Δ, ϕ in (3.6), (3.7) with Δ_h, ϕ_h, we obtain a natural discretization F_h of F. Then, path following of (1.1) beyond $(0, 2\pi^2)$ can be done as follows, see e.g. Böhmer [7], Böhmer/Mei [8] and Mei [16, 19].

Algorithm 3.1: *Path following of the nontrivial solution branch of (1.1) at $(0, 2\pi^2)$.*

Step 1) *For $\alpha_h^0 = \frac{2}{\pi}\sqrt{1/(3\,f'''(0))} + O(h^2)$, solving the approximations v_h^0 of v^0 by*

$$\begin{cases} \Delta_h v + 2\pi^2 v = -\alpha_h^0 \phi_h[1 + \dfrac{\pi^2}{3}\,f'''(0)(\alpha_h^0 \phi_h)^2], \\ (\phi_h, v)_h = 0. \end{cases} \tag{3.8}$$

$(v_h^0, \alpha_h^0, 0)$ yields an approximation for $(v^0, \alpha^0, 0)$;

Step 2) *Let δt be the step size in t direction and $t^i := i \cdot \delta t, i = 1, \ldots$ Starting from $(v_h^0, \alpha_h^0, 0)$, compute the approximations (v_h^j, α_h^j, t^j) of the nonsingular solution branch $(v_h(t^j), \alpha_h(t^j), t^j)$ of (3.6) by continuation methods, see e.g. Allgower/Georg [6];*

Step 3) *Define up approximations for the solution branch (3.5) by $(u_h^j, \lambda_h^j) := (t^j \alpha_h^j \phi_h + (t^j)^3 v_h^j,\ 2\pi^2 + (t^j)^2)$;*

Step 4) *Extend (u_h^j, λ_h^j) to the whole square $[0,1] \times [0,1]$ by group orbit operations to yield approximations of the nontrivial solution branch of (1.1) at $(0, 2\pi^2)$.*

4. Path Following beyond Corank-2 Bifurcation Points

Let $(0, \lambda_0)$ be a corank-2 bifurcation point of (1.1), i.e. (2.2). Then there is a unique pair $(m, n) \in \mathbf{N} \times \mathbf{N}, m \neq n$, such that $\lambda_0 = (m^2 + n^2)\pi^2$ and

$$N(D_u G_0^{\tilde{\Sigma}_M}) = \mathrm{span}[\phi_1, \phi_2] \quad \text{with} \quad \phi_1 := 2\sin m\pi x \sin n\pi y, \ \phi_2 := 2\sin n\pi x \sin m\pi y. \tag{4.1}$$

The problem (1.1), i.e. (2.2) has four different nontrivial solution branches at $(0, \lambda_0)$, see Allgower/Böhmer/Mei [1], Mei [17, 18]. Two of these four solution branches are conjugate and of rectangular type, i.e. with the symmetry groups Σ_{ϕ_i}, $i = 1, 2$. The other two solution branches are of triangular type with the symmetry groups $\Sigma_{\phi_1 \pm \phi_2}$, see e.g. Budden/Norbury [11, 12], Mei [19]. Based on the scaling law, we may choose m, n, such that the greatest common integer factor of m, n is 1. Hence, the possible combinations are $(m, n)=$(odd, odd) or (even, odd) or (odd, even).

Let us consider at first the rectangular solution branches. Denote the isotropy subgroups of ϕ_1, ϕ_2 by Σ_1 , $\tilde{\Sigma}_1$. Then it is easy to see

$$\Sigma_1 := \{-S_1, -S_1', (\tfrac{1}{m}, 0)S_1, (0, \tfrac{1}{n})S_1'; \cdots\}, \quad \tilde{\Sigma}_1 := \{-S_1, -S_1', (\tfrac{1}{n}, 0)S_1, (0, \tfrac{1}{m})S_1'; \cdots\}. \tag{4.2}$$

The groups $\Sigma_1, \tilde{\Sigma}_1$ are obviously conjugate and $\tilde{\Sigma}_1 = S_2 \Sigma_1 S_2$. Moreover, the groups $\Sigma_1, \tilde{\Sigma}_1$ contain $\tilde{\Sigma}_M$ in (2.1) as a proper subgroup. Thus any Σ_1- and $\tilde{\Sigma}_1$-branch of (1.11) also satisfies (2.2) (resp. (1.1)). Due to the anti-symmetries $-S_1$, $-S_1'$, the functions in X^{Σ_1} have nodal lines parallel to the x and y axes and they are also called rectangular functions. We want to consider the rectangular solution branches of (1.11) with the symmetry Σ_1 in (4.2). To this end, we choose a fundamental domain Ω_1 of $\mathbf{R}^2$ in Ω with respect to the group Σ_1:

$$\Omega_1 := [0, \frac{1}{2m}] \times [0, \frac{1}{2n}]. \tag{4.3}$$

Let $Y_1 := C^{0,s}(\bar{\Omega}_1)$ and

$$X_1 := \{u \in C^{2,s}(\bar{\Omega}_1) \mid u(0,y) = u(x,0) = 0,\ \frac{\partial u}{\partial x}(\frac{1}{2m},y) = \frac{\partial u}{\partial y}(x,\frac{1}{2n}) = 0\}. \qquad (4.4)$$

For any $u \in X^{\Sigma_1}$, it is easy to verify that $u|_{\Omega_1} \in X_1$. Similarly to Theorem 3.1, we have

Theorem 4.1: *Let Σ_1 be defined in (4.2) and Ω_1, X_1 in (4.3), (4.4). Then the space X^{Σ_1} is isomorphic to X_1 and the reduced problem $G^{\Sigma_1}(u,\lambda) = 0$ is equivalent to:*

$$Find \quad (u,\lambda) \in X_1 \times \mathbf{R}, \qquad such\ that \quad G_1(u,\lambda) := \Delta u + \lambda f(u) = 0. \qquad (4.5)$$

Since

$$N(D_u G_1(0,\lambda_0)) = \mathrm{span}[\phi_1], \qquad D_\lambda G_1(0,\lambda_0) = 0,$$

and the nondegeneracy condition (1.3) holds, $(0,\lambda_0)$ is a simple bifurcation point of (4.5) on $\{(0,\lambda); \lambda \in \mathbf{R}\}$. The nontrivial solution branch of (4.5) at $(0,\lambda_0)$ has the form:

$$(u(t),\lambda(t)) = (t\alpha(t)\phi_1 + t^3 v(t),\lambda_0 + t^2), \quad t \in \mathbf{R},\ v(t) \in R(D_u G_1(0,\lambda_0)), \qquad (4.6)$$

see Mei [16, 19]. Path following of this branch can be done on an enlarged system similarly to (3.6), (3.7) and Algorithm 3.1, see Mei [19] for the details. This solution branch is extended to the whole domain Ω by reflections and periodicities to obtain a rectangular solution branch of the original problem (1.1). The rectangular branch with the symmetry Σ_2 can be derived from the first one via the group operation S_2.

Concerning the two triangular solution branches, one sees easily that if $(m,n) = (\text{odd, odd})$, then

$$\Sigma_i := \Sigma_{\phi_1 + (-1)^i \phi_2} = \{-S_1, (-1)^i S_2, (1,0)S_1, (0,1)S_1'; \cdots\}, \quad i = 2,3. \qquad (4.7)$$

If (m,n) or $(n,m) = (\text{odd, even})$, then

$$\Sigma_i := \Sigma_{\phi_1 + (-1)^i \phi_2} = \{-S_1, (-1)^i S_2, (-1)^{i+1}(1,1)S_2'; \cdots\}, \quad i = 2,3 \qquad (4.8)$$

and these two groups are conjugate via S_1. In this case path following of one branch is sufficient.

For simplicity, we consider the case $(m,n) = (\text{odd, odd})$. The case (m,n) or $(n,m) = (\text{odd, even})$ can be treated similarly with the corresponding fundmental domains and boundary conditions. For $(m,n) = (\text{odd, odd})$ and the space X^{Σ_i} with the isotropy group Σ_i in (4.7), we choose a common fundamental domain Ω_2 of $\mathbf{R}^2$ in Ω with respect to $\Sigma_i,\ i = 2,3$:

$$\Omega_2 := \{(x,y) \mid 0 \le x \le y,\ 0 \le y \le \frac{1}{2}\}. \qquad (4.9)$$

Let

$$\begin{cases} X_2 := \{u \in C^{2,s}(\bar{\Omega}_2) \mid u(0,y) = 0,\ \frac{\partial u}{\partial x}(\frac{1}{2},y) = 0,\ \frac{\partial u}{\partial x}(x,x) = \frac{\partial u}{\partial y}(x,x)\}, \\[2mm] X_3 := \{u \in C^{2,s}(\bar{\Omega}_2) \mid u(0,y) = u(x,x) = 0,\ \frac{\partial u}{\partial x}(\frac{1}{2},y) = 0\}. \end{cases} \qquad (4.10)$$

The space X^{Σ_i} is isomorphic to X_i and the reduced problem $G^{\Sigma_i}(u, \lambda) = 0$ is equivalent to the following classical PDE on Ω_2:

$$Find \quad (u, \lambda) \in X_i \times \mathbf{R}, \qquad such\ that \quad G_i(u, \lambda) := \Delta u + \lambda f(u) = 0, \quad i = 2, 3. \qquad (4.11)$$

Now $(0, \lambda_0)$ is a simple bifurcation point of (4.11) and

$$N(D_u G_i(0, \lambda_0)) = \operatorname{span}[\phi_1 + (-1)^i \phi_2], \qquad D_\lambda G_i(0, \lambda_0) = 0, \quad i = 2, 3. \qquad (4.12)$$

Under the condition (1.3), the reduced problem (4.11) has a unique nontrivial solution branch $(u(t), \lambda(t))$ passing through $(0, \lambda_0)$ and this solution can be traced by a slightly enlarged system as in Section 3. After that, extensions of this branch to the whole domain Ω yield the corresponding triangular solution branch of (2.2), i.e. (1.1).

In conclusion, path following of the four nontrivial solution branches of (1.1) at a corank-2 bifurcation point $(0, \lambda_0)$ is transformed via reduced problems (2.2), (1.12) into path following across simple bifurcation points of small problems on subdomains of Ω, e.g. (4.5), (4.11). If $(m, n) = $ (odd, odd), two triangular and one of the two rectangular branches have to be calculated. If (m, n) or $(n, m) = $ (odd, even), only one rectangular and one triangular branch should be traced. The others follow from group operations.

Remark 4.1: Neumann boundary value problems can be treated similarly. At a corank-3 bifurcation point, scaling laws reduce the case to $(m', m') = (m, n) = (n, m) = $(odd, odd). Nevertheless, distinct nontrivial solution branches may have the identical symmetry groups, see Mei [17]. Hence, symmetries cannot simplify (1.1) to a simple bifurcation problems. Modified Lyapunov-Schmidt methods have to be employed in the path following, see e.g. Allgower/Böhmer/Mei [3], Mei [17].

5. A Numerical Example

We consider a model problem in Weber [22]:

$$\begin{cases} \Delta u + \lambda(u - u^3) = 0 & in \quad \Omega := [0, 1] \times [0, 1], \\ u = 0 & on \quad \partial\Omega. \end{cases} \qquad (5.1)$$

The solution branch at $(0, 2\pi^2)$ can be traced directly with Algorithm 3.1, see Fig. 5.1a) for a discretization of Δ with the five-points difference method with $h := 1/N$, where $N = 10$ and $t = 1.0$. At the corank-2 bifurcation point $(0, 5\pi^2)$,

$$\phi_1 := 2 \sin \pi x \sin 2\pi y, \quad \phi_2 := 2 \sin 2\pi x \sin \pi y, \qquad (5.2)$$

see (4.1). To determine the rectangular solution branch (4.6) bifurcating at $(0, 5\pi^2)$, we restrict (5.1) to the subdomain $\Omega_1 := [0, \frac{1}{2}] \times [0, \frac{1}{4}]$ with the boundary conditions

$$u(0, y) = u(x, 0) = 0, \quad \frac{\partial u}{\partial x}(\frac{1}{2}, y) = \frac{\partial u}{\partial y}(x, \frac{1}{4}) = 0 \quad for\ all \quad 0 \leq x \leq \frac{1}{2}, 0 \leq y \leq \frac{1}{4}. \qquad (5.3)$$

After discretizing the Laplace operator Δ in Ω_1 with the five-points difference scheme, we obtain a discretization of the enlarged system similarly to (3.6):

$$F_h(v, \alpha, t) := \begin{pmatrix} \Delta_h v + f(v, \alpha, t) \\ (\phi_1, v)_h \end{pmatrix} = 0, \qquad (5.4)$$

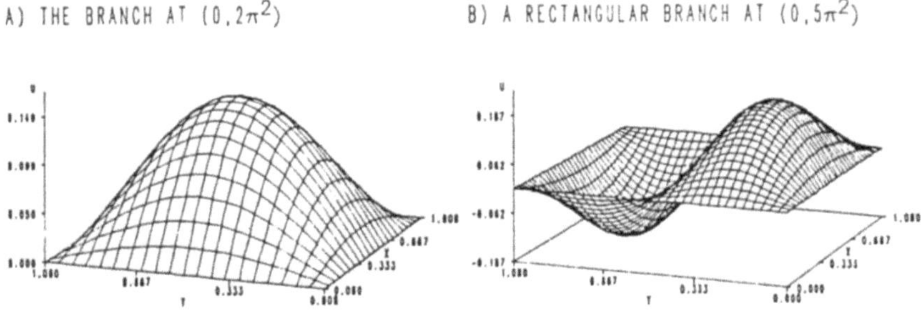

Fig. 5.1

where $v := (v_{i,j}, i = 1, \ldots, N, j = 1, \ldots, M)$ and $\mathbf{f}(v, \alpha, t) := (\alpha\phi_1(x_i, y_j) + (5\pi^2 + t^2)[v_{i,j} - (\alpha\phi_1(x_i, y_j) + t^2 v_{i,j})^3], i = 1, \ldots, N, j = 1, \ldots, M)$ are vectors of function values on the grid points $(x_i, y_j) := (i \cdot h_1, j \cdot h_2), h_1 = 1/2N, h_2 = 1/4M$, and

$$(u, v)_h := h_1 h_2 \sum_{i,j} u_{i,j} \cdot v_{i,j}.$$

At $t = 0$ the mapping F_h is defined by its limit and (5.4) has two different solutions $\pm(v_h^0, \alpha_h^0, 0)$, where $\alpha_h^0 = \frac{2}{3\sqrt{5\pi}} + O(h_1 h_2)$. Using a modified Algorithm 3.1 and the Euler-Newton continuation method, we follow the solution branch of (5.4) starting from $(v_h^0, \alpha_h^0, 0)$. Extensions of the discrete solutions to the whole domain Ω yields approximations of the rectangular solution of (5.1), see Fig. 5.1b) for the solution in the whole domain $[0, 1] \times [0, 1]$ with N=20, M=10 and t=0.5.

Path following of the Σ_2-symmetric branch of (5.1) with Σ_2 in (4.8) can be done similarly by restricting (1.1) to $\Omega_2 := \{(x, y) \in \Omega \mid 0 \leq x \leq y, 0 \leq y \leq 1 - x\}$ with the boundary conditions $u(0, y) = u(x, 1 - x) = 0$, $\frac{\partial u}{\partial x}(x, x) = \frac{\partial u}{\partial y}(x, x)$. Numerical results showed that this branch can be also followed for arbitrarily large $|t| > 0$, which indicates the global existence of this branch (cf. Healey/Kielhöfer [15]). Extending the computed solution to the whole Ω by reflections $S_2, -(1,1)S_2'$ yields approximations of the Σ_2-symmetric solution curve of (5.1).

References

[1] Allgower, E. L., Böhmer, K., Mei, Z.: A complete bifurcation scenario for the 2d-nonlinear Laplacian with Neumann boundary conditions on the unit square, in: *Bifurcation and Chaos: Analysis, Algorithms, Applications*, R. Seydel, F. W. Schneider, T. Küpper, H. Troger (Eds.), ISNM **97**, pp. 1-18, Birkhäuser Verlag, Basel 1991

[2] Allgower, E. L., Böhmer, K., Mei, Z.: On a problem decompsition for semilinear nearly symmetric elliptic problems, in: *Parallel Algorithms for Partial Differential Equations*, W. Hackbusch (Ed.), pp. 1-17, Viehweg Verlag 1991

[3] Allgower, E. L., Böhmer, K., Mei, Z.: Branch switching at a corank-4 bifurcation point of semi-linear elliptic problems with symmetry, submitted to *IMA J. Numer. Anal.*, 1991

[4] Allgower, E. L., Böhmer, K., Mei, Z.: An extended equivariant branching theory, to appear in *Math. Meth. in Appl. Sci.*, 1991

[5] Allgower, E. L., Böhmer, K., Mei, Z.: A generalized equibranching lemma with application to $D_4 \times Z_2$ symmetric elliptic problems, *Bericht des Fachbereichs Mathematik*, 9, University of Marburg, 1990

[6] Allgower, E. L., Georg, K.: *Numerical Continuation Methods: An Introduction*, Springer-Verlag, Berlin Heidelberg New York 1990

[7] Aston, P. J.: Analysis and computation of symmetry-breaking bifurcation and scaling laws using group theoretic methods, *SIAM J. Math. Anal.* 22, 181-212 (1991)

[8] Böhmer, K.: Developing a numerical Lyapunov-Schmidt method, *Bericht des Fachbereichs Mathematik*, 2, University of Marburg 1989

[9] Böhmer, K., Mei, Z.: On a numerical Lyapunov-Schmidt method, in: *Computational Solutions of Nonlinear Systems of Equations*, E. L. Allgower, K. Georg (Eds.), Lectures in Appl. Math. 26, pp. 79-98, AMS Providence, Rhode Island 1990

[10] Bossavit, A.: Symmetry, group, and boundary value problems - A progressive introduction to noncommutative harmonic analysis of partial differential equations in domains with geometrical symmetry, *Comput. Meths. in Appl. Mech. Engrg.* 56, 167-215 (1986)

[11] Budden, P., Norbury, J.: A nonlinear elliptic eigenvalue problem, *J. Inst. Math. Appl.* 24, 9-33 (1979)

[12] Budden, P., Norbury, J.: Solution branches for nonlinear equilibrium problems - bifurcation and domain perturbations, *IMA J. Appl. Math.* 28, 109-129 (1982)

[13] Crandall, M. D., Rabinowitz, P. H.: Bifurcations from simple eigenvalues, *J. Func. Anal.* 8, 321-340 (1971)

[14] Golubitsky, M., Stewart, I. N., Schaeffer, D. G.: *Singularities and Groups in Bifurcation Theory*, Vol. II, Springer-Verlag, Heidelberg Berlin New York 1988

[15] Healey, T. J., Kielhöfer, H.: Symmetry and nodal properites in global bifurcation analysis of quasi-linear elliptic equations, *Arch. Rat. Mech. Anal.* 113, 299 - 311 (1991)

[16] Mei, Z.: A numerical approximation for the simple bifurcation problems, *Numer. Func. Anal. Optimiz.* 10, 383-400 (1989)

[17] Mei, Z.: Solution branches at corank-2 bifurcation points with symmetry, in: *Bifurcations and Chaos: Analysis, Algorithms, Applications*, R. Seydel, F. W. Schneider, T. Küpper, H. Troger (Eds.), ISNM 97, pp. 251-255, Birkhäuser Verlag, Basel 1991

[18] Mei, Z.: Bifurcations of a simplified buckling problem and the effect of discretizations, *Manuscripta Mathematica* 71, 225-252 (1991)

[19] Mei, Z.: Path following across corank-2 bifurcation points of a semi-linear elliptic problem with symmetry, *Computing* 46, 492-509 (1991)

[20] Scovel, J. C., Kevrekidis, I. G., Nicolaenko, B.: Scaling laws and the prediction of bifurcations in systems modelling pattern formation, *Phys. Letter. A,* 130, 73-80 (1988)

[21] Vanderbauwhede, A.: *Local Bifurcation Theory and Symmetry*, Pitman London 1982

[22] Weber, H.: An efficient technique for the computation of stable bifurcation branches, *SIAM J. Sci. Stat. Comput.* 5, 332-348 (1984)

LINEAR STABILITY OF AXISYMMETRIC
THERMOCAPILLARY CONVECTION IN CRYSTAL GROWTH

HANS D. MITTELMANN
Department of Mathematics
Arizona State University
Tempe, AZ 85287-1804

K.-T. CHANG, D. F. JANKOWSKI
Department of Mechanical
and Aerospace Engineering
Arizona State University
Tempe, AZ 85287-6106

G. P. NEITZEL
The George W. Woodruff School
of Mechanical Engineering
Georgia Institute of Technology
Atlanta, GA 30332-0405

Abstract. Linear stability theory is applied to a basic state of thermocapillary convection occurring in a cylindrical half-zone of finite length to determine values of the Marangoni number above which instability is guaranteed. In an earlier work energy theory results had been presented. They yield Marangoni numbers below which the convection is stable under arbitrary perturbations. The least-stable mode in these results was never an axisymmetric one. Here, first results on the most unstable modes will be reported and the numerical procedure employed will be described in detail. Our results indicate that the instability is due to a subcritical *Hopf bifurcation* which *breaks* the *axisymmetry* of the basic state.

1. Introduction

The float-zone crystal-growth process is a containerless method for producing high-quality electronic material in which a rod of the material to be refined is passed through some sort of heater, producing a zone of molten material which is held in place by surface-tension forces. The requirement that surface-tension forces alone support the weight of the molten zone makes the process unsuited for use with certain materials (notably gallium arsenide) in terrestrial environments. Consequently, there has been a great deal of interest in exploring the microgravity environment of space to grow larger crystals of electronic material using the float-zone method.

Along with the reduction in weight provided by a microgravity environment comes a reduction in any convection in the melt induced by buoyancy. At one time, it was believed that buoyancy induced convection was responsible for the appearance of *striations* observed in float-zone-grown material. If this were the case, then one might also expect to produce *better* material in a microgravity environment. Associated with the float-zone process, however, is another type of convection which will not vanish in space, namely *thermocapillary convection*, driven by temperature-induced surface-tension gradients along the free surface of the melt. In fact, it has been widely speculated that the *instability* of this convective mode is responsible for the appearance of the observed *striations*. The desire to utilize the unique environment of space coupled with these observations has led to a significant body of research associated with the stability of thermocapillary convection in models of the float-zone process.

Hydrodynamic stability theory is concerned with determining the conditions under which a certain flow, called the *basic state*, will remain stable or become unstable due to the inevitable presence of unknown perturbations. In general, these perturbations are governed by nonlinear partial differential equations. Linear-stability theory assumes the perturbations to be infinitesimally small and neglects the nonlinear terms in comparison with their linear counterparts. This theory is local in nature and results in a criterion which guarantees growth of these small disturbances. Typically, an externally controllable dimensionless parameter, say R, is selected and linear theory yields a value R_L such that $R > R_L$ is a sufficient condition for instability.

Energy stability theory, on the other hand, adopts a global approach by examining the behavior of a generalized integral disturbance-energy. Unlike linear stability theory, energy-stability theory provides a value R_E such that $R < R_E$ is a sufficient condition for stability of a given basic state to disturbances of *arbitrary amplitude*. This technique is equivalent to a stability analysis utilizing a Lyapunov function. The application of either theory gives rise, in general, to an eigenvalue problem.

If R_E and R_L should coincide, a rigorous stability bound is obtained. However, this is usually not the case and the proximity of R_E to R_L is a function of the physical mechanism which gives rise to the instability. Two such mechanisms for which R_E and R_L may be expected to be relatively close to each other are buoyancy and thermocapillarity.

Experimentally, it has been shown by PREISSER, SCHWABE & SCHARMANN (1983), among others, that thermocapillary convection in a model of the float-zone process undergoes a transition from steady to oscillatory convection when a dimensionless parameter known as the *Marangoni number* exceeds a particular value, the other parameters held fixed. The geometry employed by PREISSER et al. is termed a *half-zone* because it is meant to simulate the lower half of an actual float-zone. It consists of a pair of coaxial, solid cylindrical rods, oriented vertically, with a bridge of liquid material suspended between them. The rods are heated differentially, with the upper rod being at a higher temperature than the lower one. Buoyancy, therefore, plays a stabilizing role in the experiment since the liquid is stably stratified due to temperature in the axial direction. The basic state of thermocapillary convection which results consists of a single toroidal eddy with motion on the free surface in the direction from the hot cylinder toward the cold one.

Motivated by the above work, SHEN, NEITZEL, JANKOWSKI & MITTELMANN (1990) undertook a stability analysis of a half-zone of $O(1)$ aspect ratio, employing energy-stability theory rather than linear theory. Their results, computed primarily for Prandtl number $Pr = 1$, compared favorably with the experimental results of PREISSER et al. However, their analysis had made the simplifying assumption of permitting only axisymmetric disturbances, while the oscillations observed by PREISSER et al. were clearly *non*-axisymmetric and for a material with significantly larger Prandtl number.

Computations for general disturbances were undertaken by NEITZEL, LAW, JANKOWSKI & MITTELMANN (1991). Fortunately, experiments performed by VELTEN, SCHWABE & SCHARMANN (1991) for KCl provided results for unit Prandtl number which eliminated the need to attempt the more difficult calculations for larger Prandtl numbers. These newer energy-theory results are in excellent agreement with the experimentally determined onset Marangoni numbers in order of magnitude, although the azimuthal structure emerging from the energy theory does not (and should not necessarily) agree with that observed experimentally.

The present work seeks to complete the stability picture for the finite half-zone with a non-deformable free surface by calculating linear-stability limits for this basic state. The degree of closeness of the linear- and energy-theory results provides a bound for the region of parameter space *possibly* subject to subcritical instability. To summarize the results obtained: There appears to be a substantial gap between the energy and the linear stability bounds, a result consistent with *subcritical Hopf bifurcation*. Since further both bounds do not correspond to axisymmetric modes *symmetry* of the basic state appears to be *broken*.

The following sections describe the analysis and present the numerical method used to compute the stability bounds. Some computational results are reported and compared to energy bounds. For a more complete presentation and discussion of results, see NEITZEL, CHANG, JANKOWSKI & MITTELMANN (1992).

2. Linear-stability analysis

The basic state of thermocapillary convection in a half-zone with non-deformable free surface is identical to that employed and described in SHEN et al. and NEITZEL et al. (1991). The axisymmetric Boussinesq equations are discretized in stream-function/vorticity form using finite differences on an equally spaced grid in $r - z$ plane. The computed basic state consists of a single toroidal thermocapillary eddy with flow on the free surface directed toward the bottom, cold cylinder. Isotherms become increasingly deformed from their conductive profiles as the Prandtl number of the melt is increased. Details of the numerical procedure and a discussion of the results obtained may be found in SHEN et al. and NEITZEL et al. (1991).

The stability analysis of the basic-state velocity, $U(x)$, temperature, $T(x)$, and pressure, $P(x)$, fields begins in the usual fashion by assuming there exists a solution to the Boussinesq equations of the form

$$q(x,t) = Q(x) + q'(x,t)$$

where q refers to any flow quantity (i.e., velocity, temperature or pressure), a capital letter denotes the basic state and a prime is used to denote a disturbance. Substitution of the solution into the Boussinesq equations and linearization in disturbance quantities leads to the *linearized disturbance equations* (dropping primes):

$$Re[u_t + Uu_r + uU_r + Wu_z + wU_z] = -p_r + \left[\frac{1}{r}(ru)_r\right]_r + \frac{1}{r^2}u_{\varphi\varphi} + u_{zz} - \frac{2}{r^2}v_\varphi, \quad (2.1)$$

$$Re\left[v_t + Uv_r + Wv_z + \frac{Uv}{r}\right] = -\frac{1}{r}p_\varphi + \left[\frac{1}{r}(rv)_r\right]_r + \frac{1}{r^2}v_{\varphi\varphi} + v_{zz} + \frac{2}{r^2}u_\varphi, \quad (2.2)$$

$$Re[w_t + Uw_r + uW_r + Ww_z + wW_z]$$
$$= -p_z + \frac{Gr}{Re}\theta + \left[\frac{1}{r}(rw)_r\right]_r + \frac{1}{r^2}w_{\varphi\varphi} + w_{zz}, \quad (2.3)$$

$$Ma[\theta_t + U\theta_r + uT_r + W\theta_z + wT_z] = \frac{1}{r}[r\theta_r]_r + \frac{1}{r^2}\theta_{\varphi\varphi} + \theta_{zz} \quad (2.4)$$

$$(ru)_r + v_\varphi + (rw)_z = 0. \quad (2.5)$$

In equations (2.1)–(2.5) we have scaled velocities by $\gamma\Delta T/\mu$, pressure by $\gamma\Delta T/R$, temperature by ΔT and time by $R\mu/(\gamma\Delta T)$, where $\Delta T = T_H - T_C$, γ is the (positive) rate of decrease of surface tension with respect to temperature, and μ is the coefficient of dynamic viscosity of the liquid in the zone. The disturbance temperature is denoted by θ.

The dimensionless parameters appearing in (2.1)–(2.5) are

$$\text{Reynolds number} \quad Re = \frac{\gamma R\Delta T}{\mu\nu},$$

$$\text{Grashof number} \quad Gr = \frac{g\alpha R^3 \Delta T}{\nu^2},$$

$$\text{Marangoni number} \quad Ma = \frac{\gamma R\Delta T}{\mu\kappa},$$

where $\nu = \mu/\rho$ is the kinematic viscosity, ρ is the mean density, α is the coefficient of volumetric expansion and g is the gravitational acceleration. The Prandtl number, Pr, is given by $Pr = Ma/Re$.

The boundary conditions which complete the specification of the problem are

$$u = v = w = \theta = 0, \quad z = 0, \Gamma, \tag{2.6a-b}$$

$$u = w_r + \theta_z = v_r - v/r + \theta_\varphi = \theta_r + Bi\theta = 0, \qquad r = 1, \tag{2.7a-d}$$

in addition to the requirement that all flow quantities remain bounded at $r = 0$. The quantity $\Gamma = H/R$ is the dimensionless *aspect ratio* of the zone. The additional parameter appearing in the free-surface heat-transfer condition (2.7d) is the Biot number, defined as

$$Bi = hR/k,$$

where h is a heat-transfer coefficient and k is the thermal conductivity of the liquid.

We make use of Floquet theory and normal modes to assume that all flow quantities may be decomposed as

$$q(r, \varphi, z, t) = q^*(r, z) \exp(\sigma t + \mathrm{i}m\varphi) \tag{2.8}$$

where $\sigma = \sigma_R + i\sigma_I$ is the complex growth rate and m is restricted to be an integer. Marginal stability corresponds to the condition $\sigma_R = 0$. The form (2.8) is now substituted into equations (2.1–5) and the boundary conditions. A discrete version of the resulting problem is a complex, generalized eigenvalue problem of the form

$$Ax = \sigma Bx, \tag{2.9}$$

where x is the vector of unknown velocity, temperature and pressure values at the nodes of the appropriate grid.

The corresponding eigenvalue problem from the energy-theory analysis of this basic state was, at worst, complex-Hermitian (in addition to being indefinite and sparse), whereas (2.9) has no such symmetry. An additional complication, which also existed as part of the energy-stability analysis is the fact that the basic-state velocity and temperature fields depend upon the stability parameter, Ma (equivalently, Re). For the energy-theory calculations, this required an additional level of iteration to obtain the energy limit, Ma_E. Since we formulate the problem with σ as the eigenvalue, our procedure is to fix the Prandtl, Grashof and Biot numbers as well as the azimuthal wavenumber m and calculate the eigenvalue of system (2.9) with largest *real* part, call it σ^*, for various values of the Marangoni number. The Marangoni number

Ma^* corresponding to $\sigma_R^* = 0$ is the value above which infinitesimal disturbances of azimuthal wavenumber m will grow. The linear-stability limit, $Ma_L(Pr, Gr, Bi)$, is therefore given by

$$Ma_L(Pr, Gr, Bi) = \min_m Ma^*(Pr, Gr, Bi; m). \tag{2.10}$$

3. The numerical method

As was shown above, the determination of the linear stability bounds Ma_L requires the solution of the following eigenvalue problem

$$Ax = \sigma Bx \tag{3.1a}$$

where A, B are $N \times N$ matrices, $N = k + l$, which are partitioned as

$$A = \begin{pmatrix} A_{11} & A_{12} \\ A_{21} & A_{22} \end{pmatrix}, \quad B = \begin{pmatrix} B_{11} & 0 \\ 0 & 0 \end{pmatrix} \tag{3.1b}$$

$A_{11}, B_{11} \in C^{k,k}$, $A_{22} \in C^{l,l}$, here k is the number of velocity/temperature unknowns and l the number of pressure unknowns. The matrices are derived from linearizing the boundary value problem for the Boussinesq equations at the solution (basic state) for a specific value of the parameters (Pr, Gr, Bi). If the real parts of all eigenvalues are negative, the basic state is stable, and for increasing Ma that value Ma^* has to be found where for the first time an eigenvalue, to be called the critical eigenvalue in the following, crosses the imaginary axis. Here, it is expected that this corresponds to a simple Hopf bifurcation point (HOPF 1942), that is, there will be exactly one complex-conjugate eigenvalue pair with nonzero imaginary part which crosses the imaginary axis with nonzero speed.

There is a sizable literature on the numerical computation of Hopf bifurcation points; instead of attempting to give a necessarily rather incomplete listing of works, we refer to a recent collection of articles on bifurcation problems (MITTELMANN & ROOSE 1990) in which several papers address this issue.

Generalized eigenvalue problems of the form (3.1), on the other hand, in particular if they arise from applications as is the case here, have been studied thoroughly, and many contributions have been made to their numerical solution. Again, we confine references to just one recent survey work (KERNER 1989) which also includes an extensive bibliography.

Frequently, Hopf bifurcation points are detected during a continuation process. For this, in general, all eigenvalues of (3.1) are computed, a procedure which is prohibitively expensive for large problems ($N > 10^4$). After the detection of a Hopf point its precise location may be determined through characterizing extended systems, see (SPENCE, et al. 1990), (DEDIER, et al. 1990) and the references therein. This technique has proven to be useful for problems of moderate size while it is of limited use for very large problems.

From the literature dealing with the computation of Hopf points in flow problems (CHRISTODOULOU & SCRIVEN 1988) shall be quoted exemplarily. In this work a relatively sophisticated combination of numerical techniques was applied to a nontrivial hydrodynamic stability problem. There is one feature common to the numerical approaches in the above and many other works, namely, the direct solution of the occurring linear systems of equations. This still limits the size of the problems. The dimensions of the eigenvalue problems solved are a few thousand. On the other hand, in SHEN et al. the iterative solution of these problems permitted (in-core) solution of problems of dimension $10^4 - 10^5$. From the experience with the related but different computations of the energy stability bounds in SHEN et al., there was reason to hope that, also for the determination of the linearized stability limits, a method of inverse iteration type would yield the desired results.

If it was known in advance with which imaginary part β the critical eigenvalue crosses the imaginary axis, then inverse iteration with shift $i\beta$ would allow detection of this Hopf point at least when the computation is started in the stable range, not too far from the critical parameter value and continuation in this parameter is employed. It is clear that, in order to safeguard the obtained results, computations have to be done with different values of β or, alternatively, generalizations of the numerical algorithm, as, for example, the Arnoldi method, have to be used to compute more than one eigenvalue at a time.

Let s denote the shift which is anticipated to, usually, be purely imaginary with, say, positive imaginary part β. The following form of inverse iteration was successfully applied to the present problem.

$$(A - sB)\delta x_k = (\sigma_k B - A)x_k, \tag{3.2a}$$

$$\sigma_{k+1} = \frac{(x_k + \delta x_k)_i}{\sigma_k [x_k]_i}, \tag{3.2b}$$

$$x_{k+1} = \frac{x_k + \delta x_k}{[x_k + \delta x_k]_i}, \tag{3.2c}$$

where $[x]_i$ denotes the component of the largest modulus of a vector $x \in C^n$. Here, x_0 was initially, i.e., for the first Marangoni number used, chosen as a random vector and σ_0 as 1. For a fixed shift convergence of both the eigenvalue and eigenvector approximations will in general be linear with a factor

$$\left|\frac{s - \sigma^*}{s - \sigma_n}\right| < 1 \tag{3.3}$$

where σ^*, σ_n are the nearest and the next nearest eigenvalue to s. A faster convergence for the σ_k could be obtained by also iterating with approximate left eigenvectors and using the generalized Rayleigh quotient, cf., e.g., (KERNER 1989).

The essential computational work involved in the proposed method is the solution of the linear system in (3.2a). The matrix $C = A - sB$ is complex and non-Hermitian. First, an equivalent real system of order $2N$ was formed for the matrix

$$\overline{C} = \begin{pmatrix} Re(C) & -Im(C) \\ Im(C) & Re(C) \end{pmatrix}.$$

This system was then solved using the conjugate gradient method for the normal equations (PAIGE & SAUNDERS 1982). For this the matrix $\overline{C}$ was first scaled by multiplying the matrices A_{ij} in (3.1b) with appropriate powers of the discretization parameters in order to balance $\overline{C}$. Then, additionally, a diagonal scaling was used such that the columns of $\overline{C}$ had unit norm. Different preconditionings may well lead to a more efficient solution method; several, however, were tried without yielding a substantial reduction in work. These included outer-inner iterations utilizing the partitioning in (3.1b) as, for example, in (BANK, WELFERT & YSERENTANT 1990) for the Stokes problem. Here, however, $A_{11} - sB_{11}$ is neither Hermitian nor definite. An incomplete LU decomposition was used to precondition the inner iterations. Nevertheless, a higher efficiency is to be expected from an adaptive multi-level approach such as hierarchical bases (BANK et al.). Instead, the finite difference discretization given above was used on a fixed grid producing results that should be comparable to the energy stability results of SHEN, et al.

Instead of the normal equations, the system with $\overline{C}$ could have been solved directly by suitable generalizations of conjugate gradients, as biconjugate gradients, biconjugate gradients squared, GMRES, etc. No preconditioners could easily be found which made these methods sufficiently efficient for the cases to be solved. While $\overline{C}$ will in general not be exactly singular if $i\beta \neq Im(\sigma^*)$, it will be nearly singular. It has also to be noted that the values of $Im(\sigma^*)$ were in the range $10^{-2} - 10^{-1}$.

4. Numerical results

Algorithm (3.2) was applied to the eigenvalue problem (3.1) for aspect ratio $\Gamma = 1$ and the parameter values $Pr = 1$, $Gr = 0$, and $Bi = .3$. For coarse discretizations on square grids, up until about $\delta r = \delta z = \frac{1}{20}$, the results were checked with a complex

QZ routine. Then, the approximate values of Ma^* and σ^* were known including the trend for decreasing gridsize. It was not difficult to find suitable guesses for the shift s and a Marangoni number somewhat smaller than the expected Ma^*. Continuation in Ma with a bisection or secant method on $Re(\sigma^*)$ was used to determine Ma^*.

Table 4.1 lists some of the results obtained. The critical eigenvalues were in the range .02i–.03i. The asterisk denotes that these runs were not completed since the results were too large compared to the values given. The complex linear systems (3.2a) had total dimension N where $l = (1 + \frac{1}{\delta r})^2$ in (3.1b). The number of cg iterations was less than N but $O(N)$ indicating the need for a better preconditioner. While a more complete discussion of results is given in NEITZEL et al. (1992) it should be noted that the least-stable mode from energy theory, here $m = 1$, is not the same as the most-unstable mode from the linear theory. While the linear results, as can be seen by comparing results for $\delta r = 1/49$ and $\delta r = 1/59$, have not yet fully converged, it nevertheless appears clear that the linear bound is obtained for $m = 2$ and thus

$$Ma_E \approx 1503, \qquad Ma_L \approx 3072.$$

These results may be interpreted as indicating a subcritical Hopf bifurcation with *breaking* of the *axisymmetry* of the basic state.

The computations were performed on a 4-processor Stardent 3000 in the Advanced Research Computing Facility of the Department of Mathematics at Arizona State University; the QZ computations were done on a CRAY2 at NASA Ames Research Center.

stability	gridsize	wavenumber			
theory	$\delta r = \delta z$	0	1	2	3
linear	1/49	*	*	2988	3289
linear	1/59	*	*	3072	*
energy	1/69	6245	1503	2633	3545

Table 4.1. Critical Marangoni numbers for different grids and wavenumbers

Acknowledgments

This work was supported by the Microgravity Science and Applications Division of NASA under grant NAG-3-1221 and, in part (HDM), by the US Air Force Office of Scientific Research under grant AFOSR-90-0080.

REFERENCES

BANK, R. E., WELFERT, B. D. & YSERENTANT, H. (1990) A class of iterative methods for solving saddle point problems. *Numer. Math.* **56**, 645-666.

CHRISTODOULOU, K. N. & SCRIVEN, L. E. (1988) Finding leading modes of a viscous free surface flow: An asymmetric generalized eigenproblem. *J. Sci. Comp.* **3**, 355-405.

DEDIER, B., ROOSE, D. & VAN ROMPAY, P. (1990) Interaction between fold and Hopf curves leads to new bifurcation phenomena. 171-186 in MITTELMANN & ROOSE.

HOPF, E. (1942) Abzweigung einer periodischen Lösung von einer stationären Lösung eines Differentialsystems. *Bericht der Math.-Phys. Klasse der Sächsischen Akademie der Wissenschaften zu Leipzig* **94**.

KERNER, W. (1989) Large-scale complex eigenvalue problems. *J. Comp. Phys.* **85** 1-85.

MITTELMANN, H. D. & ROOSE, D. (eds.) (1990), Continuation techniques and bifurcation problems. *J. Comp. Appl. Math.* **26** and ISNM 92, Birkhäuser-Verlag, Basel.

NEITZEL, G. P., CHANG, K.-T., JANKOWSKI, D. F. & MITTELMANN, H. D. (1992) Linear-stability theory of thermocapillary convection in a model of float-zone crystal growth. Proceedings of the AIAA meeting, January 1992, Reno, Nevada.

NEITZEL, G. P., LAW, C. C., JANKOWSKI, D. F. & MITTELMANN, H. D. (1991) Energy stability of thermocapillary convection in a model of the float-zone crystal-growth process. Part 2. Non-axisymmetric disturbances, to appear in *Phys. Fluids A*.

PAIGE, C. C. & SAUNDERS, M. A. (1982) LSQR: An algorithm for sparse linear equations and sparse least squares. *ACM Trans. Math. Software* **8**, 43-71.

PREISSER, F., SCHWABE, P. & SCHARMANN, A. (1983) Steady and oscillatory thermocapillary convection in liquid columns with free cylindrical surface. *J. Fluid Mech.* **126**, 545-567.

SHEN, Y., NEITZEL, G. P., JANKOWSKI, D. F. & MITTELMANN, H. D. (1990) Energy stability of thermocapillary convection in a model of the float-zone crystal-growth process. *J. Fluid Mech.* **217**, 639-660.

SPENCE, A., CLIFFE, K.A. & JEPSON, A. D. (1990) A note on the calculation of paths of Hopf bifurcations. 125-131 in MITTELMANN & ROOSE.

VELTEN, R., SCHWABE, D. & SCHARMANN, A. (1991) The periodic instability of thermocapillary convection in cylindrical liquid bridges, *Phys. Fluids A* **3**, 267-279.

An Indirect Approach to Computing Origins of Hopf Bifurcation and Its Application to Problems with Symmetry

Gerd Pönisch

In the paper an indirect method for numerically computing TB-turning points of nonlinear equations depending on two parameters is presented. The method allows the efficient approximation of the second order partial derivatives ocurring in the algorithm. Moreover, the method is adapted to G-equivariant equations. In this case the computations can be restricted to the symmetric part of the turning point curve.

1 Introduction

Let us consider an evolution equation of the form

$$\frac{dy}{dt} \;=\; F(y,\lambda,\gamma)\,, \quad F : D \subset R^n \times R^1 \times R^1 \to R^n\,, \tag{1}$$

where λ, γ are real parameters.

The aim of this paper is to detect branches of Hopf bifurcation points. One way consists of searching for Hopf points in the one-parameter stationary problem

$$F(y,\lambda,\gamma^\star) \;=\; 0\,, \quad \gamma^\star \in R^1 \text{ fixed,} \tag{2}$$

This requires using a continuation process in which one has to compute all eigenvalues of the Jacobian $F_y(y,\lambda,\gamma^\star)$ at each continuation point (y,λ) and looking whether there is a pair of conjugate complex eigenvalues crossing the imaginary axis. For large dimension, such a procedure is not practical. But within a two-parameter problem, one can first search for a turning point with respect to λ on the branch of steady state solutions satisfying (2) for a fixed $\gamma^\star \in R^1$. Since simple turning points are structurally stable in a one-parameter problem, branches of simple turning points exist in a generic two-parameter problem. Such a branch can be parametrized by γ and traced by a path following algorithm, and one can look for Takens-Bogdanov points. It has been proven in [3] that a branch of Hopf bifurcation points emanates from a degenerated Hopf point which lies on a branch of turning points, i.e., a branch of Hopf points bifurcates from a turning point branch and the branching is one-sided. Such a steady state solution of (1) is called an *origin for Hopf bifurcation* or TB-turning point, see [3].

Now we want to give a definition of a TB-turning point which allows an efficient numerical determination. Assume that

(A1) $\qquad\qquad (y^\star, \lambda^\star, \gamma^\star)$ is a solution of $F(y, \lambda, \gamma) = 0$

such that the Jacobian $F_y(y^\star, \lambda^\star, \gamma^\star)$ has a one-dimensional nullspace $\mathcal{N}$ given by

(A2) $\qquad\qquad \mathcal{N}\,(F_y(y^\star, \lambda^\star, \gamma^\star)) = span\{\varphi^\star\}\,,\ \|\varphi^\star\| = 1\,.$

Then the range $\mathcal{R}$ of this Jacobian can be represented as

$$\mathcal{R}(F_y(y^\star, \lambda^\star, \gamma^\star)) = \{z \in R^n : (\psi^\star)^T z = 0\}$$

where $\psi^\star$ satisfies

$$\mathcal{N}(F_y(y^\star, \lambda^\star, \gamma^\star)^T) = span\{\psi^\star\}\,,\ \|\psi^\star\| = 1\,.$$

Assume further that $(y^\star, \lambda^\star)$ is a *simple turning point* with respect to λ, i.e., that

(A3) $\qquad\qquad F_\lambda(y^\star, \lambda^\star, \gamma^\star) \notin \mathcal{R}(F_y(y^\star, \lambda^\star, \gamma^\star))$

and

(A4) $\qquad\qquad F_{yy}(y^\star, \lambda^\star, \gamma^\star)\varphi^\star\varphi^\star \notin \mathcal{R}(F_y(y^\star, \lambda^\star, \gamma^\star))$

hold. The zero eigenvalue of $F_y(y^\star, \lambda^\star, \gamma^\star)$ can have Riesz index two, i.e., the Jacobian $F_y(y^\star, \lambda^\star, \gamma^\star)$ can have a double eigenvalue zero and a one-dimensional nullspace. A necessary condition is

(A5) $\qquad\qquad (\psi^\star)^T \varphi^\star = 0\,.$

Each point $(y^\star, \lambda^\star, \gamma^\star)$ satisfying (A1-A5) is called a TB-turning point with respect to λ. Consequently, a TB-turning point can be characterized by the augmented system

$$F(y, \lambda, \gamma) = 0 \tag{3a}$$
$$F_y(y, \lambda, \gamma)\varphi = 0,\ \|\varphi\| = 1 \tag{3b}$$
$$F_y(y, \lambda, \gamma)^T\psi = 0,\ \|\psi\| = 1 \tag{3c}$$
$$\psi^T\varphi = 0. \tag{3d}$$

Since the system (3) of $3n + 3$ equations for $3n + 2$ variables is overdetermined, one can employ the fact that the Jacobian $F_y(y^\star, \lambda^\star, \gamma^\star)$ has a generalized eigenvector $\xi^\star$ satisfying

$$F_y(y^\star, \lambda^\star, \gamma^\star)\xi^\star = \varphi^\star\,,\quad (\varphi^\star)^T\xi^\star = 0\,.$$

Hence, we replace (3c)-(3d) by

$$F_y(y, \lambda, \gamma)\xi = \varphi\,,\quad \varphi^T\xi = 0\,.$$

This characterization has been used by Roose [9] for an efficient direct method.

2   An Indirect Method

The turning point curve can be parametrized by the second parameter γ in the neighborhood of a TB-turning point $(y^\star, \lambda^\star, \gamma^\star)$. Then the vectors φ and ψ defined by (3b) and (3c), resp., can also be parametrized by γ, and a necessary condition for a TB-turning point is

$$\psi(\gamma)^T \varphi(\gamma) = 0 .\tag{4}$$

The monitoring of the scalar product of $\psi(\gamma)$ and $\varphi(\gamma)$ requires low additional costs during the continuation process along the turning point curve. In the neighborhood of a TB-turning point equation (4) can be used to compute a sequence $\{\gamma_k\}$ which converges to $\gamma^\star$.

In order to compute $\psi(\gamma)$ and $\varphi(\gamma)$ by standard software from a nonsingular system of linear equations we reformulate the given problem. Denoting the couple $(y, \lambda) \in R^n \times R^1$ by $x \in R^{n+1}$, a turning point (x, γ) satisfies the nonlinear system

$$T(x,\gamma) := \begin{pmatrix} F(x,\gamma) \\ f(x,\gamma) \end{pmatrix} = 0 , \quad T : D \in R^{n+1} \times R^1 \to R^{n+1} ,\tag{5}$$

where the scalar function $f : D \to R^1$ is defined by

$$f(x,\gamma) := (e^{n+1})^T B(x,\gamma,r)^{-1} e^{n+1} \quad \text{with} \quad B(x,\gamma,r) := \begin{bmatrix} F_x(x,\gamma) \\ r^T \end{bmatrix} ,$$

where $e^{n+1} := (0,\ldots,0,1)^T \in R^{n+1}$, and $r \in R^{n+1}$ is chosen such that

$$r^T v^\star \neq 0 , \quad v^\star := \begin{pmatrix} \varphi^\star \\ 0 \end{pmatrix}$$

holds, see [7]. Note that the $(n+1)\times(n+1)$ matrix $B(x,\gamma,r)$ is nonsingular on a neighborhood of $(x^\star, \gamma^\star)$. Since, as a consequence of (A4),

$$f_x(x^\star,\gamma^\star)v^\star \neq 0,\tag{6}$$

the Jacobian $T_x(x^\star, \gamma^\star)$ is nonsingular too, see [7] again.

Thus, by the Implicit Function Theorem, the curve of simple turning points can be parametrized by γ in a neighborhood of each simple turning point. Along the branch

$$C := \{(x(\gamma),\gamma) \in D : \ T(x(\gamma),\gamma) = 0, \ \gamma \in J \subset R^1\}$$

the vectors $v(\gamma) \in R^{n+1}$ and $u(\gamma) \in R^{n+1}$ are defined by

$$B(x(\gamma),\gamma,r)v(\gamma) = e^{n+1} \quad \text{and} \quad B(x(\gamma),\gamma,r)^T u(\gamma) = e^{n+1},\tag{7}$$

resp. Because of the relations

$$\begin{aligned} v(\gamma) &= (\alpha\varphi(\gamma)^T, 0)^T , \quad \alpha := 1/(\varphi(\gamma)^T, 0)r \\ u(\gamma) &= (\beta\psi(\gamma)^T, 0)^T , \quad \beta := 1/\psi(\gamma)^T F_\lambda(y(\gamma),\lambda(\gamma),\gamma) \end{aligned}\tag{8}$$

the necessary condition (4) for a TB-turning point can be described by

$$h(\gamma) = u(\gamma)^T v(\gamma) = 0 \ , \quad h : J \subset R^1 \to R^1 \ . \tag{9}$$

This description is numerically advantageous because the $(n+1)$–dimensional vectors $u(\gamma)$ and $v(\gamma)$ are calculated from linear systems with the nonsingular matrix $B(x(\gamma),\gamma,r)$. Hence the scalar function h can be stably computed along $\mathcal{C}$.

Obviously, at a TB-turning point $(x(\gamma^\star),\gamma^\star)$ there holds

$$h(\gamma^\star) = 0 \ .$$

In order to use (9) for the computation of a TB-turning point we suppose

(A6) $$\dot{h}(\gamma^\star) \neq 0 \ ,$$

where the derivative $\dot{h}$ of h is given by

$$\dot{h}(\gamma) = -u(\gamma)^T \big[d(\gamma) B(x(\gamma),\gamma,r)^{-1} \ + \ B(x(\gamma),\gamma,r)^{-1} d(\gamma) \big] v(\gamma) \tag{10}$$

with

$$d(\gamma) := \left[\begin{array}{c} F_{xx}(x(\gamma),\gamma)\dot{x}(\gamma) + F_{x\gamma}(x(\gamma),\gamma) \\ o^T \end{array} \right] \ \in \ L(R^{n+1})$$

and

$$\dot{x}(\gamma) = -T_x(x(\gamma),\gamma)^{-1} T_\gamma(x(\gamma),\gamma) \ .$$

We summarize these considerations with the following procedure.

Algorithm 1 (A procedure for computing TB-turning points)

input $\gamma_0 \in R^1, r \in R^{n+1}$

for $k := 0, 1, \ldots do$

begin S1: *Compute x^k as approximation to $x(\gamma_k)$ from the defining system*

$$T(x,\gamma_k) = 0$$

with the accuracy $\|T(x^k,\gamma_k)\| \leq \varepsilon_k \ , \ \varepsilon_k > 0 \ .$

S2: *Compute $h(\gamma_k)$ and $\dot{h}(\gamma_k)$ according to (9),(10) with x_k instead of $x(\gamma_k)$.*

S3: *Set $\gamma_{k+1} := \gamma_k - h(\gamma_k)/\dot{h}(\gamma_k)$.*

end

The substep S1 can be performed by a standard solver for nonlinear equations or by a direct method for computing turning points. We prefer the turning point method [7] which is based on the same principle described above. Hence, the main emphasis lies on an efficient computation of $\dot{h}(\gamma_k)$.

The computation of $\dot{h}(\gamma)$ according to (10) causes the main computational costs. At first we want to compute $\dot{x}(\gamma)$ from

$$F_x(x(\gamma),\gamma)\dot{x}(\gamma) + F_\gamma(x(\gamma),\gamma) = 0 \tag{11a}$$
$$f_x(x(\gamma),\gamma)\dot{x}(\gamma) + f_\gamma(x(\gamma),\gamma) = 0. \tag{11b}$$

Using the LU decomposition of $B(x(\gamma),\gamma,r)$, which is already available from the computation of $v(\gamma)$ and $u(\gamma)$ according to (7), one can write the general solution of (11a) in the form

$$\dot{x}(\gamma) = \kappa v(\gamma) + p(\gamma), \quad \kappa \in R^1, \tag{12}$$

where $p(\gamma)$ is a particular solution of (11a), say

$$p(\gamma) := -B(x(\gamma),\gamma,r)^{-1} \begin{pmatrix} F_\gamma(x(\gamma),\gamma) \\ 0 \end{pmatrix}. \tag{13}$$

Since (6) holds in a neighborhood of each turning point, by (11b) the free coefficient κ is determined as

$$\kappa := -\big(f_x(x(\gamma),\gamma)p(\gamma) + f_\gamma(x(\gamma),\gamma)\big) \,/\, f_x(x(\gamma),\gamma)v(\gamma), \tag{14}$$

where

$$f_x(x,\gamma)p + f_\gamma(x,\gamma) = -u(\gamma)^T \begin{pmatrix} F_{xx}(x,\gamma)pv(\gamma) + F_{x\gamma}(x,\gamma)v(\gamma) \\ 0 \end{pmatrix}. \tag{15}$$

Inspecting (10) and (15) it is seen that two directional derivatives of the form

$$F_{xx}(x,\gamma)wv + \delta F_{x\gamma}(x,\gamma)v \tag{16}$$

with given $w, v \in R^{n+1}$ and $\delta \in \{-1,0,1\}$ have to be computed. The computational costs can be drastically reduced if we use efficient approximations instead of the explicit second order partial derivatives F_{xx} and $F_{x\gamma}$. For reasons of numerical stability, it is desirable to work with normalized directions $w \in R^{n+1}$ and $v \in R^{n+1}$. Hence, we define

$$\begin{aligned} a(x,\gamma,v,w,\delta,\mu) \;:=\; & \frac{\|v\|\|w\|}{\mu^2}[F(x + \mu\bar{w} + \mu\bar{v}, \gamma + \delta\mu) \\ & - F(x + \mu\bar{w}, \gamma + \delta\mu) - F(x + \mu\bar{v}, \gamma) + F(x,\gamma)], \end{aligned} \tag{17}$$

where $\bar{v} := v/\|v\|$, $\bar{w} := w/\|w\|$, $\delta \in \{-1,0,1\}$ and $\mu \neq 0 \in R^1$.
Assume that $F : D_0 \subset R^{n+1} \times R^1 \to R^n$ is twice Lipschitz-continuously differentiable on a neighborhood of $(x,\gamma) \in \text{int}(D_0)$. Then $a(x,\gamma,v,w,\delta,\mu)$ is an $O(\mu)$-approximation of (16) for sufficiently small $\mu \in R^1 \setminus \{0\}$, see [8].
Using the described approximation for the second order directional derivatives of F with stepsize μ_k at each iteration step k, the convergence properties of the resulting indirect method are given by

Theorem 2 *Let $F : D \subset R^{n+1} \times R^1 \to R^n$ be twice Lipschitz-continuously differentiable on some neighborhood $D_0 \subset D$ of (x^*, γ^*). Assume that the assumptions (A1 - A6) hold. Then, for sufficiently good starting values (x^0, γ_0) and for sufficiently small $\bar{\varepsilon} > 0$ and $\bar{\mu} > 0$, Algorithm 1 generates a well defined sequence $\{(x^k, \gamma_k)\}$ which remains in D_0, provided that the sequences $\{\mu_k\}$ and $\{\varepsilon_k\}$ are choosen according to*

$$0 < \mu_k \leq \bar{\mu} \quad and \quad 0 \leq \varepsilon_k \leq \bar{\varepsilon} ,$$

resp. If, additionally,

$$\lim_{k \to \infty} \varepsilon_k = 0 \tag{18}$$

then the sequence $\{(x^k, \gamma_k)\}$ converges to (x^*, γ^*), where the inequality

$$\max\{\|x^{k+1} - x^*\| , |\gamma_{k+1} - \gamma^*|\} \leq c \max\{\varepsilon_k , \mu_k|\gamma_k - \gamma^*| , |\gamma_k - \gamma^*|^2\} \tag{19}$$

holds with some $c > 0$. In case of

$$0 < \varepsilon_k \leq \alpha\, h(\gamma_k)^2 \leq \bar{\varepsilon} , \quad 0 < \mu_k \leq \beta \|T(x^k, \gamma_k)\| \leq \bar{\mu}$$

with $\alpha, \beta > 0$ the R-convergence order is at least 2. If, in particular, $\varepsilon_k \equiv 0$ for all $k \geq k_0 \geq 0$, the parameter sequence $\{\gamma_k\}$ converges at least Q-quadratically to γ^*.

This theorem follows from a more general theorem which is proven in [6].

In Algorithm 1 the main computational costs are produced by the substeps S1 and S2. The following scheme gives the essential costs of substep S2, at first, and, in parenthesis, the costs of one loop of substep S1, where the direct turning point method [7] is used:

1	(1)	Evaluation of the derivatives $F_x(.,.)$ and $F_\gamma(.,.)$	$O(n^2)$
9	(5)	Evaluations of the function vector $F(.,.)$	$O(n)$
1	(1)	LU decomposition of the $(n+1) \times (n+1)$-matrix $B(.,.,r)$	$O(n^3)$
4	(2)	Back substitutions	$O(n^2)$

Note that, in general, one or two loops are sufficient to compute the corresponding turning points to a desired accuracy. This experience is confirmed by numerical tests. For example, in order to compute the four TB-turning points in the model of the tubular reactor [4] the following numbers of loops are required, where the given starting values γ_0 are chosen according to $|\gamma_0 - \gamma^*| \geq 0.1$:

Starting values γ_0	1.08	1.50	2.50	0.70
Loops at Algorithm 1 for $|\gamma_k - \gamma^*| \leq 10^{-10}$	6	3	5	6
Loops at substep S1 with $\varepsilon_k = 10^{-10}$	12	6	6	10

The computational costs can be reduced further if it is possible to employ the symmetries of the given problem.

3 G-equivariant systems

In the following we assume that $F : R^n \times R^1 \times R^1 \to R^n$ is G-equivariant where G is a finite group operating on R^n. Let $\vartheta_1, \ldots, \vartheta_m$ be the system of all real irreducible representations of G. Each irreducible representation ϑ_i of dimension n_i is a mapping

$$\vartheta_i : G \to Aut(R^{n_i})$$

from G in the group $Aut(R^{n_i})$ of automorphisms of R^{n_i} which can be described by orthogonal matrices $\vartheta_i(s) \in Aut(R^{n_i})$. Particularly, let $\vartheta_1 : G \to Aut(R^1)$ be the trivial representation of G. The representation $\vartheta : G \to Aut(R^n)$ of G corresponding to F is equivalent to

$$\bigoplus_{i=1}^{m} \left(\bigoplus_{j=1}^{c_i} \vartheta_i \right)$$

where $c_i \in N$ denotes the multiplicity of ϑ_i in ϑ. One usually writes

$$\vartheta = \sum_{i=1}^{m} c_i \vartheta_i \ .$$

Then there exists a unique canonical orthogonal decomposition

$$R^n = \bigoplus_{i=1}^{m} Y_i$$

such that $Y_i \subset R^n$ is the direct sum of c_i irreducible subspaces $Y_{ij} \subset R^n, j = 1, \ldots, c_i$, of type ϑ_i, see [2, 10]. Hence, there is an orthogonal matrix $M = [M_1, \ldots, M_m] \in L(R^n)$ such that the column vectors of $M_i = [M_{i1}, \ldots, M_{in_i}]$, $i = 1, \ldots, m$ are a basis of Y_i where each M_{ij} consists of c_i basis vectors. Such a matrix M consisting of the ordered basis vectors of the symmetry adapted basis in R^n with respect to ϑ is called symmetry adapted transformation matrix [1]. The colums of M can be determined numerically. An efficient implementation is given by the program SYMCON [1] where symbolic and numerical computations are performed. Hence, a G-invariant vector $y \in Y_1 \subset R^n$ is related to

$$\hat{y} = (\hat{y}_1, \ldots, \hat{y}_{c_1}, 0, \ldots, 0)^T = (\tilde{y}^T, 0, \ldots, 0)^T = M^T y \tag{20}$$

by the coordinate transformation belonging to M. Via M the so-called symmetrical normal form of F is defined by

$$g(\hat{y}, \lambda, \gamma) = M^T F(M\hat{y}, \lambda, \gamma), \quad g : R^n \times R^1 \times R^1 \to R^n. \tag{21}$$

In [1] it is proven, that the Jacobian

$$g_{\hat{y}}(\hat{y}, \lambda, \gamma) = M^T F_y(y, \lambda, \gamma) M \tag{22}$$

has block diagonal form for every $\hat{y} \in R^n$ with $\hat{y}_j = 0$, $j = c_1 + 1, \ldots, n$, i.e.,

$$g_{\hat{y}}(\hat{y}, \lambda, \gamma) = \begin{pmatrix} A_1 & & & \\ & A_2 & & \\ & & \ddots & \\ & & & A_m \end{pmatrix} \, , \quad A_i = \begin{pmatrix} D_i & & \\ & \ddots & \\ & & D_i \end{pmatrix}$$

where the blocks A_i are themselves block diagonal matrices consisting of n_i identical block-matrices $D_i \in L(R^{c_i})$, $i = 1, \ldots, m$. Moreover, for $\hat{y}$ as in (20)

$$g_j(\hat{y}, \lambda, \gamma) = 0 \, , \quad j = c_1 + 1, \ldots, n \tag{23}$$

holds for all $\tilde{y} \in R^{c_1}$. Consequently, instead of the whole system we can restrict ourselves to the so-called symmetry reduced system

$$\tilde{g}(\tilde{y}, \lambda, \gamma) = 0 \, , \quad \tilde{g} : R^{c_1} \times R^1 \times R^1 \to R^{c_1} \, , \tag{24}$$

given by

$$\tilde{g}_i(\tilde{y}, \lambda, \gamma) := g_i(\hat{y}, \lambda, \gamma) \, , \quad i = 1, \ldots, c_1 \, , \tag{25}$$

for analyzing G-invariant solutions. Both blocks of the corresponding Jacobian

$$\tilde{g}_{\tilde{w}}(\tilde{w}, \gamma) := A := \begin{bmatrix} A_1 \vdots A_\lambda \end{bmatrix}$$

with $\tilde{w} := (\tilde{y}, \lambda) \in R^{c_1} \times R^1$, $A_1 := \tilde{g}_{\tilde{y}}(\tilde{y}, \lambda, \gamma) \in L(R^{c_1})$, $A_\lambda := \tilde{g}_\lambda(\tilde{y}, \lambda, \gamma) \in R^{c_1}$ are extracted from the whole Jacobian $g_w(w, \gamma)$ with the following block structure

$$g_w(w, \lambda, \gamma) = \begin{pmatrix} A_1 & & & & A_\lambda \\ & A_2 & & & \\ & & \ddots & & \\ & & & A_m & \end{pmatrix} \tag{26}$$

where w is defined by $w := (\hat{y}^T, \lambda)^T = (\tilde{y}^T, 0, \ldots, 0, \lambda)^T \in R^{n+1}$, cf. [1]. Furthermore, the null vectors $\hat{\varphi}(\gamma) \in R^n$ and $\hat{\psi}(\gamma) \in R^n$ such that

$$\mathcal{N}\big(g_{\hat{y}}(\hat{y}(\gamma), \lambda(\gamma), \gamma)\big) = span\{\hat{\varphi}(\gamma)\} \, , \quad \mathcal{N}\big(g_{\hat{y}}(\hat{y}(\gamma), \lambda(\gamma), \gamma)^T\big) = span\{\hat{\psi}(\gamma)\}$$

are given by $\quad \hat{\varphi}(\gamma) := M^T \varphi(\gamma) \quad$ and $\quad \hat{\psi}(\gamma) := M^T \psi(\gamma) \quad$ for $\gamma \in J$. The block structure (26) of g_w implies

$$\hat{\varphi}(\gamma) := (\hat{\varphi}_1, \ldots, \hat{\varphi}_{c_1}, 0, \ldots, 0)^T \quad \text{and} \quad \hat{\psi}(\gamma) := (\hat{\psi}_1, \ldots, \hat{\psi}_{c_1}, 0, \ldots, 0)^T.$$

Hence, $\quad \mathcal{N}(A_1(\gamma)) = span\{\tilde{\varphi}(\gamma)\}$, $\quad \mathcal{N}(A_1(\gamma)^T) = span\{\tilde{\psi}(\gamma)\}\quad$ holds where $\tilde{\varphi}$ and $\tilde{\psi}$ denotes the first c_1 components of $\hat{\varphi}$ and $\hat{\psi}$ given by $\quad \tilde{\varphi}(\gamma) := C\hat{\varphi}(\gamma) \in R^{c_1}\quad$ and $\tilde{\psi}(\gamma) := C\hat{\psi}(\gamma) \in R^{c_1}$, resp., with the cutting matrix $\quad C := \left[I_{c_1} \vdots 0 \right] \in L(R^n, R^{c_1})$. Because of

$$\hat{\psi}^T g_\lambda(\hat{y}, \lambda, \gamma) = \hat{\psi}^T M^T F_\lambda(M\hat{y}, \lambda, \gamma), \quad \hat{\psi}^T g_{\hat{y}\hat{y}}(\hat{y}, \lambda, \gamma)\hat{\varphi}\hat{\varphi} = \hat{\psi}^T M^T F_{yy}(M\hat{y}, \lambda, \gamma) M\hat{\varphi}M\hat{\varphi}$$

for $(\hat{y}, \lambda, \gamma) \in R^n \times R^1 \times R^1$ with $(M\hat{y}, \lambda, \gamma) \in \mathcal{C}$, the assumptions (A3) and (A4) imply

$$\tilde{\psi}(\gamma)^T \tilde{g}_\lambda(\tilde{y}, \lambda, \gamma) \neq 0 , \quad \tilde{\psi}(\gamma)^T \tilde{g}_{\tilde{y}\tilde{y}}(\tilde{y}, \lambda, \gamma)\tilde{\varphi}(\gamma)\tilde{\varphi}(\gamma) \neq 0 . \tag{27}$$

Moreover, the necessary condition (A5) for $(M\hat{y}^\star, \lambda^\star, \gamma^\star)$ to be a TB-turning point changes into

$$(\tilde{\psi}^\star)^T \tilde{\varphi}^\star = \tilde{\psi}(\gamma^\star)^T \tilde{\varphi}(\gamma^\star) = 0 . \tag{28}$$

Hence, the technique for computing a TB-turning point described in section 2 can be applied to the corresponding problem of smaller dimension.

In case of smooth G-invariant solutions the block structure (26) of $g_w(w, \gamma)$ allows the characterization of the symmetric part of the turning point branch $\mathcal{C}$ by

$$\tilde{\mathcal{C}} := \{(\tilde{w}(\gamma), \gamma) \in R^{c_1+1} \times R^1 : \tilde{T}(\tilde{w}(\gamma), \gamma) = 0\},$$

where $\tilde{T} : R^{c_1+1} \times R^1 \to R^{c_1+1}$ is defined by

$$\tilde{T}(\tilde{w}, \gamma) := \begin{pmatrix} \tilde{g}(\tilde{w}, \gamma) \\ \tilde{f}(\tilde{w}, \gamma) \end{pmatrix}$$

with $\tilde{f}(\tilde{w}, \gamma) := (e^{c_1+1})^T \tilde{B}(\tilde{w}, \gamma, \tilde{r})^{-1} e^{c_1+1} \in R^1$. Here $\tilde{B}$ denotes the matrix

$$\tilde{B}(\tilde{w}, \gamma, \tilde{r}) := \left[\begin{array}{c} \tilde{g}_{\tilde{w}}(\tilde{w}, \gamma) \\ \tilde{r}^T \end{array} \right] = \left[\begin{array}{c} A(\tilde{w}, \gamma) \\ \tilde{r}^T \end{array} \right] \in L(R^{c_1+1}, R^{c_1+1})$$

where the bordering direction $\tilde{r} \in R^{c_1+1}$ is obtained from $r \in R^{n+1}$ by $\tilde{r} := \hat{C}\hat{M}^{-1}\quad$ with

$$\hat{M} := \left[\begin{array}{c} M \vdots 0 \\ (e^{n+1})^T \end{array} \right] \in L(R^{n+1}) , \quad \hat{C} := \left[\begin{array}{c} C \vdots 0 \\ (e^{n+1})^T \end{array} \right] \in L(R^{n+1}, R^{c_1+1}) .$$

Note that, because of (27) and

$$\tilde{r}^T \begin{pmatrix} \tilde{\varphi} \\ 0 \end{pmatrix} \neq 0 ,$$

the matrix $\tilde{B}(\tilde{w}, \gamma, \tilde{r})$ is nonsingular for $(\tilde{w}, \gamma) \in \tilde{\mathcal{C}}$ on the neighborhood of $(\tilde{w}(\gamma^\star), \gamma^\star)$.

In analogy to (7) we define

$$\tilde{u}(\gamma) := \tilde{B}(\tilde{w}(\gamma), \gamma, \tilde{r})^{-T} e^{c_1+1} , \quad \tilde{v}(\gamma) := \tilde{B}(\tilde{w}(\gamma), \gamma, \tilde{r})^{-1} e^{c_1+1} .$$

Then, the necessary condition for $(\tilde{w}(\gamma^*), \gamma^*) \in \tilde{C}$ to be a G-invariant TB-turning point is given by $\tilde{h}(\gamma^*) = 0$, where $\tilde{h}: J \to R^1$ is defined by $\tilde{h}(\gamma) := \tilde{u}(\gamma)^T \tilde{v}(\gamma)$. Following the technique outlined in section 2, Algorithm 1 can be applied substituting T, x, h by $\tilde{T}$, $\tilde{w}$, $\tilde{h}$. In this way the second order partial derivatives can also be approximated efficiently via the corresponding directional approximations. Hence, Theorem 2 holds, too. Finally, we remark that the computational costs are very low in contrast to the original problem. The dimension reduces to c_1 in the table given in section 2. In addition, the transformation matrix M has to be computed.

Acknowledgement

The author would like to thank E. L. Allgower and H. Schwetlick for their helpful comments.

References

[1] Gatermann, K., Hohmann, A.: *Symbolic exploitation of symmetry in numerical pathfollowing.* Preprint **SC 90 - 11**, Konrad - Zuse - Zentrum für Informationstechnik, Berlin, 1990; appeared in Impact of Computing in Science and Engineering.

[2] Golubitsky, M., Stewart, T., Schaeffer, D.: *Singularities and Groups in Bifurcation Theory.* Vol. 2, Springer, Berlin, 1988.

[3] Guckenkeimer, J., Holmes, P.: *Nonlinear Oscilation, Dynamical Systems and Bifurcation of Vector Fields.* Applied Math. Science **42**, Springer, New York, 1983.

[4] Heinemann, R.F., Poore, A.B.: *Multiplicity, stability and oscilatory dynamics of tubular reactors.* Chem. Engng. Sci. **36**(1981), pp. 1411 - 1419.

[5] Jepson, A.D., Keller, H.B.: *Steady state and periodic solution paths: Their bifurcations and computations.* pp. 219 - 246 in: *Numerical Methods for Bifurcation Problems* (Küpper, T., Mittelmann, H.D.; eds.) ISNM **70**, Birkhäuser, Basel, 1984.

[6] Pönisch, G.: *An indirect method for computing origins for Hopf bifurcation in two-parameter problems.* Computing **46**(1991), pp. 307 - 320.

[7] Pönisch, G., Schwetlick, H.: *Computing turning points of curves implicitly defined by nonlinear equations depending on a parameter.* Computing **26**(1981), pp. 101 - 126.

[8] Pönisch, G., Schwetlick, H.: *Some types of inverse problems for nonlinear equations having coupled turning points.* Preprint **07 - 01 - 90**, Technical University of Dresden 1990.

[9] Roose, D.: *Numerical computation of origins for Hopf bifurcation in two-parameter problems.* pp. 268 - 276 in: *Bifurcation: Analysis, Algorithms, Applications.* (Küpper, T., Seydel, R., Trogner, H.; eds.) ISNM **79**, Birkhäuser, Basel, 1987.

[10] Werner, B.: *Eigenvalue problems with the symmetry of a group and bifurcations.* in: *Continuation and Bifurcation: Numerical Techniques and Applications.* (Roose, D., De Dier, B. Spence, A.; eds.) NATO ASI SERIES C Vol. **313**, 1990.

Gerd Pönisch: Institut of Numerical Mathematics, Technical University of Dresden, Mommsenstr. 13, D-O-8027 Dresden, Federal Republic of Germany

International Series of Numerical Mathematics, Vol. 104, © 1992 Birkhäuser Verlag Basel

A Version of GMRES for Nearly Symmetric Linear Systems

R. Sebastian
Dept. of Mathematics
Philipps University Marburg

Abstract

We present a version of GMRES as a solver for systems of linear equations in large sparse continuation problems. Following a solution curve for a nonlinear problem with an Euler-Newton method requires the solution of linear systems. For many problems the coefficient matrices of these systems can be written as the sum of a symmetric and a rank-one unsymmetric matrix. First we show that for this specific class of systems the orthogonalization process in GMRES does not require the full effort as for general matrices and that it is possible to reduce the computational effort to about twice the effort for symmetric matrices. Secondly, a few numerical results are given for a discretization of the problem $\Delta u + \lambda f(u) = 0$ on the square $\Omega = [0, 1]^2$ and its restriction to $D_4 \times Z_2$-fixed-point spaces.

1   Introduction

In this paper we study the systems of linear equations arising from the approximation of solution branches of nonlinear eigenvalue problems. Suppose that a nonlinear problem

$$H(u, \lambda) = 0,$$

is given , where u denotes a discrete function , $u \in \mathbb{R}^n$, λ a real parameter and $H : \mathbb{R}^n \times \mathbb{R} \to \mathbb{R}^n$ satisfies certain smoothness conditions. Following a solution curve for $H(u, \lambda) = 0$ using an Euler-Newton-continuation method involves the solution of systems of linear equations of the form

$$A_g x = b,$$

where

$$A_g = \left(\begin{array}{cc} D_u H & D_\lambda H \\ c^T & d \end{array} \right) \in \mathbb{R}^{(n+1)\times(n+1)},$$

$b \in \mathbb{R}^{n+1}$, $c \in \mathbb{R}^n$, $d \in \mathbb{R}$ and $(D_u H, D_\lambda H) \in \mathbb{R}^{n \times (n+1)}$ is the Jacobian matrix of H evaluated at a point $(u, \lambda) \in \mathbb{R}^{n+1}$.

In many cases the matrix $D_u H$ will be symmetric. The matrix A_g however in general will be unsymmetric and is assumed here to be indefinite.

Linear conjugate gradient methods to solve unsymmetric systems of equations are Arnoldi's method and GMRES,see [8]. Since for unsymmetric matrices the amount of work in the orthogonalization process of GMRES depends quadratically on the number of iterations, but only linearly for symmetric matrices, a considerable reduction of the computational effort can be expected by exploiting the symmetry of the matrix $D_u H$. Methods to exploit the symmetry of $D_u H$ are for example Incomplete Orthogonalization Methods or the combination of GMRES(m) and block elimination. Incomplete Orthogonalization Methods (see e.g. [7]) reduce the computational effort for the orthogonalization process. However they often require a higher number of iterations than the original algorithm. The combination of GMRES(m) and block elimination (see e.g. [4]) results in the solution of two symmetric systems of equations instead of one unsymmetric system.

The purpose of this paper is to show how GMRES can be modified for matrices which can be written as the sum of a symmetric and a rank-one unsymmetric matrix. The remainder of the paper is organized as follows. In section 2 we briefly recall the GMRES(m) algorithm. In section 3 we outline the structure of the Hessenberg matrix computed in GMRES for the specific class of matrices and show how this structure can be used to reduce the number of scalar products and linear combinations in the orthogonalization process. Finally some numerical results are reported in section 4 for the discrete problem $\Delta u + \lambda \sin u = 0$ on $\Omega = [0, 1]^2$ and its restriction to $\Gamma := D_4 \times Z_2$-fixed-point spaces.

2   The GMRES(m) algorithm

Given a regular matrix $A \in \mathbb{R}^{n \times n}$, $b \in \mathbb{R}^n$ and x_0 as an initial approximation for the solution of the linear system

$$Ax = b,$$

GMRES computes at each step an approximate solution $x_k = x_0 + z_k \in \mathbb{R}^n$ which satisfies the following minimization condition:

$$\|b - Ax_k\|_2 = \min_{z \in K_k} \|b - A(x_0 + z)\|_2 = \min_{z \in K_k} \|r_0 - Az\|_2 = \|r_0 - Az_k\|_2$$

with

$$K_k := K[r_0, A, k] = \mathrm{span}[r_0, Ar_0, \dots, A^{k-1}r_0] \qquad \text{and} \qquad r_0 := b - Ax_0.$$

The algorithm involves the computation of an l_2-orthonormal basis for the Krylov-subspace K_k and a Hessenberg matrix H_k which can be considered as an approximation of the matrix A and is used to approximate the solution of $Ax = b$. The restart version of GMRES(m) is described as follows:

Algorithm 2.1 : GMRES(m)

1. Start: Choose x_0, $r_0 := b - Ax_0$, $v_1 := r_0/\|r_0\|_2$, $\beta := \|r_0\|_2$.

2. For j=1,2,...,m :   (Determine l_2-orthonormal basis $v_1, \dots, v_{m+1}$ for K_{m+1})

$$
\begin{aligned}
h_{ij} &:= (Av_j, v_i), \quad i = 1, \dots, j \\
\hat{v}_{j+1} &:= Av_j - \sum_{i=1}^{j} h_{ij} v_i \\
h_{j+1,j} &:= \|\hat{v}_{j+1}\|_2 \\
v_{j+1} &:= \frac{\hat{v}_{j+1}}{h_{j+1,j}}
\end{aligned}
$$

3. Compute approximate solution : $x_m := x_0 + z_m$,

$$z_m := V_m y_m, \quad \text{where } y_m \text{ minimizes } \|\beta e_1 - H_m y\|_2 = \|r_0 - Az_m\|_2, y \in \mathbb{R}^m.$$

4. Restart :

$$r_m := b - Ax_m$$

If $\|r_m\|_2 \leq \varepsilon$, then stop, otherwise $x_0 := x_m$, continue with 1.

Practically, the orthogonalization in step 2 would be done with a modified Gram-Schmidt process for reasons of numerical stability . It is possible to combine step 2 and step 3 such that for each j, an estimate of the current residual norm is available without computing $x_j := x_0 + z_j$ and the algorithm can be stopped if the residual norm is sufficiently small.

The main part of the computational work is done in step 2. For each $j = 1, \dots, m$ the elements of the Hessenberg matrix $H_m := (h_{ij})_{i,j=1}^{m}$ are computed as the scalar products $(Av_j, v_i), i = 1, \dots, j$. The new basis vector v_{j+1} is computed as a linear combination of $(j + 1)$ vectors.

For symmetric matrices, the matrix H_m would be a tridiagonal matrix and would require only 3 scalar products, and a linear combination of 3 vectors for each j . Hence the amount of work grows only linearly depending on the number of iterations m in step 2 of the GMRES algorithm.

In the next section we will show that for matrices of the specific form $A_g = A_s + U_s$, where A_s is symmetric and U_s is a rank-one matrix, only a fixed number of scalar products and linear combinations for each j have to be computed to perform the GMRES(m) algorithm as before. In this modified version the amount of work in step 2 also grows only linearly depending on m .

3   GMRES(m) for matrices $A_g = A_s + U_s$

Here it is assumed that a system of linear equations has to be solved with the special form :

$$A_g x = b$$
$$A_g = \begin{pmatrix} A & d \\ c^T & \alpha \end{pmatrix}$$
$$A \in \mathbb{R}^{(n-1)\times(n-1)} \text{ symmetric,}$$
$$c, d \in \mathbb{R}^{n-1}, \quad \alpha \in \mathbb{R},$$
$$A_g \text{ nonsingular,} \quad A_g \in \mathbb{R}^{n\times n}.$$

At first, A_g can be rewritten as

$$A_g = \begin{pmatrix} A & d \\ c^T & \alpha \end{pmatrix} = \underbrace{\begin{pmatrix} A & c \\ c^T & \alpha \end{pmatrix}}_{=:A_s} + \underbrace{\begin{pmatrix} 0_{(n-1)\times(n-1)} & d-c \\ 0_{n-1}^T & 0 \end{pmatrix}}_{=:U_s}.$$

Recall that for the matrix $H_m = (h_{ij})_{i,j=1}^m$ from the GMRES(m) algorithm and the orthonormal basis V_m of the Krylov-subspace K_k the following property holds:

$$\begin{aligned} H_m &= V_m^T A_g V_m \\ &= V_m^T A_s V_m + V_m^T U_s V_m. \end{aligned}$$

The second term in the above sum is a rank-one matrix. Define

$$V_m = (v_1, \ldots, v_m) = \begin{pmatrix} v_{11} & \cdots & v_{1m} \\ \vdots & & \vdots \\ v_{n1} & \cdots & v_{nm} \end{pmatrix} \text{ and } s := \begin{pmatrix} d-c \\ 0 \end{pmatrix}.$$

Then

$$\begin{aligned} V_m^T U_s V_m &= V_m^T \begin{pmatrix} 0_{(n-1)\times(n-1)} & d-c \\ 0_{n-1}^T & 0 \end{pmatrix} V_m = \begin{pmatrix} & v_1^T s \\ 0_{m\times(n-1)} & \vdots \\ & v_m^T s \end{pmatrix} V_m \\ &= \begin{pmatrix} v_1^T s v_{n1} & \cdots & v_1^T s v_{nm} \\ \vdots & & \vdots \\ v_m^T s v_{n1} & \cdots & v_m^T s v_{nm} \end{pmatrix}. \end{aligned}$$

Hence H_m also is the sum of a symmetric and a rank-one matrix :

$$
\begin{aligned}
H_m &= V_m^T A_g V_m \\
&= \begin{pmatrix} v_1^T A_s v_1 & \cdots & v_1^T A_s v_m \\ \vdots & & \vdots \\ v_m^T A_s v_1 & \cdots & v_m^T A_s v_m \end{pmatrix} + \begin{pmatrix} v_1^T s v_{n1} & \cdots & v_1^T s v_{nm} \\ \vdots & & \vdots \\ v_m^T s v_{n1} & \cdots & v_m^T s v_{nm} \end{pmatrix}.
\end{aligned}
$$

Since H_m is a Hessenberg matrix we get that

$$
v_i^T A_s v_j + v_i^T s v_{nj} = 0 \quad i > j+1, \quad j = 1, \ldots, m. \tag{1}
$$

Then by the symmetry of A_s the following equations hold for the elements h_{ji} above the first upper diagonal :

$$
\begin{aligned}
h_{ji} &= v_j^T A_g v_i = v_j^T A_s v_i + v_j^T s v_{ni} = v_i^T A_s v_j + v_j^T s v_{ni} \\
&= v_j^T s v_{ni} - v_i^T s v_{nj} \qquad \text{because of (1)} \\
&\quad \text{for } i > j+1, \quad j = 1, \ldots, m.
\end{aligned} \tag{2}
$$

Overall, H_m has the following structure for the matrix $A_g = A_s + U_s$.

$$
H_m = \begin{pmatrix}
v_1^T A_g v_1 & v_1^T A_g v_2 & & & & & \\
v_2^T A_g v_1 & v_2^T A_g v_2 & v_2^T A_g v_3 & & & & \\
& v_3^T A_g v_2 & v_3^T A_g v_3 & \ddots & & & \\
& & & \ddots & \ddots & \ddots & \\
& & & & \ddots & \ddots & \ddots \\
& & & & \ddots & \ddots & v_{m-1}^T A_g v_{m-1} \\
& & & & & v_m^T A_g v_{m-1} & v_m^T A_g v_m
\end{pmatrix}
$$

$$
+ \begin{pmatrix}
0 & 0 & v_1^T s v_{n3} & v_1^T s v_{n4} & \cdots & \cdots & v_1^T s v_{nm} \\
& \ddots & 0 & v_2^T s v_{n4} & \cdots & \cdots & v_2^T s v_{nm} \\
& & \ddots & \ddots & \ddots & & \vdots \\
& & & \ddots & \ddots & \ddots & \vdots \\
& & & & \ddots & \ddots & v_{m-2}^T s v_{nm} \\
& & & & & \ddots & 0 \\
& & & & & & 0
\end{pmatrix}
$$

$$
- \begin{pmatrix}
0 & 0 & v_3^T s v_{n1} & v_4^T s v_{n1} & \cdots & \cdots & v_m^T s v_{n1} \\
& \ddots & 0 & v_4^T s v_{n2} & \cdots & \cdots & v_m^T s v_{n2} \\
& & \ddots & \ddots & \ddots & & \vdots \\
& & & \ddots & \ddots & \ddots & \vdots \\
& & & & \ddots & \ddots & v_m^T s v_{n,(m-2)} \\
& & & & & \ddots & 0 \\
& & & & & & 0
\end{pmatrix}
$$

It can be seen immediately that for the computation of the elements in H_m above the first upper diagonal, only the scalar products $v_j^T s$ need to be computed. Hence in this case, the number of scalar products needed to determine H_m is 4 for each $j = 1, \ldots, m$, rather than $j + 1$ for each $j = 1, \ldots, m$, compared to a fully unsymmetric matrix.

The next step is to show how the calculation of the basis vectors $\hat{v}_{j+1}$ can be done more efficiently. Recall that for the elements h_{ji} we get, see (2) :

$$h_{ji} = v_j^T s v_{ni} - v_i^T s v_{nj} \text{ for } i > j + 1, \ j = 1, \ldots, m.$$

The vectors $\hat{v}_{k+1}$ in GMRES(m) are determined as:

$$
\begin{aligned}
\hat{v}_{k+1} &= Av_k - \sum_{i=1}^{k} h_{ik} v_i \\
&= Av_k - h_{kk} v_k - h_{k-1,k} v_{k-1} - \sum_{i=1}^{k-2} (v_i^T s v_{nk} - v_k^T s v_{ni}) v_i \\
&= Av_k - h_{kk} v_k - h_{k-1,k} v_{k-1} - V_{k-2} \begin{pmatrix} v_1^T s v_{nk} & - & v_k^T s v_{n1} \\ v_2^T s v_{nk} & - & v_k^T s v_{n2} \\ & \vdots & \\ v_{k-2}^T s v_{nk} & - & v_k^T s v_{n,k-2} \end{pmatrix} \\
&= Av_k - h_{kk} v_k - h_{k-1,k} v_{k-1} - V_{k-2} \left[\begin{pmatrix} v_1^T s \\ v_2^T s \\ \vdots \\ v_{k-2}^T s \end{pmatrix} v_{nk} - v_k^T s \begin{pmatrix} v_{n1} \\ v_{n2} \\ \vdots \\ v_{n,k-2} \end{pmatrix} \right].
\end{aligned}
$$

Thus the $\hat{v}_{k+1}$ can be determined such that in each step, $j = 3, \ldots, m$

$$
V_{k-2} \begin{pmatrix} v_1^T s \\ \vdots \\ v_{k-2}^T s \end{pmatrix} \qquad \text{and} \qquad V_{k-2} \begin{pmatrix} v_{n1} \\ \vdots \\ v_{n,k-2} \end{pmatrix}
$$

are obtained as appropriate multiples of

$$
V_{k-3} \begin{pmatrix} v_1^T s \\ \vdots \\ v_{k-3}^T s \end{pmatrix}, \qquad V_{k-3} \begin{pmatrix} v_{n1} \\ \vdots \\ v_{n,k-3} \end{pmatrix} \qquad \text{and} \qquad v_{k-2}.
$$

Therefore the term $\sum_{i=1}^{k-2} h_{ik} v_i$ requires only the linear combination of 3 vectors respectively, rather than a linear combination of $k - 2$ vectors.

This shows that for the specific matrices $A_g = A_s + U_s$ the orthogonalization process in GMRES(m) can be carried out with an effort of linear order with respect to the number of iterations m. Overall, the number of scalar products in the orthogonalization process of GMRES is $3 \cdot m$ for symmetric matrices, $(m + 1)^2/2$ for unsymmetric matrices, and $4 \cdot m$ for matrices of the type A_g. The number of linear combinations is $3 \cdot m$ for symmetric matrices, about $(m + 1)^2/2$ for unsymmetric matrices and $6 \cdot m$ for matrices of the type A_g.

4 Numerical Results

Consider the semi-linear elliptic boundary value problem

$$G(u, \lambda) = \Delta u + \lambda \sin(u) = 0 \text{ on } \Omega = [0, 1]^2,$$

with Dirichlet boundary conditions $u = 0$ on $\partial\Omega$.

There are four solution curves branching off the trivial solution curve $(0, \lambda \in \mathbb{R})$ at the bifurcation point $\lambda_{14} = (1^2 + 4^2)\pi^2 = 17\pi^2$. For each of these curves the functions u are fixed under the elements of a symmetry subgroup of $\Gamma := D_4 \times Z_2$, see [1], [6].

The discretization of the problem with the finite difference method results in a problem of the form $H(u^h, \lambda) = 0$, where u^h denotes a discrete function, $u^h \in \mathbb{R}^n$, defined on a uniform grid $\Omega^h \subset \Omega$. For the numerical experiments the standard five-point-star discretization was used to approximate the Laplacian operator. Tracing a solution curve with an Euler-Newton-continuation method, see [3], leads to unsymmetric linear systems with coefficient matrices of the form

$$A_g = \begin{pmatrix} A & d \\ c^T & \alpha \end{pmatrix},$$

where $A = \Delta^h$ is symmetric, $d = D_\lambda H$ evaluated in a point (u^h, λ) and (c^T, α) is a vector of local parametrization.

Let S_1, S_1' denote the reflections leaving the middle parallels of the square fixed. The solution curve in the Σ-fixed-point space, Σ generated by $S_1, -S_1'$, was traced. By the Γ-equivariance of H, the problem was restricted to the fundamental domain of Σ, for details of the symmetry subgroup Σ and the restriction see [1], [2] and [9]. A uniform meshsize of $h = 1/(k+1)$, $k = 49$, was used, such that the dimension of the fixed point space was $n_\Sigma = \dim X_N^\Sigma = 600$, $u^h \in X_N^\Sigma$. The linear equation systems had coefficient matrices $A_g \in \mathbb{R}^{(n_\Sigma+1)\times(n_\Sigma+1)}$.

For all the experiments given below , the residual tolerance of the nonlinear problems for the Newton corrector was $1 \cdot 10^{-6}$. The residual tolerance for the approximations obtained by GMRES(m) was $1 \cdot 10^{-7}$. A stepsize of $h_i = 1.0$ was kept for the predictor step. For this stepsize two Newton iterations were needed to perform the corrector step. The linear systems were solved without preconditioning with the original GMRES(m) algorithm and the modified version described above. The experiments were carried out on a DEC VaxStation 3100 with 16 MB RAM. The following tables show the number of iterations and the CPU-time needed to solve the linear systems for different (not consecutive) parameter values λ.

Results are given for different dimensions m of the Krylov-subspaces. Note that theoretically, both algorithms require the same number of iterations, but that round-off errors produce different results. While the number of iterations is almost identical for a maximal dimension of the Krylov-subspace of $m = 100$, the values differ for for the cases $m = 70, 80$ and 90 where a restart is necessary. For the parameter values where two corrector steps had to be performed the modified version required less iterations than the original GMRES(m) algorithm in the first step and more in the second step.

It can also be seen that the choice of the parameter m is important for the convergence behaviour. It is remarkable that for both versions, the choice of $m = 70$ yielded better results than the choice $m = 80$ for the first parameter values. And for the original version, a higher number of iterations was needed in the case $m = 70$ compared to $m = 100$, but the performance of $m = 70$ was better in terms of CPU-time, since the computational effort grows quadratically with the parameter value m. At $\lambda = 173.04$ for example, restarting the algorithm after $m = 70$ iterations and performing 30 more iterations is more efficient than the choice $m = 100$ where 91 iterations without a restart are needed to to obtain an acceptable approximation.

Overall, a comparison of the CPU-times shows that the modification of the orthogonalization process (step 2 of Algorithm 2.1), as outlined in section 3, results in a considerable reduction of computational effort.

Parameter values, iterations and CPU-times for GMRES(m) :

λ	m = 70		m = 80		m = 90		m = 100		
	1.Step	2.Step	1.Step	2.Step	1.Step	2.Step	1.Step	2.Step	
173.04	100	69	110	69	92	70	91	69	Itn
	20.79s	16.57s	25.46s	16.60s	27.41s	17.03s	27.99s	16.62s	CPU
177.75	96	63	91	62	87	62	87	62	Itn
	19.89s	14.04s	22.54s	13.59s	25.51s	13.60s	25.62s	13.62s	CPU
183.03	95	56	84	57	83	56	83	56	Itn
	19.73s	11.30s	22.00s	11.63s	23.36s	11.26s	23.45s	11.28s	CPU
188.70	92	45	82	54	82	43	82	43	Itn
	19.22s	7.63s	21.92s	10.57s	22.77s	6.97s	22.92s	6.99s	CPU
195.50	64	51	62	41	61	41	61	41	Itn
	14.45s	9.50s	13.55s	6.40s	13.15s	6.41s	13.20s	6.42s	CPU

Parameter values, iterations and CPU-times for modified version :

λ	m = 70		m = 80		m = 90		m = 100		
	1.Step	2.Step	1.Step	2.Step	1.Step	2.Step	1.Step	2.Step	
173.04	75	73	85	76	91	71	91	69	Itn
	4.16s	3.98s	4.77s	4.20s	5.17s	3.91s	5.15s	3.82s	CPU
177.75	74	67	83	66	87	62	87	62	Itn
	4.03s	3.69s	4.54s	3.62s	4.80s	3.40s	4.82s	3.50s	CPU
183.03	73	62	81	59	83	56	83	56	Itn
	4.00s	3.39s	4.48s	3.25s	4.59s	3.08s	4.59s	3.09s	CPU
188.70	71	55	81	55	82	43	82	43	Itn
	3.87s	3.00s	4.47s	3.02s	4.53s	2.36s	4.56s	2.35s	CPU
195.50	71	53	62	41	61	41	61	41	Itn
	3.88s	2.87s	3.38s	2.24s	3.37s	2.31s	3.37s	2.32s	CPU

References

[1] Allgower,E., Böhmer,K., Mei,Z. : *A generalized equibranching lemma with applications to $D_4 \times Z_2$ symmetric elliptic problems*, Part I, Bericht Nr.9 (1990), Fachbereich Mathematik, Universität Marburg.

[2] Allgower,E., Böhmer,K., Mei,Z. : On a Problem Decomposition for Semilinear Nearly Symmetric Elliptic Problems , in: *Parallel Algorithms for PDE's*, W. Hackbusch (Ed.), Notes on Numerical Fluid Mechanics **31**, Vieweg Verlag, Braunschweig, 1991, pp.1-17.

[3] Allgower,E., Georg,K. : *Numerical Continuation Methods, An introduction*, Springer-Verlag, Berlin Heidelberg New York 1990.

[4] Chan,T.F., Saad,Y. : Iterative Methods for Solving Bordered Systems with Applications to Continuation Methods, *SIAM J. Sci. Stat. Comp.* **6** (1985) 438-451.

[5] Golub,G.H., Van Loan,Ch.F. : *Matrix Computations*, The John Hopkins University Press, Baltimore, MD, 1983.

[6] Mei,Z. : Path following around Corank-2 Bifurcation Points of a Semi-Linear Elliptic Problem with Symmetry , *Computing* **46** (1991) 492-508.

[7] Saad,Y. : Krylov Subspace Methods for Solving Large Unsymmetric Linear Systems, *Math. Comp* **37** (1981) 105-125.

[8] Saad,Y., Schultz,M.H. : A Generalized Minimal Residual Algorithm for Solving Nonsymmetric Linear Systems, *SIAM J. Sci. Stat. Comp.* **7** (1988) 857-869.

[9] Sebastian,R. : *Anwendung von verallgemeinerten konjugierten Gradientenverfahren auf Verzweigungsprobleme mit Symmetrien*, Diplomarbeit, 1991, Department of Mathematics, University of Marburg.

International Series of Numerical Mathematics, Vol. 104, © 1992 Birkhäuser Verlag Basel

Hopf/Steady-state Mode Interaction for a Fluid Conveying Elastic Tube with $\mathbf{D}_4$-symmetric Support

A. Steindl

1 Introduction

The dynamics of tubes conveying fluid has been studied intensively over a long period of time from quite different points of view. While the linear models were used to calculate the critical flow rate, nonlinear investigations determined the post-critical behaviour of the pipes. The transition from planar to spatial models with rotational symmetry showed that planar oscillations and rotating solutions were possible. By adding a rotational symmetric elastic support of appropriate stiffness more degenerate bifurcations could be generated, leading to a large variety of interacting branches of solutions. All solutions are quite easily classified and visualized by their symmetry properties.

In order to get rid of the artificial rotational symmetry of the support and to gain further insight into problems with less symmetry we started our investigations for a support with discrete symmetry $\mathbf{D}_4$, which is generated by 4 identical springs located at the endpoints of a square.

The case $n = 4$ is a special subcase of $\mathbf{D}_n$-symmetric problems with rich solution structure. It is also computationally attracting, since the main solution types (primary branches) can be distinguished by the cubic terms in the bifurcation equations.

Nevertheless the Center Manifold reduction is nontrivial, since the consideration of twisting deformation due to damping forces leads to quadratic terms in the equations of motion, which have to be eliminated from the final bifurcation equations.

While trying to include all necessary terms, we attempted to keep the mechanical model as simple as possible: We assume a linear visco-elastic material law and neglect extension of the tube, shearing deformations and rotatory inertia terms.

2 Mechanical model and equations of motion

The tube is treated as an inextensible thin elastic rod of uniform annular cross-section. It is clamped at the upper end $s = 0$ and free at the lower end $s = \ell$. At the location $s = \xi\ell$ the tube is supported by 4 identical linearly visco-elastic springs, arranged along the horizontal coordinate axes. The fluid enters the pipe at the upper end and flows with constant speed U through the pipe. The problem may be expressed in

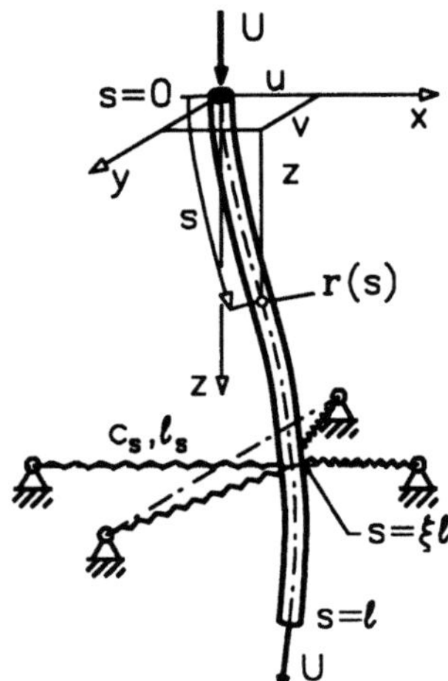

Figure 1: Fluid conveying elastic tube with $\mathbf{D}_4$-symmetric elastic support. One spring is arranged along the positive x-axis. The position of the centerline is denoted by $\mathbf{r}(s)$.

nondimensional terms by defining the following quantities:

$$\tilde{s} = s/\ell \qquad \tilde{r} = r/\ell \qquad \tilde{\Omega} = \ell\Omega \qquad \omega = \sqrt{\frac{EJ}{m_T + m_F}}\frac{1}{\ell^2},$$

$$\tilde{t} = \omega t, \qquad \beta = \frac{m_F}{m_F + m_T}, \qquad \tilde{q} = \frac{\ell^3 q}{EJ}, \qquad \tilde{T} = \frac{\ell T}{EJ},$$

$$\widetilde{F} = \frac{\ell^2 F}{EJ}, \qquad \tilde{\alpha}_1 = \alpha_1\omega, \qquad \tilde{\alpha}_3 = \alpha_3\omega, \qquad \tilde{\alpha}_s = \alpha_s\omega,$$

$$\tilde{\alpha}_e = \frac{\ell^4 \omega \alpha_e}{EJ}, \qquad \tilde{f}_s = \frac{\ell^2 f_s}{EJ}, \qquad c = \frac{4\ell^3 c_s}{2EJ}, \qquad \gamma = \frac{\ell^3 g(m_T + m_F)}{EJ},$$

$$\gamma_3 = \frac{GJ_T}{EJ}, \qquad \varepsilon_s = \ell/\ell_s, \qquad \rho = \sqrt{\frac{m_F}{EJ}}\ell U, \quad \widetilde{A} = \ell A,$$

After dropping the tildes the nondimensional differential equations and boundary conditions read (derivatives with respect to arclength s are denoted by $(\cdot)'$ and time derivatives by $(\dot{\cdot})$)

$$\begin{aligned}
\mathbf{r}' &= \mathbf{B}e_3, \\
\mathbf{B}' &= \mathbf{B} \cdot \mathbf{A}, \\
\mathbf{T}' &= \mathbf{T} \times \mathbf{\Omega} + \mathbf{F} \times e_3, \\
\mathbf{F}' &= \mathbf{F} \times \mathbf{\Omega} + \mathbf{B}^T \left(\ddot{\mathbf{r}} + 2\sqrt{\beta}\rho\dot{\mathbf{r}}' + \rho^2 \mathbf{r}'' - \gamma e_3 \right), \\
\mathbf{r}(0) &= \mathbf{0}, \\
\mathbf{B}(0) &= \mathbf{E}, \\
\mathbf{F}(\xi_+) - \mathbf{F}(\xi_-) &= -\mathbf{B}(\xi)\mathbf{f}_s, \\
\mathbf{F}(1) &= \mathbf{0}, \\
\mathbf{T}(1) &= \mathbf{0},
\end{aligned} \qquad (1)$$

with

$$
\begin{aligned}
T_1 &= \kappa_1 + \alpha_1 \dot\kappa_1, \\
T_2 &= \kappa_2 + \alpha_1 \dot\kappa_2, \\
T_3 &= \gamma_3(\kappa_3 + \alpha_3 \dot\kappa_3).
\end{aligned}
\tag{2}
$$

Here r denotes the position of the centerline, the orthogonal matrix $\mathbf{B}$ gives the orientation of the cross-sections, T and F are the resultant couple and force vectors in body coordinates. The skewsymmetric matrix $\mathbf{A}$ and its associated vector Ω contain the bending and twisting deformations κ_i. The restoring force due to the support is denoted by f_s and e_3 is the unit vector in z-direction. The parameter c denotes the stiffness of the support, ρ measures the flow rate and the α_i are damping coefficients. The derivation of (1) will be described in a forthcoming paper [6].

Due to the circular cross-section of the tube and the symmetric arrangement of the support the differential equations (1) are equivariant under the representation of $\mathbf{D}_4$

$$
\begin{aligned}
r &\mapsto \mathbf{D}r, & \mathbf{B} &\mapsto \mathbf{D} \cdot \mathbf{B} \cdot \mathbf{D}^T \\
T &\mapsto (\det \mathbf{D})\mathbf{D}T, & F &\mapsto \mathbf{D}F,
\end{aligned}
\tag{3}
$$

where $\mathbf{D} \in \mathbf{D}_4$ is a (possibly improper) rotation matrix

$$
\mathbf{D} = \mathbf{R}_3(\zeta) =
\begin{pmatrix}
\cos\zeta & -\sin\zeta & 0 \\
\sin\zeta & \cos\zeta & 0 \\
0 & 0 & 1
\end{pmatrix}
\quad \text{or} \quad
\mathbf{D} = \mathbf{diag}(1,-1,1) \cdot \mathbf{R}_3(\zeta),
\tag{4}
$$

with $\zeta = k\pi/2$.

3 Linearization

The straight downhanging configuration

$$
\begin{aligned}
r &= se_3, & \mathbf{B} &\equiv \mathbf{E}, \\
T &\equiv 0, & F &= (1-s)\gamma e_3
\end{aligned}
\tag{5}
$$

is a solution of (1) for all parameter values. It is called the "trivial solution" and is invariant under $\mathbf{D}_4$. Due to the symmetry the linearization of (1) decouples into 2 identical bending equations

$$
\ddot{x}_i + \alpha_e \dot{x}_i + x_i'''' + \alpha_i \dot{x}_i'''' + 2\sqrt{\beta}\rho \dot{x}_i' + \rho^2 x_i'' - (\gamma(1-s)x_i')' = 0
\tag{6}
$$

for $i = 1,2$ and one simple equation for the twisting angle χ:

$$
\gamma_3(\chi'' + \alpha_3 \dot\chi'') = 0.
\tag{7}
$$

Also the boundary and jump conditions separate:

$$
\begin{aligned}
x_i(0) &= 0, \\
x_i'(0) &= 0, \\
(x_i'' + \alpha_1 \dot{x}_i'')(1) &= 0, \\
(x_i''' + \alpha_1 \dot{x}_i''')(1) &= 0, \\
(x_i''' + \alpha_1 \dot{x}_i''')\big|_{\xi_-}^{\xi_+} + c(x_i + \alpha_s \dot{x}_i)(\xi) &= 0, \\
\chi(0) &= 0, \\
\gamma_3(\chi' + \alpha_3 \dot\chi')(1) &= 0.
\end{aligned}
\tag{8}
$$

Since every solution of (7) decays exponentially, the linear stability investigation may be restricted to one bending equation, corresponding to an equivalent planar problem. But every critical eigenvalue of the planar problem will appear twice in the full 3-dimensional system.

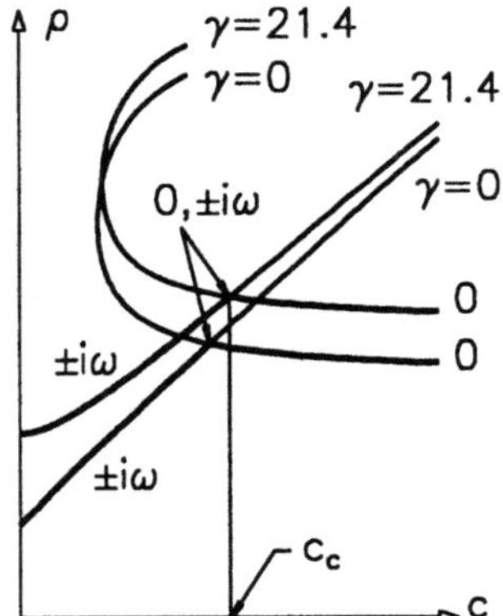

Figure 2: Stability boundary in (c, ρ)-space for $\xi = 0.5$. The material parameters are chosen according to pipe 1 in [7]. Inclusion of gravity ($\gamma \neq 0$) increases the critical flow rate ρ. For stiff support ($c > c_c$) the loss of stability occurs by a zero eigenvalue, for soft support ($c < c_c$) purely imaginary eigenvalues lead to flutter instabilities. At $c = c_c$ a Steady-state/Hopf mode interaction occurs.

Fig. 2 shows a typical stability chart in (c, ρ)-space for $\xi = 0.5$. For stiff support the trivial state loses its stability by a zero eigenvalue, whereas for soft support generally purely imaginary eigenvalues occur. For a certain critical stiffness c_c a codimension 2 bifurcation occurs with simultaneous appearance of zero and imaginary eigenvalues. By varying the location ξ of the support it is possible to find double zero eigenvalues; also selfintersections of the Hopf bifurcation boundary have been observed, leading to Hopf-Hopf-mode interactions. ([8])

4   Nonlinear bifurcation equations

In order to derive the bifurcation equations in the neighbourhood of the codimension 2 bifurcation point, we perform Center Manifold reduction onto the critical eigenspace and simplify the expressions by Normal Form theory.

The Center Manifold reduction requires some information about the remaining eigenvalues of the linearized system. It is quite easy to calculate the limiting behaviour of the spectrum for high spatial modes: There exist 2 families of eigenvalues, one of which converges rapidly to $-\infty$, whereas the second one accumulates at $\lambda = -1/\alpha$ due to the appearance of material damping in the highest order derivatives. The location of the intermediate eigenvalues has to be calculated numerically: An approximation of the eigenvalues by finite differences and subsequent "exact" computation shows that for material data according to pipe 1 in [7] only two pairs of stable complex eigenvalues are found at the critical parameter values; all further eigenvalues are real and strictly negative.

According to the separation of the linear system into bending and twisting equations the approximation of the Center Manifold splits into 2 parts:

In order to obtain cubic terms in the bifurcation equations, a trivial ansatz suffices for the stable components in the *bending* system. It remains to show that there exists a sufficiently smooth Center Manifold:

Since the highest order derivatives are the same as in the panel flutter problem (page 355ff in [4]), the linear equations generate a contractive semiflow on $(\mathbf{H}^2 \times \mathbf{L}^2)^2$.

We augment the system (1) by differential equations for the velocities and accelerations and define a continous operator

$$\mathbf{T}(\Omega, \dot{\Omega}, \ddot{\Omega}) \mapsto T$$

by successively solving the initial value problems for $\mathbf{B}$, $\dot{\mathbf{B}}$, $\ddot{\mathbf{B}}$, r, $\dot{r}$, $\ddot{r}$, F and T. Inserting $T = \mathbf{T}(\Omega, \dot{\Omega}, \ddot{\Omega})$ into the material law (2) restates (1) as a smooth system of integro-differential equations for the bending functions Ω_i.

Because the variables corresponding to *twisting* deformations appear quadratically in the bending equations, these have to be removed by the following steps: By (1_3) and (2) the torque T_3 satisfies the quadratic differential equation

$$T_3' = \kappa_1 T_2 - \kappa_2 T_1 = \alpha_1(\kappa_1 \dot{\kappa}_2 - \kappa_2 \dot{\kappa}_1), \tag{9}$$

with boundary condition $T_3(1) = 0$. It has to be integrated for every distinct pair of eigenfunctions. Now the material law (2_3) can be regarded as ordinary differential equation for κ_3:

$$\gamma_3(\kappa_3 + \alpha_3 \dot{\kappa}_3) = T_3, \tag{10}$$

which is solved as usual by power series expansion. Finally the twisting angle χ can be calculated from (1_2).

The Normal Form reduction introduces the additional phase-shift symmetry $\mathbf{S}^1$ into the equations of motion. Since the bifurcation equations also inherit the $\mathbf{D}_4$-symmetry of the original system, the bifurcation equations are equivariant wrt. the group $\mathbf{D}_4 \times \mathbf{S}^1$. If we choose complex coordinates z_1, z_2, z_3 according to [8], the symmetry transformations operate as follows:

$$
\begin{aligned}
\zeta \in \mathbf{Z}_4 : \quad & (z_1, z_2, z_3) \mapsto (e^{i\zeta} z_1, e^{-i\zeta} z_2, e^{i\zeta} z_3), \\
\kappa \in \mathbf{Z}_2 : \quad & (z_1, z_2, z_3) \mapsto (z_2, z_1, \bar{z}_3), \\
\theta \in \mathbf{S}^1 : \quad & (z_1, z_2, z_3) \mapsto (e^{i\theta} z_1, e^{i\theta} z_2, z_3).
\end{aligned}
\tag{11}
$$

It's a straightforward task to find those nonlinear functions $g(z)$, which commute with any element $\sigma \in \mathbf{D}_4 \times \mathbf{S}^1$: If the terms are written in multi-index notation

$$g_j(z) = \sum_m g_{j,m} z^m = \sum_m g_{j,m_1 m_2 m_3 m_4 m_5 m_6} z_1^{m_1} \bar{z}_1^{m_2} z_2^{m_3} \bar{z}_2^{m_4} z_3^{m_5} \bar{z}_3^{m_6}$$

equivariance under $\mathbf{Z}_4$ requires

$$m_1 - m_2 - m_3 + m_4 + m_5 - m_6 \equiv 1 \quad (\text{mod } 4) \quad \text{for } j = 1 \text{ and } 3. \tag{12}$$

Equivariance wrt. $\mathbf{S}^1$ yields the conditions

$$
\begin{aligned}
m_1 - m_2 + m_3 - m_4 &= 1 && \text{for } j = 1, \tag{13} \\
m_1 - m_2 + m_3 - m_4 &= 0 && \text{for } j = 3. \tag{14}
\end{aligned}
$$

Nr.	Orbit repr.	Σ	$\Sigma_x = Fix(\Sigma)$	$\dim \Sigma_x$	
(0)	$(0,0,0)$	$\mathbf{D}_4 \times \mathbf{S}^1$	$(0,0,0)$	0	Trivial solution
(1)	$(0,0,b)$	$\mathbf{Z}_2(\kappa) \times \mathbf{S}^1$	$(0,0,x)$	1	Buckling mode 1
(2)	$(0,0,\sqrt{-ib})$	$\mathbf{Z}_2(\kappa\zeta) \times \mathbf{S}^1$	$(0,0,\sqrt{-i}x)$	1	Buckling mode 2
(3)	$(a,a,0)$	$\mathbf{Z}_2(\kappa) \oplus \mathbf{Z}_2^c$	$(w,w,0)$	2	Standing wave 1
(4)	$(a,ia,0)$	$\mathbf{Z}_2(\kappa\zeta) \oplus \mathbf{Z}_2^c$	$(w,iw,0)$	2	Standing wave 2
(5)	$(a,0,0)$	$\tilde{\mathbf{Z}}_4 = \{(-\zeta,\zeta)\}$	$(w,0,0)$	2	Rotating wave
(6)	(a,a,b)	$\mathbf{Z}_2(\kappa)$	(w,w,x)	3	(1) $\parallel$ (4)
(7)	(a,a,ib)	$\mathbf{Z}_2(\kappa\pi,\pi)$	(w,w,iy)	3	(1) $\perp$ (4)
(8)	$(a,ia,\sqrt{-ib})$	$\mathbf{Z}_2(\kappa\zeta)$	$(w,iw,\sqrt{-i}x)$	3	(2) $\parallel$ (5)
(9)	$(a,ia,\sqrt{ib})$	$\mathbf{Z}_2(\kappa\zeta\pi,\pi)$	$(w,iw,\sqrt{i}x)$	3	(2) $\perp$ (5)
(10)	$(a,0,b)$	1	$(w_1,\mathcal{O}(w_1,w_3),w_3)$	6	modulated wave
(11)	$(0,0,w)$	$\mathbf{S}^1$	$(0,0,w)$	2	asym. buckling
(12)	$(a,b,0)$	$\mathbf{Z}_2^c$	$(w_1,w_2,0)$	4	modulated wave
(13)	(a,a,w)	1		6	
(14)	(a,ia,w)	1		6	

Table 1: Solutions of (16) and their isotropy subgroups. The solutions (11) and (12) correspond to degenerate bifurcations.

The 2nd component of $g(z)$ is calculated by exploiting the flip symmetry:

$$g_2(z) = g_1(\kappa \cdot z). \tag{15}$$

To 3rd order we obtain the unfolded equivariant bifurcation equations

$$\begin{aligned}
g_1(z) &= (\lambda + i\omega - i\tau_1)z_1 + (A_1 z_1\bar{z}_1 + A_2 z_2\bar{z}_2 + A_3 z_3\bar{z}_3)z_1 + \\
&\quad + A_4 z_2 z_3^2 + A_5 \bar{z}_1 z_2^2 + A_6 z_2\bar{z}_3^2, \\
g_2(z) &= (\lambda + i\omega - i\tau_1)z_2 + (A_2 z_1\bar{z}_1 + A_1 z_2\bar{z}_2 + A_3 z_3\bar{z}_3)z_2 + \\
&\quad + A_4 z_1\bar{z}_3^2 + A_5 z_1^2\bar{z}_2 + A_6 z_1 z_3^2, \\
g_3(z) &= (\mu - i\tau_3)z_3 + (A_7 z_1\bar{z}_1 + A_8 z_2\bar{z}_2 + A_9 z_3\bar{z}_3)z_3 + \\
&\quad + A_{10} z_1\bar{z}_2\bar{z}_3 + A_{11}\bar{z}_1 z_2\bar{z}_3 + A_{12}\bar{z}_3^3,
\end{aligned} \tag{16}$$

with $A_j = c_j + id_j$ for $i = 1\ldots 12$, $A_8 = \overline{A}_7$ and $d_9\ldots d_{12} = 0$.

Zeroes of (16) correspond to stationary states of the pipe ($\tau_j = 0$), periodic oscillations ($\tau_1 \neq 0, \tau_3 = 0$) or quasiperiodic motions. The solutions can be classified by their symmetry properties. Table 1 lists the different solutions and their isotropy subgroups Σ.

4.1 Branching equations and stability

According to [1] the equations (16) are solved by restricting to the invariant subspaces. The calculation of the stability properties requires the eigenvalue computation for the linearization of (16) at the solution branches. By isotypic decomposition of the 6-dimensional critical eigenspace into invariant subspaces for the isotropy subgroups the matrices are block-diagonalized.

By the symmetry the various solution branches correspond to orbits of different solutions, which are conjugate to each other and behave identically. For example buckling in x-direction is just a representative for buckling in any vertical plane containing a spring. It is of course different from buckling in diagonal direction, which is denoted by another solution orbit with different branching and stability behaviour. In the following description of the solution branches we will therefore refer to special representations, rather than to orbits.

The stability of the branches is determined by the eigenvalues of the linearized equations, evaluated at the solutions. Eigenvalues with negative real parts indicate stability. For the 2×2 matrices below the stability is more easily calculated from their determinant and trace: The trace has to be negative and the determinant positive for a stable solution.

Generically we expect to find the following solution branches in the neighborhood of the bifurcation point:

(0) $z = 0$: Trivial solution.

Real parts of eigenvalues:

$$\lambda \ (4 \text{ times}), \ \mu \ (\text{twice}).$$

(1) $z = (0,0,b)$: Stationary buckling in x-direction.
Amplitude equation

$$\mu + (c_9 + c_{12})b^2 = 0.$$

Real parts of eigenvalues:

$$\lambda + c_3 b^2 \pm (c_4 + c_6)b^2,$$
$$2(c_9 + c_{12})b^2,$$
$$-4c_{12}b^2.$$

(2) $z = (0,0,\sqrt{-i}b)$: Stationary buckling in diagonal direction.
Amplitude equation

$$\mu + (c_9 - c_{12})b^2 = 0.$$

Real parts of eigenvalues:

$$\lambda + c_3 b^2 \pm (c_4 - c_6)b^2,$$
$$2(c_9 - c_{12})b^2,$$
$$4c_{12}b^2.$$

(3) $z = (a,a,0)$: Planar oscillation in x-direction.
Amplitude equation

$$\lambda + (c_1 + c_2 + c_5)a^2 = 0.$$

Real parts of eigenvalues:

$2(c_1 + c_2 + c_5)a^2,$

Eigenvalues of complex matrix $(A_1 - A_2 - 3A_5)a^2 w + (A_1 - A_2 + A_5)a^2\overline{w}$:
$\quad \cdot$ trace: $2(c_1 - c_2 - 3c_5)a^2,$
$\quad\quad$ det: $8a^4(|A_5|^2 - \mathrm{Re}((A_1 - A_2)\overline{A}_5)),$
$\mu + 2c_7 a^2 \pm (c_{10} + c_{11})a^2.$

(4) $z = (a, ia, 0)$: Planar oscillation in diagonal direction.
Amplitude equation

$$\lambda + (c_1 + c_2 - c_5)a^2 = 0.$$

Real parts of eigenvalues:

$2(c_1 + c_2 - c_5)a^2,$
Eigenvalues of complex matrix $(A_1 - A_2 + 3A_5)a^2 w + (A_1 - A_2 - A_5)a^2\overline{w}$:
     trace: $2(c_1 - c_2 + 3c_5)a^2,$
       det: $8a^4(|A_5|^2 + \mathrm{Re}((A_1 - A_2)\overline{A}_5)),$
$\mu + 2c_7 a^2 \pm (c_{10} - c_{11})a^2.$

(5) $z = (a, 0, 0)$: Discrete analogue of a "Rotating wave".
Amplitude equation

$$\lambda + c_1 a^2 = 0.$$

Real parts of eigenvalues:

$2c_1 a^2,$
Eigenvalues of complex matrix $(A_2 - A_1)a^2 w + A_5 a^2\overline{w}$:
     trace: $2(c_2 - c_1)a^2,$
       det: $4a^4(|A_2 - A_1|^2 - |A_5|^2),$
$\mu + c_7 a^2.$

(6) $z = (a, a, b)$ Buckling and oscillation in x-direction.
Amplitude equations

$$\begin{aligned}
\lambda + (c_1 + c_2 + c_5)a^2 + (c_4 + c_6)b^2 &= 0, \\
\mu + (2c_7 + c_{10} + c_{11})a^2 + (c_9 + c_{12})b^2 &= 0.
\end{aligned}$$

Stability properties are determined by the matrices

$$M_0 = 2 \begin{pmatrix} (c_1 + c_2 + c_5)a^2 & (c_3 + c_4 + c_6)ab \\ (2c_7 + c_{10} + c_{11})ab & (c_9 + c_{12})b^2 \end{pmatrix}$$

and

$$M_1 = \begin{pmatrix} (c_1 - c_2 - c_5)a^2 - (c_4 + c_6)b^2 & 2d_5 a^2 + (d_4 + d_6)b^2 & -(d_4 - d_6)ab \\ (d_1 - d_2 - d_5)a^2 - (d_4 + d_6)b^2 & -2c_5 a^2 - (c_4 + c_6)b^2 & (c_4 - c_6)ab \\ 2d_7 ab & (c_{10} - c_{11})ab & -(c_{10} + c_{11})a^2 - 2c_{12}b^2 \end{pmatrix}$$

(7) $z = (a, a, ib)$: Oscillation in x-direction and buckling in y-direction.
Amplitude equations

$$\lambda + (c_1 + c_2 + c_5)a^2 - (c_4 + c_6)b^2 = 0,$$
$$\mu + (2c_7 - c_{10} - c_{11})a^2 + (c_9 + c_{12})b^2 = 0.$$

The stability properties are computed by changing the signs of A_4, A_6, A_{10} and A_{11} in the matrices M_i of case (6).

(8) $z = (a, ia, \sqrt{-ib})$: Buckling and planar oscillation in the same diagonal direction.
Amplitude equations

$$\lambda + (c_1 + c_2 - c_5)a^2 + (c_4 - c_6)b^2 = 0,$$
$$\mu + (2c_7 + c_{10} - c_{11})a^2 + (c_9 - c_{12})b^2 = 0.$$

The stability properties are computed by changing the signs of A_5, A_6, A_{11} and A_{12} in the matrices M_i of case (6).

(9) $z = (a, ia, \sqrt{ib})$: Buckling and planar oscillation in diagonal direction. The oscillation plane is perpendicular to the buckling direction.
Amplitude equations

$$\lambda + (c_1 + c_2 - c_5)a^2 - (c_4 - c_6)b^2 = 0,$$
$$\mu + (2c_7 - c_{10} + c_{11})a^2 + (c_9 - c_{12})b^2 = 0.$$

The stability properties are computed by changing the signs of A_5, A_4, A_{10} and A_{12} in the matrices M_i of case (6).

(10) $z \approx (a, 0, b)$: Modulated rotating wave. It is generated by a Hopf bifurcation from the rotating wave solution.
Amplitude equations

$$\lambda + c_1 a^2 + c_3 b^2 = 0,$$
$$\mu + c_7 a^2 + \tilde{c}_9 b^2 = 0.$$

where

$$\tilde{c}_9 = c_9 + c_{10}\tilde{c}_4 + c_{11}\tilde{c}_6,$$
$$\tilde{c}_4 = \mathrm{Re}(A_4/[3c_1 - c_2 + i(d_1 - d_2 - 2d_7)]),$$
$$\tilde{c}_6 = \mathrm{Re}(A_6/[3c_1 - c_2 + i(d_1 - d_2 + 2d_7)]).$$

These coefficients follow from Center manifold computations at the secondary Hopf bifurcation point. They also determine the stability of the branch in the invariant subspace close to the bifurcation point.

The solutions (11) and (12) correspond to degenerate simple stationary and Hopf bifurcations. Their treatment would require higher order terms in the bifurcation equations. The solutions (13) and (14) indicate tertiary branches of solutions.

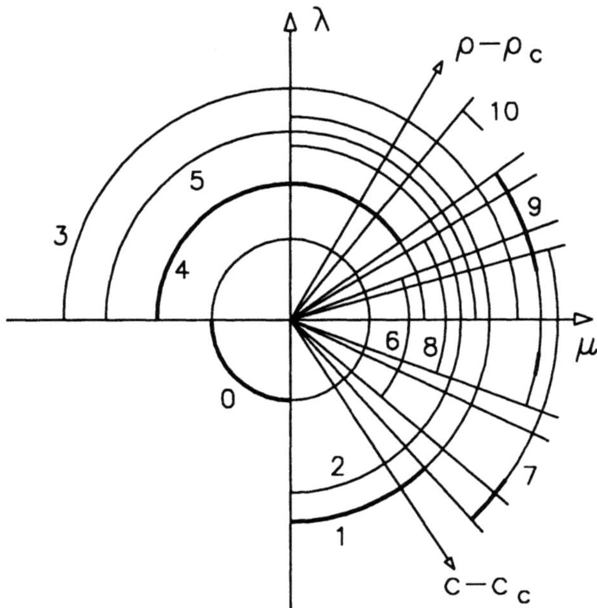

Figure 3: Stratification of the (ρ, c)-parameter plane into domains of existence of the different solution types. Stable solution orbits are indicated by bold arcs. The solution (9) has 2 disjoint regions of stability.

The (local) domain of existence in (ρ, c)-parameter space of the different solutions for parameter values corresponding to pipe 1 in [7] is shown in Fig. 3 by circular arcs. Bold arcs indicate stable solutions.

Fig. 3 indicates a complex picture of solution branches: If we fix the stiffness c sligthly above the critical value c_c and increase the flow rate ρ, the pipe first buckles in x-direction. Later it starts an out of plane oscillation about the stationary state, until that solution becomes unstable. A little bit later we encounter an out of plane oscillation about a diagonally buckled configuration with changing stability properties. Finally the pipe performs a pure planar oscillation in diagonal direction.

Concluding remarks

By the methods developed in [1] it is a straightforward task to derive and solve the bifurcation equations for a Hopf/Steady state mode interaction. It should be worth to apply these methods to different bifurcations (eg. Takens-Bogdanov bifurcation, Hopf/Hopf-mode interaction) and symmetry groups. Especially the group $\mathbf{D_3}$ should exhibit quite different behaviour: For example we cannot expect pure oscillation modes in that case.

For experimental reasons small imperfections should be taken into account, which do not preserve the full original symmetry.

Address: Alois Steindl, Institut für Mechanik, Techn. Univ. Vienna, Wiedner Hauptstraße 8–10, A–1040 Wien.

References

[1] M. Golubitsky, I. Stewart, and D. Schaeffer. *Singularities and Groups in Bifurcation Theory*, volume 51 and 69 of *Applied Math. Sciences*. Springer-Verlag, New York – Heidelberg – Berlin, 1985, 1988.

[2] P. Holmes and J Marsden. Bifurcation to divergence and flutter in flow induced oscillations; an infinite dimensional analysis. *Automatica*, 14:367–384, 1978.

[3] T. S. Lundgren, P. R. Sethna, and A. K. Bajaj. Stability Boundaries for Flow Induced Motions of Tubes with an Inclined Nozzle. *J. Sound and Vibrations*, 64:553–571, 1979.

[4] J. E. Marsden and Th. J.R. Hughes. *Mathematical Foundations of Elasticity*. Prentice-Hall, Inc., 1983.

[5] R. Seydel. A continuation algorithm with step control. In *Numerical methods for bifurcation problems*. ISNM 70. Birkhäuser, 1984.

[6] A. Steindl. Bifurcations of a fluid conveying elastic tube with D_n-symmetric support. In preparation.

[7] Y. Sugiyama, Y. Tanaka, T. Kishi, and H. Kawagoe. Effect of a Spring Support on the Stability of Pipes Conveying Fluid. *J. Sound and Vibration*, 100:257–270, 1985.

[8] H. Troger and A. Steindl. *Nonlinear Stability and Bifurcation Theory: An Introduction for Engineers and Applied Scientists*. Springer-Verlag, Wien – New York, 1991.

International Series of Numerical Mathematics, Vol. 104, © 1992 Birkhäuser Verlag Basel

TEST FUNCTIONS FOR BIFURCATION POINTS AND HOPF POINTS IN PROBLEMS WITH SYMMETRIES.

Bodo Werner

AMS (MOS) subject classification: 65H10, 20C30, 58F14.

Key words: *test functions for bifurcation points, interaction of bifurcation points, computation and detection of bifurcation points, Hopf bifurcation, symmetry of a group, subspace breaking bifurcation point.*

1 Introduction

Following a path $\mathcal{C} := \{(x(s), \lambda(s)) : |s| < \delta\}$ of equilibria of $\dot{x} = g(x, \lambda)$, $\lambda \in \mathbb{R}$, $x \in X \subset \mathbb{R}^N$, **test functions** t are real functions defined on a neighborhood of $\mathcal{C}$, being monitored during continuation and strictly changing sign in bifurcation points $(x_0, \lambda_0) = (x(0), \lambda(0))$ of interest (Def.2.1, see also SEYDEL [10]).

Test functions can be used for the detection, computation and (after releasing a second parameter) for path following of the corresponding bifurcation points (see KHIBNIK [8]).

In sec. 3 and 4 we review some test functions for stationary and Hopf bifurcation without symmetries. We will introduce a new test function for *real* and *imaginary* (see WERNER-JANOVSKY [13]) Hopf points near TB-points. This test function has some similarity with that in GRIEWANK-REDDIEN [5] for generalized turning points.

In sec. 5 we show that block diagonalization for problems with the symmetry of a group Γ (WERNER [12], STORK-WERNER [11], IKEDA-MUROTA [7], GATERMANN-HOHMANN [3], HEALEY-TREACEY [6]) can recast the problem of finding test functions for Γ-*simple (symmetry breaking) bifurcation points* onto the non-symmetric case.

In sec. 6 we consider problems with a subspace-invariance ($g(X_+, \lambda) \subset X_+ \subset X$). We have in mind $\mathbb{Z}_2$-symmetric problems or problems with a more general symmetry of a group Γ (here $X = X^\Sigma$ and $X_+ = X^\Gamma$ are fixed point spaces, and Σ is a bifurcation subgroup , see DELLNITZ-WERNER [1]).

We are concerned with the interaction of general **t-points** (essentially defined by $g(x, \lambda) = 0$, $t(x, \lambda) = 0$) with subspace breaking bifurcation points (SB-points). In Th.6.1 we give conditions under which such an interaction leads to secondary t-points (on the asymmetric branch of equilibria). Th.6.1 generalizes the analysis in WERNER-JANOVSKY [13] concerning $\mathbb{Z}_2$-TB points.

In sec. 7 (Th.7.1) we apply Th.6.1 to analyse the occurrence of secondary symmetry breaking bifurcation points of problems with the symmetry of a group Γ.

Our *interaction theorem* Th.7.1 says that the interaction of two Γ-simple bifurcation points of different symmetry types ϑ and η on a Γ-symmetric branch leads to secondary Σ-simple bifurcation points of symmetry type ϱ, if Σ is a bifurcation subgroup for ϑ and if ϱ satisfies an interaction condition (7.21).

2 Test functions

Let $g : \mathbb{R}^N \times \mathbb{R} \to \mathbb{R}^N$ be a smooth mapping and let X be a subspace of $\mathbb{R}^N$ invariant under $g(\cdot, \lambda)$ (in sec.6, 7 we set $X = X^\Sigma$, the fixed point space corresponding to a subgroup

Σ of Γ).

Definition 2.1 *Let U be an (relatively) open subset of $X \times \mathbb{R}$ and $t : U \to \mathbb{R}$ be continously differentiable. A common zero (x_0, λ_0) of g and t is called a* **t-point** *of g. A t-point (x_0, λ_0) is* **regular** *iff (x_0, λ_0) is a regular zero of*

$$G : U \to X \times \mathbb{R}, \quad G(x, \lambda) := (g(x, \lambda), t(x, \lambda)) \tag{2.1}$$

e.g. if $DG(x_0, \lambda_0)$ is regular.

Observe that *regularity* can be forced by restricting g (and t) to the invariant subspace X, see Rem. 3.1.

To be a regular t-point, (x_0, λ_0) has to be a *regular* root of g with respect to X ($Dg(x_0, \lambda_0)$ has full rank, g being restricted to $X \times \mathbb{R}$).

If (x_0, λ_0) is a regular root of g with respect to X, there is locally a unique (smooth) branch of equilibria

$$\mathcal{C} := \{(x(s), \lambda(s)) : |s| < \delta\} \subset U \subset X \times \mathbb{R}, \quad \delta > 0$$

through $(x_0, \lambda_0) = (x(0), \lambda(0))$. We state the following theorem which can easily been proved.

Theorem 2.2 *(x_0, λ_0) is a regular t-point of g iff (x_0, λ_0) is a regular root of g with respect to X and if*

$$\tau(0) = 0, \quad \tau'(0) \neq 0, \quad \text{where } \tau(s) := t(x(s), \lambda(s)). \tag{2.2}$$

Denoting a tangent vector of $\mathcal{C}$ through (x_0, λ_0) by $w_0 \neq 0$, we can assume that either $w_0 = (v_\lambda, 1) \in X \times \mathbb{R}$ or that $w_0 = (\varphi_0, 0)$. In the first (λ-*regular*) case, $\mathcal{C}$ can be parametrized by $s = \lambda$, in the second case, (x_0, λ_0) is a *turning point* of g (defined by $g(x_0, \lambda_0) = 0, g_x^0 \varphi_0 = 0, \varphi_0 \neq 0, g_\lambda \notin im(g_x^0)$).

Here and in the following we use the abbreviations

$$g_x^0 = \frac{\partial g}{\partial x}(x_0, \lambda_0), \quad g_\lambda^0 = \frac{\partial g}{\partial \lambda}(x_0, \lambda_0), \quad g_{x\lambda}^0 := \frac{\partial^2 g}{\partial x \partial \lambda}(x_0, \lambda_0),$$

and analogously $t_x^0, t_\lambda^0, t_{x\lambda}^0$, etc.

Locally $t(x, \lambda) = 0$ defines a manifold of codimension 1 in U transversally intersected by $\mathcal{C}$ in (x_0, λ_0), provided that (x_0, λ_0) is a regular t-point of g. (2.2) is the transversality condition and can also be expressed by $\nabla t(x_0, \lambda_0) \cdot w_0 \neq 0$ or in the case of λ-regular (x_0, λ_0) by

$$d_\lambda := t_x^0 v_\lambda + t_\lambda \neq 0, \quad \text{where} \quad g_x^0 v_\lambda + g_\lambda^0 = 0. \tag{2.3}$$

For certain classes of bifurcation points (x_0, λ_0), one can associate a function t, for which (x_0, λ_0) is a regular t-point. In this case we call t a **test function** for the bifurcation point class. Th.2.2 shows that the bifurcation point under consideration can be detected by a strict sign change of the test function during path following of $\mathcal{C}$. Morover the *minimally extended* system $G(x, \lambda) = 0$ (2.1) can be used for the precise location of the bifurcation point either by some direct or some indirect method.

Since bifurcation often depends on (the eigenvalues of) the Jacobian $A(s) := g_x(x(s), \lambda(s))$, test functions t being considered here, have the form

$$t(x, \lambda) = T(g_x(x, \lambda)),$$

where $T(A)$ is defined on a neighborhood of $A_0 = g_x(x_0, \lambda_0)$.

Again, locally near A_0, $T(A) = 0$ defines a manifold of codimension 1 in the matrix space $\mathcal{L}(X)$ and (2.2) describes the transversal crossing of this manifold by the path $A(s)$ of the Jacobians, see also [13] for this kind of view. Usually, certain *eigenvalue crossing conditions* are reflected by (2.2).

We will use the name *test function* for t and for T as well.

3 Test functions for steady state bifurcation points.

For steady state bifurcation points (x_0, λ_0) the Jacobian g_x^0 is singular. Hence

$$t(x, \lambda) := \det(g_x(x, \lambda)), \quad T(A) := \det(A) \tag{3.4}$$

seems the most obvious choice for a test function ($X = \mathbb{R}^N$). But the regularity condition for t-points requires the bifurcation points to be *quadratic* turning points ($\lambda''(0) \neq 0$).

Remark 3.1 *It is well known that also simple bifurcation points can be detected by a strict sign change of the determinant. Nevertheless our regularity condition for a test function is so strong that - for $X = \mathbb{R}^N$ - the determinant is not a test function for simple bifurcation points (to be more precise: (2.2) holds, but DG^0 in Def. 2.1 is not regular).*

But in the case of subspace-breaking bifurcation points, see sec. 6, the equilibria of C lie in an invariant subspace $X \neq \mathbb{R}^N$ and there emanates a branch of equilibria not in X from (x_0, λ_0). Then (3.4) is a test function according to the obove Def.2.1, see sec. 5.

Another choice which can replace the somehow unpopular determinant has been suggested by [5]: Given *generic* vectors $l, r \in \mathbb{R}^N$ there are unique $u \in \mathbb{R}^N, T(A) \in \mathbb{R}$ solving the bordered linear system

$$\begin{pmatrix} A & r \\ l^T & 0 \end{pmatrix} \begin{pmatrix} u \\ T(A) \end{pmatrix} = \begin{pmatrix} 0 \\ 1 \end{pmatrix}, \tag{3.5}$$

if $\operatorname{rank}(A) \geq N - 1$. Now by $T(A)$, a test function for quadratic turning points (or in the sense of Rem. 3.1 for simple bifurcation points) is defined.

Both test functions (3.4) and (3.5) aim at the same bifurcation points and have the same theoretical, but different numerical properties. The derivative of the test function in (3.5) can be obtained much cheaper than that of (3.4). On the other hand (3.4) is a globally valid test function while in (3.5) the vectors l, r must be chosen suitably.

For other related choices of test functions see [10].

4 Test functions for Hopf points.

Hopf points are essentially defined by the fact that a pair of complex conjugate eigenvalues of the Jacobians $A(s)$ crosses the imaginary axis with nonzero speed.

In practice, the detection of Hopf points is usually based on $T(A) :=$ real part of some eigenvalue of A, while certain extended *Hopf systems*, for instance

$$H(A, \xi, \omega) := \left((A^2 + \omega^2 I)\xi, \, l^T \xi, \, \xi^T \xi - 1 \right) = 0, \tag{4.6}$$

320   B. Werner

are used for computation, where $l \in \mathbb{R}^N$ is a suitable vector ([9]).

Another interesting test function is used in the package LOCBIF by KHIBNIK, KUZNETSOV, LEVITIN, NIKOLAEV [8], namely $T(A) = \text{res}(p_A(\mu), p_A(-\mu))$. Here $\text{res}(p, q)$ is the *resultant* of the two polynomials p, q being zero iff p and q have at least one common (complex) root, and p_A is the characteristic polynomial of A. By this test function also *imaginary Hopf points* in the sense of [13] are detected (two *real* eigenvalues of A sum up to zero).

Both possibilities are expensive in the way that all eigenvalues or the characteristic polynomial of the Jacobians are involved. The advantage is the robustness which makes these test functions very reliable for low dimensional cases. Other methods for the detection of Hopf points were recently discussed in [2].

4.1 A NEW TEST FUNCTION

A slight modification of (4.6), replacing ω^2 by the *Hopf number* ν and using different normalizing conditions, leads to

$$H(A, \xi, \nu) := \left((A^2 + \nu I)\xi, \; l^T \xi, \; l^T A \xi - 1 \right) = 0. \tag{4.7}$$

Regular zeros of (4.7) correspond to *real Hopf matrices* A ($\nu = \omega^2 > 0$) or to *Takens-Bogdanov (TB-) matrices* ($\nu = 0$) or to *imaginary Hopf matrices* ($\nu = \omega^2 < 0$, hence ω is imaginary) ([13]).

To end up with a test function $T(A)$ for Hopf matrices A, we border the matrix $A^2 + \nu I$ using vectors $l, r \in \mathbb{R}^N$ and consider the following bordered linear system:

$$\begin{pmatrix} A^2 + \nu I & Ar & r \\ l^T A & 0 & 0 \\ l^T & 0 & 0 \end{pmatrix} \begin{pmatrix} \xi \\ \alpha \\ \beta \end{pmatrix} = \begin{pmatrix} 0 \\ 1 \\ 0 \end{pmatrix}. \tag{4.8}$$

If the coefficient matrx $H(A, \nu)$ in (4.8) is regular, there are unique solutions $\xi \in \mathbb{R}^N, \alpha, \beta \in \mathbb{R}$. Comparing (4.7) and (4.8), a Hopf matrix with Hopf number ν is characterized by *two* implicitely definied equations

$$\alpha(A, \nu) = 0, \quad \beta(A, \nu) = 0.$$

If one can eleminate ν by solving $\beta(A, \nu(A)) = 0$, we end up with

$$T(A) := \alpha(A, \nu(A)). \tag{4.9}$$

The elemination can be performed numerically by solving for fixed A the scalar nonlinear equation $\beta(A, \nu) = 0$ by a Newton like method with a given starting value ν_0.

Hence our candidate (4.9) for a test function for Hopf points has some shortcomings in comparison with the comparable test function (3.5) for stationary bifurcation points. A scalar nonlinear equation has to be solved involving the solutions of bordered linear systems (4.8). Moreover, a first estimate ν_0 for the Hopf number must be given (different ν_0 lead to different test functions).

It follows from the following theorem that (4.9) defines a (local) test function for Hopf points.

Theorem 4.1 *Let A_0 be a Hopf matrix with Hopf number ν_0. For almost all bordering vectors $l, r \in \mathbb{R}^N$ there is a neighborhood U_0 of A_0 in the matrix space $\mathbb{R}^{N,N}$ and a number $\delta > 0$ such that for $A \in U_0$, $|\nu - \nu_0| < \delta$, $H(A, \nu)$ is regular, $T(A)$ (and $\nu(A)$) in (4.9) are well defined and $T(A) = 0$ is equivalent with A being a Hopf matrix with Hopf number $\nu(A)$. $(T(A) = 0$ is a local description of the Hopf manifold $\mathcal{M}_H$ in [13]).*

A proof of this theorem and discussions of details of the algorithm involved in evaluation and differentiation of our test function (4.1) is beyond the scope of this paper.

Being interested in Hopf points close to TB-points one should set $\nu_0 = 0$. Then also imaginary Hopf points are detected, and - by continuation - one can reach the more important real Hopf points.

In the following we utilize the analytical consequence of Th.4.1 saying that for each Hopf bifurcation point (x_0, λ_0) (including TB-points and imaginary Hopf points) there exists a test function such that (x_0, λ_0) is a regular t-point of g in the sense of Def.2.1.

5   Test functions for Γ-simple bifurcation points

For 1-parameter problems without symmetries generically only turning points and Hopf points occur with associated algebraically simple eigenvalues. But if symmetries are present, the situation changes dramatically ([4]). Assume that

$$g(\gamma x, \lambda) = \gamma g(x, \lambda), \quad x \in \mathbb{R}^N, \lambda \in \mathbb{R}, \gamma \in \Gamma, \tag{5.10}$$

where Γ is a compact Lie group of orthogonal $N \times N$-matrices acting on $\mathbb{R}^N$.

Along a branch of Γ-*symmetric* equilibria $C^\Gamma = \{(x(s), \lambda(s)) : |s| < \delta\}$ $(\gamma x(s) = x(s)$ for all $\gamma \in \Gamma)$, the Jacobians $A(s) := \frac{\partial g}{\partial x}(x(s), \lambda(s))$ *have the symmetry of the group* Γ *in the sense that*

$$\gamma A(s) = A(s)\gamma \text{ for all } \gamma \in \Gamma. \tag{5.11}$$

We denote by $C(\Gamma)$ the space of all real $N \times N$-matrices, having the symmetry of Γ.

The basic fact for understanding the eigenvalue properties of $A(s)$ and hence bifurcation for g is the *block diagonalization theorem*, see Th.2.3 in [12]. There is a *symmetry adapted basis* of $\mathbb{R}^N$, such that **all** matrices $A \in C(\Gamma)$ have certain block diagonal structure with respect to this basis. To be more concrete, to each irreducible representation of Γ, say ϑ, there corresponds an isotypic component X_ϑ which is invariant under all $A \in C(\Gamma)$ yielding a rough block structure of A. But the corresponding ϑ-blocks have again a block diagonal form. It consists of a number d_ϑ of *equal* blocks $A_\vartheta \in \mathbb{R}^{c_\vartheta, c_\vartheta}$, where d_ϑ depends on the dimension of the irreducible representation ϑ. The size c_ϑ of A_ϑ is related to the multiplicity of ϑ_ϑ in Γ, see [12] for more details. Eigenvalues of A_ϑ are called ϑ-*eigenvalues*.

Definition 5.1 *A point $(x_0, \lambda_0) = (x(0), \lambda(0)) \in C$ is called a Γ-semisimple* **steady state**, *resp. a Hopf* **bifurcation point** *of* **symmetry type** ϑ *if an algebraically simple eigenvalue, resp. a pair of algebraically simple conjugate complex eigenvalues of the subblocks $A_\vartheta(s)$ of the Jacobians $A(s)$ crosses the imaginary axis with nonzero speed for $s = s_0 = 0$. If no other eigenvalues of $A(s)$ are lying on the imaginary axis for $s = s_0 = 0$, (x_0, λ_0) is called a* **Γ-simple bifurcation point of symmetry type** ϑ.

If $A(s_0)$ has a vanishing ϑ-eigenvalue, resp. a pair of imaginary ϑ-eigenvalues, (x_0, λ_0) will be called a **potential** *steady state resp. a Hopf bifurcation point of symmetry type ϑ.*

322   **B. Werner**

We cannot discuss here the whole variety of (*symmetry breaking*) bifurcations of branches of equilibria, resp. periodical orbits from a Γ-simple bifurcation point (x_0, λ_0). But we briefly present the idea of a *bifurcation subgroup* [1].

$\Sigma \subset \Gamma$ is a bifurcation subgroup associated with an irreducible representation ϑ (of real type) if the corresponding fixed point space X^Σ has a one-dimensional space in common with every irreducible subspace of X_ϑ. Then if we restrict ourselves to states $x \in X^\Sigma$ Γ-simple bifurcation points of symmetry type ϑ are reduced to *simple* (symmetry breaking) steady state or Hopf bifurcation points.

Each test function T_S, resp. T_H for stationary, resp. for Hopf bifurcation points from the preceding sections defines a test function for Γ-semisimple bifurcation points setting in Def.2.1

$$t(x, \lambda) = T_\vartheta(g_x(x, \lambda)), \qquad T_\vartheta(A) := T_S(A_\vartheta) \quad \text{or} \quad T_\vartheta(A) := T_H(A_\vartheta), \qquad (5.12)$$

and

$$X := X^\Gamma := \{x \in \mathbb{R}^N : \gamma x = x \text{ for all } \gamma \in \Gamma\}.$$

The regularity condition in Def.2.1 reflects the nonzero speed of the eigenvalue crossing, not the seperation condition in Def.5.1. If this condition is violated, we have an interaction of bifurcation points of different symmetry types, see sec.7.

6   Interaction of t-points and subspace breaking bifurcation points.

This section is linked directly with sec.2. It is connected with sec.5 by setting $X_+ = X^\Gamma$, $X = X^\Sigma$, where Σ is a bifurcation subgroup.

Assume that - besides X - there is another subspace X_+ of X invariant under $g(\,.\,, \lambda)$. **Subspace breaking bifurcation points (SB-points)** (x_0, λ_0) are certain *simple λ-singular points* of g defined by

$$g(x_0, \lambda_0) = 0, \quad \text{kernel}(g_x^0) = \text{span}(\varphi_0), \quad \text{range}(g_x^0) = (\text{span}(\psi_0))^\perp, \quad \varphi_0, \psi_0 \neq 0$$

with the additional properties

$$x_0 \in X_+, \varphi_0 \notin X_+, \qquad c_\lambda := \psi_0^T(g_{xx}^0 v_\lambda + g_{x\lambda}^0)\varphi_0 \neq 0, \qquad (6.13)$$

where $v_\lambda \in X_+$ is defined by $g_x^0 v_\lambda + g_\lambda^0 = 0$.

The first two conditions in (6.13) ensure the unique existence of a smooth *symmetric* branch $C^s \subset X_+ \times \mathbb{R}$ through (x_0, λ_0) which can be parametrized by $s = \lambda$. The transversality condition $c_\lambda \neq 0$ guarantees a unique *asymmetric* branch C^a emanating from C^s in (x_0, λ_0) *breaking the subspace* X_+ in the sense that the tangent vector of C^a in (x_0, λ_0) points out of $X_+ \times \mathbb{R}$.

Let $t : U \subset X \times \mathbb{R} \to \mathbb{R}$ be a candidate for a test function for a certain class of bifurcation points having nothing in common with the above SB-point, for example Hopf points.

We are interested in t-points on the symmetric and the asymmetric branch as well. But the degenerate situation we start with, is the coincidence of a t-point with a SB-point (a kind of mode interaction). We call (x_0, λ_0) a **regular t-SB-point** iff (x_0, λ_0) is a SB-point and a *regular* t-point of g_+, the restriction of g to $X_+ \times \mathbb{R}$, compare with Rem. 3.1.

The regularity condition for the **t**-point is again $d_\lambda = t_x^0 v_\lambda + t_\lambda^0 \neq 0$ (2.3), where now $v_\lambda \in X_+$, while $c_\lambda \neq 0$ is the essential SB-point condition.

Assume now there is a second parameter α unfolding the interaction of a **t**-point and a SB-point. We replace $g(x, \lambda)$ by $g(x, \lambda, \alpha)$, $t(x, \lambda)$ by $t(x, \lambda, \alpha)$ and $G(x, \lambda, \alpha)$ by $G(x, \lambda)$ (see (2.1) for the definition of G). We introduce

$$c_\alpha := \psi_0^T(g_{xx}^0 v_\alpha + g_{x\alpha}^0)\varphi_0, \quad d_\alpha := t_x^0 v_\alpha + t_\alpha^0, \tag{6.14}$$

where $v_\alpha \in X_+$ is defined by $g_x^0 v_\alpha + g_\alpha^0 = 0$.

For the following theorem observe that the subspace $\tilde{X}_+ := X_+ \times \mathbb{R}$ is invariant under $G(\,.\,, \alpha)$. Hence the name *SB-point of G* with respect to α makes sense.

Theorem 6.1 *Let $y_0 := (x_0, \lambda_0)$ be a regular* **t**-*SB-point of $g(\,.\,, \alpha_0)$ with respect to λ. Let*

$$D := Det \begin{pmatrix} c_\lambda & c_\alpha \\ d_\lambda & d_\alpha \end{pmatrix} \neq 0. \tag{6.15}$$

Then (y_0, α_0) is a SB-point of G with respect to α.

PROOF: It is easy to see that $\Phi_0 := (\varphi_0, 0)$ spans the kernel and $\Psi_0 := (\psi_0, 0)$ the cokernel of G_y^0. The desired SB-condition is

$$C_\alpha := \Psi_0^T(G_{yy}^0 V_\alpha + G_{y\alpha}^0)\Phi_0 \neq 0,$$

where $V_\alpha := (V_\alpha^x, V_\alpha^\lambda) \in X_+ \times \mathbb{R}$ is defined by $G_y^0 V_\alpha + G_\alpha^0 = 0$. Computing $V_\alpha^x = -d_\alpha v_\lambda/d_\lambda + v_\alpha$, $V_\alpha^\lambda = -d_\alpha/d_\lambda$ we obtain $C_\alpha = \psi_0^T(g_{xx}^0 V_\alpha^x + g_{x\alpha}^0)\varphi_0 = c_\alpha - d_\alpha c_\lambda/d_\lambda \neq 0$, because of $D \neq 0$. ∎

Treating α as a bifurcation parameter and applying the SB-point analysis to G, one concludes that there is a branch

$$C_t^s = \{(x_t^s(\alpha), \lambda_t^s(\alpha)) : |\alpha - \alpha_0| < \varepsilon\}, \quad \varepsilon > 0 \tag{6.16}$$

of *symmetric* **t**-points lying on the symmetric branches $C^s(\alpha)$ of steady state solutions of $g(\,.\,, \alpha) = 0$ and a branch

$$C_t^a = \{(x_t^a(\tau), \lambda_t^a(\tau), \alpha(\tau)) : |\tau| < \varepsilon)\}, \quad \varepsilon > 0 \tag{6.17}$$

of *asymmetric* **t**-points on the asymmetric branches $C^a(\alpha)$ of steady state solutions.

Moreover there is also a curve of SB-points on $C^s(\alpha)$,

$$C_{SB} = \{(x_{SB}(\alpha), \lambda_{SB}(\alpha)) : |\alpha - \alpha_0| < \varepsilon\}, \quad \varepsilon > 0.$$

Computing $(\lambda_t^s)'(\alpha_0) = -d_\alpha/d_\lambda$, $(\lambda_{SB})'(\alpha_0) = -c_\alpha/c_\lambda$, the non-degeneracy condition $D \neq 0$ tells us that - varying α -, the symmetric **t**-points overtake the SB-points at the **t**-SB-point for $\alpha = \alpha_0$ with different λ-speeds.

Remark 6.2 *1. Without further assumptions, SB-points are transcritical simple bifurcation points as opposed to sub- or supercritical bifurcation points (pitchforks) which come up for problems with $\mathbb{Z}_2$-symmetry. It can be shown in that case that the SB-point in Th.6.1 is a $\mathbb{Z}_2$-pitchfork, if the test function* **t** *is invariant.*

2. The symmetric **t**-*points are regular with respect to X_+, but Th.6.1 does not claim that the asymmetric* **t**-*points are regular with respect to X. Hence the latter may correspond only to potential bifurcation points, see Def.5.1.*

7 Applications

If in Th.6.1 a t-point is a Hopf point, there are two possibilities depending on the Hopf number ν_0 at the regular t-SB-point (x_0, λ_0). If $\nu_0 = 0$, we have the Takens-Bogdanov singularity and Th.5.5 in [13] is a special case of Th.6.1.

If $\nu_0 \neq 0$ we have a steady state - Hopf mode interaction in a SB-point, where it is well known that this mode interaction leads to Hopf points on the secondary asymmetric branch in accordance to Th.6.1. But observe that without further non-degeneracy conditions, the asymmetric t-points may be *potential* Hopf points only in the sense that their regularity properties, in the form of eigenvalue crossing conditions may fail.

7.1 INTERACTION OF Γ-SIMPLE BIFURCATION POINTS OF DIFFERENT SYMMETRY TYPE.

Now we turn to the case where $g(x, \lambda, \alpha)$ satisfies the group equivariance condition (5.10). We are interested in the case, where two Γ-simple bifurcation points $y_i(\alpha) := (x_i(\alpha), \lambda_i(\alpha)) \in C^\Gamma(\alpha), i = 1, 2$, of different symmetry types ϑ and η coalesce for $\alpha = \alpha_0$.

We assume that y_1 is a steady state bifurcation point and that ϑ is a nontrivial absolutely irreducible with bifurcation subgroup $\Sigma \neq \Gamma$, while y_2 (and $\eta \neq \vartheta$) is not yet specified.

Concerning the eigenvalues of the Jacobians $A(s)$ for fixed $\alpha = \alpha_0$ either two real eigenvalues of symmetry types ϑ and η cross the imaginary axis at zero (without splitting) or there is a simultaneous crossing of the imaginary axis by one real ϑ- and by two conjugate complex η-eigenvalues.

Let $C^\Sigma(\alpha)$ be the branches emanating in $y_1(\alpha)$ with the symmetry of Σ. We set $X := X^\Sigma$, $X_+ := X^\Gamma$, $C^s := C^\Gamma$, $C^a := C^\Sigma$. Then $y_1(\alpha)$ can be shown to be SB-points, where the symmetry of Γ breaks into the subsymmetry of Σ. The question now is, whether $y_2(\alpha)$ can be characterized as t-points with a suitable test function $t(x, \lambda, \alpha)$ defined on an open subset U not only of $X_+ \times \mathbb{R}^2$ but also of $X \times \mathbb{R}^2$ such that $y_0 := y_1(\alpha_0) = y_2(\alpha_0)$ is a *regular* t-SB-point.

Using the η-block in the symmetry adapted block diagonalization of the Jacobians $A(x, \lambda, \alpha) = \frac{\partial g}{\partial x}(x, \lambda, \alpha), x \in X^\Gamma$, the testfunctions (5.12),

$$t(x, \lambda, \alpha) := T_\eta(A(x, \lambda, \alpha)), \quad x \in X^\Gamma, \tag{7.18}$$

seem to be good candidates. Disregarding for the moment that t is defined only for $x \in X_+$ and not for all $x \in X$, the non-degeneracy condition $D \neq 0$ in (6.15) can be easily reformulated by

$$\lambda_1'(\alpha_0) \neq \lambda_2'(\alpha_0). \tag{7.19}$$

The problem now is the extension of (7.18) to $x \in X = X^\Sigma$. The answer is based on the way in which the Γ-irreducible representations ϑ and η (being considered now as representations of Σ) decompose into Σ-irreducible representations $\varrho_1, ..., \varrho_r$.

In the language of representations there are *multiplicities* $0 \leq c_j, d_j \in \mathbb{N}_0$,

$$\vartheta = c_1\varrho_1 + c_2\varrho_2 + \cdots + c_r\varrho_r, \quad \eta = d_1\varrho_1 + d_2\varrho_2 + \cdots + d_r\varrho_r. \tag{7.20}$$

Since Σ is a bifurcation subgroup for ϑ we have $c_1 = 1$ assuming that ϱ_1 is the trivial representation. Our condition can be formulated as

$$d_1 = 0; \quad \exists j > 1 : c_j = 0, d_j = 1. \tag{7.21}$$

We say that the nontrivial irreducible representation of Σ, $\varrho := \varrho_j$ **satisfies the interaction condition (7.21)**.

Now t in (7.18) can be extended to $x \in X^\Sigma$ by blockdiagonalizing of $A(x, \lambda, \alpha)$ *with respect to the subgroup* Σ, taking the ϱ-block A_ϱ and setting in (5.12)

$$t(x, \lambda, \alpha) := T_\varrho(A(x, \lambda, \alpha)), \quad x \in X^\Sigma. \tag{7.22}$$

The interaction condition (7.21) allows us to characterize the degenerate point (y_0, α_0) as a *regular* **t-***SB-point* with test function t defined by (7.22). $d_1 = 0$ implies that y_0 is still a SB-point for $\alpha = \alpha_0$, while the regularity of the t-point follows from $d_j = 1, c_j = 0$. Hence Th.6.1 applies.

The symmetric t-points in (6.16) coincide with $y_2(\alpha)$, while the asymmetric t-points in (6.17) now are potential (Σ-breaking) Σ-simple bifurcation points of symmetry type ϱ.

Theorem 7.1 *Assume that two Γ-simple bifurcation points $y_i(\alpha) := (x_i(\alpha), \lambda_i(\alpha)) \in C^\Gamma(\alpha), i = 1, 2$, of different symmetry types ϑ and η coalesce for $\alpha = \alpha_0$ in y_0 (with y_0 still Γ-semisimple).*

Let y_1 be a steady state bifurcation point and ϑ be a nontrivial absolutely irreducible representation with bifurcation subgroup $\Sigma \neq \Gamma$.

If $\varrho = \varrho_j$ satisfies the interaction condition (7.21) with ϑ and η decomposed with repect to Σ as in (7.20), then there is a branch of potential (Σ-breaking) Σ-simple bifurcation points of symmetry type ϱ

$$C_\varrho^\Sigma = \{(x_\varrho^\Sigma(\tau), \lambda_\varrho^\Sigma(\tau), \alpha(\tau)) : |\tau| < \varepsilon)\}, \quad \varepsilon > 0$$

on the Σ-branch $C^\Sigma(\alpha)$ $((x_\varrho^\Sigma(\tau), \lambda_\varrho^\Sigma(\tau)) \in C^\Sigma(\alpha(\tau)))$.

7.2 Examples

Let $\Gamma = D_4 = \{I, R, R^2, R^3, S_1, S_1', S_2, S_2'\}$ be the dihedral group of the square, where R corresponds to the rotation about $\pi/2$, and S_1, S_1' resp. S_2, S_2' correspond to reflections along the midaxes, resp. the diagonals of a square.

Let ϑ be the 1-dimensional irreducible representation defined by $R \mapsto -1, S_2 \mapsto 1$ and let η be the (unique) 2-dimensional irreducible representation of D_4. Then $\Sigma = \Sigma_D = \{I, R^2, S_2, S_2'\}$ is the unique bifurcation subgroup for ϑ being isomorphic to the symmetry group of a rectangle. Σ has four 1-dimensional irreducible representations $\varrho_1, ..., \varrho_4$. Let ϱ_1 be the trivial irreducible representation and ϱ_4 be defined by $S_2, S_2' \mapsto -1$. Then (7.20) takes the form

$$\vartheta = \varrho_1, \quad \eta = \varrho_2 + \varrho_3,$$

hence both, $\varrho = \varrho_2$ and $\varrho = \varrho_3$ satisfy the interaction condition (7.21).

The result of Th.7.1 can be demonstrated by the computations for the 4-box Brusselator in [1]. Look at the bifurcation diagram with steady state (Fig.13, [1]) and with Hopf bifurcation points (Fig.14, [1]). Primary bifurcation points labeled with 1 are D_4-simple bifurcation points of symmetry type ϑ, while those labeled with 2 are of symmetry type η. The bifurcation points labeled with the number 3 are secondary Σ-simple bifurcation points of symmetry type ϱ_2, see also the bifurcation graph in Fig.15. Varying $\alpha = B$, it is possible to explain at least two of them in Fig.13 and one of them in Fig.14 as the result of

the interaction of the steady state bifurcation point 1 and the steady state (Fig.13) or Hopf
(Fig.14) bifurcation point 2 according to Th.7.1. The conjugate Σ-simple bifurcation points
of symmetry type ϱ_3 are not shown in Fig.13 and 14.

A second example is another dihedral group

$$\Gamma = D_6 = \{I, R.R^2, R^3, R^4, R^5, S_1, S_1', S_1'', S_2, S_2', S_2''\},$$

the group of the hexagon.

Let ϑ be the 1-dimensional irreducible representation, given by $R \mapsto -1, S_1 \mapsto 1$. The
bifurcation subgroup is

$$\Sigma = \Sigma_{d_1} := \{I, R^2, R^4, S_1, S_1', S_1''\}$$

being isomorphic to the dihedral group D_3. If η is one of the two 2-dimesional irreducible
representations of D_6, then η is still D_3-irreducible. If ϱ is the (only) 2-dimensional irreducible
representation of D_3, then ϱ satisfies the interaction condition (7.21). Hence an interaction of
two D_6-simple bifurcation points of symmetry type ϑ and η leads to secondary bifurcations
of symmetry type ϱ on the D_3-branch.

These results can be verified by numerical computations for the hexagonal lattice dome in
[11].

The interaction condition is not always true. Interchanging the role of ϑ and η in the last
two examples, the interaction condition does not hold.

As the last example we briefly mention $\Gamma = O(2)$, $\vartheta = \varrho_l$ (with bifurcation subgroup
$\Sigma = D_l$) and $\eta = \varrho_{l+1}, l \in \mathbb{N}$, see [4], p.446f. Since ϱ_{l+1} considered as a representation
of D_l is still irreducible, ϱ satisfies the interaction condition (7.21). Hence, by Th.7.1, an
interaction of a steady state $O(2)$-simple bifurcation point of type ϱ_l with any (steady state
or Hopf) $O(2)$-simple bifurcation point of type ϱ_{l+1} leads to secondary D_l-simple (steady
state or Hopf) bifurcation points of type ϱ on the secondary D_l-branch. This has been also
shown in [4], Ch.XX (mode interactions with $O(2)$ symmetry).

References

[1] M.Dellnitz, B.Werner. *Computational methods for bifurcation problems with symmetries
- with special attention to steady state and Hopf bifurcation points.* J. of Comp. and Appl.
Math. **26**, 97-123, 1989.

[2] T.J.Garrat, G.Moore, A.Spence. *Two methods for the numerical detection of Hopf bifurcations.* In: **Bifurcation and Chaos** (R.Seydel, F.W.Schneider, T.Küpper, H.Troger,
eds.), ISNM **97**, Birkhäuser, 1991.

[3] K.Gatermann, H.Hohmann. *Symbolic Exploitation of symmetry in Numerical Pathfollowing.* Konrad-Zuse-Institut, Report, 1990.

[4] M.Golubitsky, I.Stewart, D.Schaeffer. **Singularities and Groups in Bifurcation
Theory**, Vol. 2, Springer 1988.

[5] A.Griewank, G.W.Reddien. *Characterization and computation of generalized turning
points.* SIAM J. Numer. Anal. **21**, 176-185, 1984.

[6] T.Healey, J.A.Treacy. *Exact block diagonalization of large eigenvalue problems for structures with symmetry.* Intern. J. for Num. Meth. in Eng. **30**, 1990.

[7] K.Ikeda, K.Murota. *Bifurcation analysis of symmetric structures using block diagonalization.* Comp. Meth. Appl. Mech. Eng. **86**, 215-243, 1991

[8] A.Khibnik. *LINLBF: A program for continuation and bifurcation analysis of equilibria up to codimension three.* In: **Continuation and Bifurcations: Numerical Techniques and Applications**, D. Roose, B. de Dier, A. Spence (eds.), NATO ASI Series C, Vol. **313**, 283-296, Kluwer Academic Publishers, Dordrecht, 1990.

[9] D.Roose, V.Hlavacek. *A direct method for the computation of Hopf bifurcation points.* SIAM J. Appl. Math. **45**, 879-894, 1985.

[10] R.Seydel. *On detecting stationary bifurcations.* Int. J. of Bifurcation and Chaos **1**, 1-5, 1991.

[11] P.Stork, B.Werner. *Symmetry adapted block diagonalization in equivariant steady state bifurcation problems and its numerical applications.* To appear in *Advances in Mathematics, China* 1991.

[12] B.Werner. *Eigenvalue problems with the symmetry of a group and bifurcations.* In: **Continuation and Bifurcations: Numerical Techniques and Applications**, D. Roose, B. de Dier, A. Spence (eds.), NATO ASI Series C, Vol. **313**, 71-88, Kluwer Academic Publishers, Dordrecht, 1990.

[13] B.Werner, V.Janovsky. *Computation of Hopf branches bifurcating from Takens-Bogdanov points for problems with symmetries.* In: **Bifurcation and Chaos** (R.Seydel, F.W.Schneider, T.Küpper, H.Troger, eds.), ISNM **97**, 377-388, Birkhäuser, 1991.

Bodo Werner
Institut für Angewandte Mathematik
Universität Hamburg
Bundesstr.55
W 2000 Hamburg 13

ZAMP

Zeitschrift für angewandte Mathematik und Physik
Journal of Applied Mathematics and Physics
Journal de Mathématiques et de Physique appliquées

ISSN 0044-2275

Editors:
U. Kirchgraber, Zürich
(Differential Equations/
Applied Mathematics)
M. Renardy, Blacksburg
(Differential Equations/
Fluid Mechanics)
I.L. Ryhming, Lausanne
(Fluid Mechanics)
M. Sayir, Zürich
(Mechanics of Solids)

Assistant Editor:
K. Nipp, Zürich

Editorial Office:
ZAMP, P.O.Box 274
CH-4024 Basel
Switzerland

Co-editors:
F. Bark, Stockholm

F.G. Blottner,
Albuquerque
S.R. Bodner, Haifa
F.H. Busse, Bayreuth
G.-Q. Chen, Chicago
T.K. Fannelöp, Zürich
J. Glimm, Stony Brook
M. Hiller, Duisburg
T. Maxworthy,
   Los Angeles
P.A. Monkewitz,
   Los Angeles
A. Moser, Zürich
P.M. Naghdi, Berkeley
N. Rott, Palo Alto
N.M. Temme,
   Amsterdam
H. Thomann, Zürich
J. Waldvogel, Zürich

Selected and forthcoming articles:
S.S. Antman and R.S. Marlow, Transcritical buckling
of columns – H. Brauchli, Mass–orthogonal
formulation of equations of motion for multibody
systems – L. Brüll and U. Pallaske, On differential
algebraic equations with discontinuities – F.H. Busse
and M. Kropp, Buoyancy driven instabilities in rotating
layers with parallel axis of rotation – B. Galperin and
G.L. Mellor, The effects of streamline curvature and
spanwise rotation on near–surface, turbulent
boundary layers – J.J. Keller and Y.-P. Chyou, On the
hydraulic lock-exchange problem – P. Podio–Guidugli
and M.E. Gurtin, On the formulation of mechanical
balance laws for structured continua – I.W. Stewart,
Density conservation for a coagulation equation.

Subscription information:
1992 subscription, volume 43(6 issues)
sFr. 668.–/DM 788.–/US$ 455.–
(plus postage & handling)
Single copy:
sFr. 128.–/DM 154.–/US$ 92.–
1/92 all prices are object to changes without notice

The *Journal of Applied Mathematics and Physics*
(ZAMP) publishes papers of high quality in Fluid
Mechanics , Mechanics of Solids and Differential
Equations / Applied Mathematics. The readers of
ZAMP will find not only articles in their own special
field but also original work in neighbouring domains.
This will lead to an exchange of ideas; concepts and
methods which have proven to be successful in one
field may well be useful to other areas. ZAMP attempts
to publish articles reasonably fast. A longer paper is
published in the section „Original Papers", a shorter
one may appear under „Brief Reports". Publication
of contributions to the „Brief Reports" section is
particularly rapid. The journal includes a „Book
Review" section and provides information on activities
(such as upcoming symposia, meetings, special
courses) which are of interest to its readers.

**You may order through your bookseller or
subscription agency, or directly from the
publisher:**

Birkhäuser Verlag AG
P.O. Box 133
CH-4010 Basel / Switzerland

Birkhäuser

Birkhäuser Verlag AG
Basel · Boston · Berlin

MIX
Papier aus verantwortungsvollen Quellen
Paper from responsible sources
FSC® C105338

If you have any concerns about our products,
you can contact us on
ProductSafety@springernature.com

In case Publisher is established outside the EU,
the EU authorized representative is:
Springer Nature Customer Service Center GmbH
Europaplatz 3, 69115 Heidelberg, Germany

Printed by Libri Plureos GmbH
in Hamburg, Germany